Lecture Notes in Mathematics

1936

Editors:
J.-M. Morel, Cachan
F. Takens, Groningen
B. Teissier, Paris

Editors *Mathematical Biosciences Subseries:*
P.K. Maini, Oxford

Pierre Magal · Shigui Ruan (Eds.)

Structured Population Models in Biology and Epidemiology

With Contributions by:

P. Auger · M. Ballyk · R. Bravo de la Parra
W.-E. Fitzgibbon · S.A. Gourley · D. Jones · M. Langlais
R. Liu · M. Martcheva · T. Nguyen-Huu · J.-C. Poggiale
E. Sánchez · H.L. Smith · H.R. Thieme · G.F. Webb · J. Wu

 Springer

Editors

Pierre Magal
University of Le Havre
25 rue philippe Lebon
76058 Le Havre
France
pierre.magal@uni-lehavre.fr

Shigui Ruan
Department of Mathematics
University of Miami
Coral Gables, FL 33124-4250
USA
ruan@math.miami.edu

ISBN: 978-3-540-78272-8   e-ISBN: 978-3-540-78273-5
DOI: 10.1007/978-3-540-78273-5

Lecture Notes in Mathematics ISSN print edition: 0075-8434
                             ISSN electronic edition: 1617-9692

Library of Congress Control Number: 2008921368

Mathematics Subject Classification (2000): 34A35, 34G20, 35B40, 35K57, 92D25, 92D30

Cover design: WMXDesign GmbH, Heidelberg

Printed on acid-free paper

9 8 7 6 5 4 3 2 1

springer.com

Preface

In this new century mankind faces ever more challenging environmental and public health problems, such as pollution, invasion by exotic species, the emergence of new diseases or the emergence of diseases into new regions (West Nile virus, SARS, Anthrax, etc.), and the resurgence of existing diseases (influenza, malaria, TB, HIV/AIDS, etc.). Mathematical models have been successfully used to study many biological, epidemiological and medical problems, and nonlinear and complex dynamics have been observed in all of those contexts. Mathematical studies have helped us not only to better understand these problems but also to find solutions in some cases, such as the prediction and control of SARS outbreaks, understanding HIV infection, and the investigation of antibiotic-resistant infections in hospitals.

Structured population models distinguish individuals from one another according to characteristics such as age, size, location, status, and movement, to determine the birth, growth and death rates, interaction with each other and with environment, infectivity, etc. The goal of structured population models is to understand how these characteristics affect the dynamics of these models and thus the outcomes and consequences of the biological and epidemiological processes. There is a very large and growing body of literature on these topics. This book deals with the recent and important advances in the study of structured population models in biology and epidemiology. There are six chapters in this book, written by leading researchers in these areas.

In Chap. 1, Population Models Structured by Age, Size, and Spatial Position, Glenn Webb systematically introduces population models with age, size, and spatial structure. The theory of semigroups of linear and nonlinear operators in Banach spaces is used to analyze these models and many numerical simulations are included to illustrate the theoretical results, as an aid to readers. A brief and historical introduction on age structured models is given in the Introduction. Section 1.1 focuses on population models structured by two factors, size and spatial position. Basic assumptions and definitions, illustrative examples, and simulations are presented. Fundamental theorems are stated and proved. Results are then established for population models structured by age and spatial position in Sect. 1.2. In Sect. 1.3 population models structured by all three factors, namely, age, size, and spatial position, are studied. As an example, a model of tumor growth with cell age corresponding to the cell cycle, cell size corresponding to cell growth and division, and cell position in space corresponding to cell motility and migration is discussed.

An important class of spatial models deals with metapopulations. A metapopulation is a group of populations of the same species that occupy separate patches and are connected by dispersal. Spatially implicit metapopulation models with discrete patch-size structure lead to infinite systems of ordinary differential equations, as do host-macroparasite models which distinguish hosts by their parasite loads, and prion proliferation. In Chap. 2, Infinite ODE Systems Modeling Size-structured Metapopulations, Macroparasitic Diseases, and Prion Proliferation, Maia Martcheva and Horst Thieme develop a general theory on the properties of the solution semiflows generated by infinite systems of ordinary differential equations. The chapter consists of 14 sections. In Sect. 2.2, homogeneous linear Kolmogorov systems are discussed. Section 2.3 addresses semilinear systems. Sections 2.4–2.11 focus exclusively on metapopulation models. Boundedness of solutions of general metapopulation models is studied in Sect. 2.4, while Sect. 2.5 focuses on how the absence of migration or colonization of empty patches can cause extinction. In Sects. 2.6–2.11, a more specific metapopulation model is studied. Topics addressed include the existence of compact attractors, stability of equilibria, metapopulation persistence, etc. Finally, results are applied to analyze specific metapopulation models, specific host-macroparasite models, and prion proliferation models in Sects. 2.12, 2.13, and 2.14, respectively.

In the classical diffusive epidemic models introduced in Chap. 1, the spatial domains are usually assumed to be homogenous. However, in reality that is not the case – one needs to consider heterogeneous environments. In Chap. 3, Simple Models for the Transmission of Microparasites between Host Populations Living on Non-coincident Spatial Domains, William E. Fitzgibbon and Michel Langlais focus on modeling the direct and indirect transmission of a microparasite between host populations living on non-coincident spatial domains. The goal is to provide a mathematical approach to model the environmentally driven transmission of microparasites between host populations living on distinct spatial domains. A key assumption is the possibility for the microparasite to persist in the environment once it is released by infective individuals. Besides criss-cross transmission, indirect transmission occurs through contacts between susceptible hosts and the contaminated part of the environment. A critical example is the indirect contamination of human populations by animals when the parasite is benign for animals but lethal to humans. This is actually the case for Hantaviruses for which rodents are the main reservoir population, and more generally for many emerging infectious diseases. Various deterministic models can be developed, ranging from unstructured populations – basic ODE systems – to spatially structured multi-patch and reaction–diffusion models to handle heterogeneous environments.

Age structures, as introduced in Chap. 1, are also important to consider both from a chronological point of view – juveniles vs. adults – and from an epidemiological point of view – variable virulence.

The emergence of diseases that are transmitted by vectors, such as the West Nile virus and Dengue fever, raises challenging issues not only epidemiologically but also mathematically. In Chap. 4, Spatiotemporal Patterns of Disease Spread: Interaction of Physiological Structure, Spatial Movements, Disease Progression and Human Intervention, Stephen A. Gourley, Rongsong Liu, and Jianhong Wu are concerned about the effects of demographic and disease ages and spatial movements of hosts on the spatiotemporal spread patterns of certain diseases, with special emphasis on West Nile virus. They start with a short review of some standard models for the transmission dynamics of vector-born diseases in homogeneous environments and populations. In Sect. 4.2 they focus on structured vector-borne diseases with particular reference to West Nile virus. As the juvenile and adult birds (the hosts) have different spatial movement behaviors, the main model is of the McKendrick–von Foerster type for age-structured bird populations. Various sufficient conditions are given for the system to evolve to the disease-free state and for the stability of this equilibrium. Age-structured control measures are considered in Sect. 4.3, and appropriate mathematical models are derived and used to asses the effectiveness of culling as a tool to eradicate WNV. The results show that eradication of WNV is possible by culling the mosquitoes at either the immature or the mature phase, even though the size of the mosquito population is oscillating and stays above a certain level. The interaction of individual movement and physiological status is considered in Sect. 4.4 with an application to the spatial spread of WNV. Finally, patchy models for the spatial spread of WNV are formulated and analyzed in Sect. 4.5.

To describe the complex dynamics of ecological systems, mathematical models frequently have a large number of variables. In Chap. 5, Aggregation of Variables and Applications to Population Dynamics, attention is directed to aggregation of variables in population and community dynamics. Pierre Auger, Rafael Bravo de la Parra, Jean-Christophe Poggiale, Eva Sanchez, and T. Nguyen-Huu introduce aggregation methods for different types of models, such as ordinary differential equations, discrete equations, delay differential equations, and partial differential equations based on the method of separation of time scales. In Sect. 5.2 the aggregation of variables for ODE systems is developed, geometric singular perturbation theory and normally hyperbolic manifolds are introduced, along with slow and fast systems, aggregation and emergence, and illustrative examples are given. Section 5.3 addresses aggregation methods for discrete models. In Sect. 5.4 the method of aggregation is also applied to partial differential equations and delay equations. Section 5.5 presents several applications in population and community dynamics.

In 1983, R. Freter and collaborators developed a simple chemostat-based model of competition between two bacterial strains, one of which is capable of wall-growth, to illuminate the role of bacterial wall attachment on the phenomenon of colonization resistance in the mammalian gut. In Chap. 6, The Biofilm Model of Freter: A Review, Mary Ballyk, Don Jones, and Hal Smith have re-formulated the model in the setting of a tubular flow reactor, extended the interpretation of the model as a biofilm model, and provided both mathematical analysis and numerical simulations of solution behavior. In Sects. 6.1 and 6.2, the original Freter model is introduced and then is generalized and re-formulated as a chemostat-based model. In Sect. 6.3, the one-dimensional thin tube flow reactor model with biofilm is proposed and analyzed, and special cases such as an advection dominated flow reactor and mobile wall-adherent cells, are considered. The three-dimensional flow reactor models are studied in Sect. 6.4. Section 6.5 focuses on mixed culture models, the associated eigenvalue problems are considered, and simulations are carried out.

This book can be used for various purposes. It is suitable as a textbook for a mathematical biology course or a summer school at the advanced undergraduate and graduate level. It can also be used as a reference book by researchers looking for either interesting and specific problems to work on or useful techniques and discussions of some particular problems. Since the book contains the most recent developments in some fields of mathematical biology and epidemiology, we hope that researchers at all levels will find the book inspiring and useful for their research and study.

September 2007

Pierre Magal
Shigui Ruan

Contents

List of Contributors

P. Auger
IRD UR Géodes
Centre IRD de l'Ile de France
32, Av. Henri Varagnat
93143 Bondy Cedex, France
pierre.auger@bondy.ird.fr

M. Ballyk
Department of Mathematical
Sciences, New Mexico State
University, Las Cruces
NM 88003-8001, USA
mballyk@nmsu.edu

R. Bravo de la Parra
Departamento de Matemáticas
Universidad de Alcalá
28871 Alcalá de Henares
Madrid, Spain
rafael.bravo@uah.es

W.-E. Fitzgibbon
Departments of Engineering
Technology and Mathematics
University of Houston, Houston
TX 77204-3476, USA
fitz@uh.edu

S.A. Gourley
Department of Mathematics
University of Surrey
Guildford, Surrey, GU2 7XH, UK
s.gourley@surrey.ac.uk

D. Jones
Department of Mathematics
and Statistics
Arizona State University, Tempe
AZ 85287-1804, USA
dajones@math.asu.edu

M. Langlais
UMR CNRS 5466 MAB and INRIA
Futurs Anubis, case 26
Université Victor Segalen
Bordeaux 2, 146 rue Léo Saignat
F 33076, Bordeaux Cedex, France
langlais@sm.u-bordeaux2.fr

R. Liu
Department of Mathematics
Purdue University, 150
University Street, West Lafayette
IN 47907-2067, USA
rliu@math.purdue.edu

M. Martcheva
Department of Mathematics
University of Florida
358 Little Hall
P.O. Box 118105, Gainesville
FL 32611-8105, USA
maia@math.ufl.edu

T. Nguyen-Huu
Laboratoire de Microbiologie
Géochimie et Ecologie Marines
UMR 6117, OSU, Case 901
Campus de Luminy
13288 Marseille Cedex 9, France
nguyen@com.univ-mrs.fr

J.-C. Poggiale
Laboratoire de Microbiologie
Géochimie et Ecologie Marines
UMR 6117, OSU, Case 901
Campus de Luminy
13288 Marseille Cedex 9, France
Jean-Christophe.Poggiale@com.
univ-mrs.fr

E. Sánchez
Departamento de Matemáticas
E.T.S.I. Industriales, U.P.M.
c/ José Gutiérrez Abascal
2, 28006 Madrid, Spain
esanchez@etsii.upm.es

H.L. Smith
Department of Mathematics and
Statistics, Arizona State University
Tempe, AZ 85287-1804, USA
halsmith@asu.edu

H.R. Thieme
Department of Mathematics and
Statistics, Arizona State University
P.O. Box 871804, Tempe
AZ 85287-1804, USA
thieme@math.asu.edu

G.F. Webb
Department of Mathematics
Vanderbilt University
Nashville, TN 37240, USA
glenn.f.webb@vanderbilt.edu

J. Wu
Laboratory for Industrial
and Applied Mathematics
Department of Mathematics
and Statistics, York University
Toronto, ON, Canada M3J 1P3
wujh@mathstat.yorku.ca

1

Population Models Structured by Age, Size, and Spatial Position

G.F. Webb

Department of Mathematics, Vanderbilt University, Nashville, TN 37240, USA
glenn.f.webb@vanderbilt.edu

Summary. Population models incorporating age, size, and spatial structure are analyzed. The methods use the theory of semigroups of linear and nonlinear operators in Banach spaces. An illustration is given of a model of tumor growth, with cell age corresponding to the cell cycle, cell size corresponding to cell growth and division, and cell position in space corresponding to cell motility and migration.

> *I know that in the study of material things number, order, and position are the threefold clue to exact knowledge: and that these three, in the mathematician's hands furnish the first outlines for a sketch of the Universe.*
> D'Arcy Thompson, Growth and Form (1917)

1.1 Introduction

Mathematical models of populations incorporating age structure, or other structuring of individuals with continuously varying properties, have an extensive history. The earliest models of age structured populations, due to Sharpe and Lotka in 1911 [104] and McKendrick in 1926 [92] established a foundation for a partial differential equations approach to modeling continuum age structure in an evolving population. At this early stage of development, the stabilization of age structure in models with linear mortality and fertility processes was recognized, although not rigorously established [85], [86]. Rigorous analysis of these linear models was accomplished later in 1941 by Feller [51], in 1963 by Bellman and Cooke [20], and others, using the methods of Volterra integral equations and Laplace transforms. Many applications of this theory have been developed in demography: Coale [34], Inaba [75], Keyfitz [77], Pollard [97], biology: Arino [12], Ayati [14], Bell and Anderson [19], Cushing [36], Gyllenberg [66], Von Foerster [117], and epidemiology: Busenberg and Cooke [26], Castillo-Chavez and Feng [30], Feng, Huang,

Castillo-Chavez [52], Feng, Li, Milner [53], Hoppensteadt [69], Kermack and McKendrick [76], to name only a few.

A new impetus of research in age structured models arose with the pioneering work of Gurtin and MacCamy in 1974 [64] for nonlinear age structured models. Their technology, which utilized a nonlinear Volterra integtral equations approach, established the existence, uniqueness, and convergence to equilibrium of solutions to nonlinear versions of the Sharpe–Lotka–McKendrick model. A rapid expansion of research in nonlinear models ensued in both theoretical developments and biological applications. A comprehensive treatment of this approach is given by Iannelli [73]. The increasingly complex mathematical issues involved in nonlinearities in age structured models led to the development of new technologies, and one of the most useful of these has been the method of semigroups of linear and nonlinear operators in Banach spaces. This functional analytic approach was developed by many researchers, including [17, 24, 33, 42, 44, 61, 62, 74, 88, 89, 98–100, 109–115, 122].

In the semigroup approach, an evolving age structured population is viewed as a dynamical system in a state space such as $Y = L^1((0, a_1); R)$, where $a_1 \leq \infty$ is the maximum age of individuals. The initial state at time $t = 0$ is a given age distribution $\phi(a), a \in (0, a_1)$, where $\phi \in Y$. The age distribution at a later time $t > 0$ is given by $(S(t)\phi)(a)$, where $S(t), t \geq 0$ is a linear or nonlinear semigroup of operators in Y. The function $p(a, t) = (S(t)\phi)(a)$ is viewed as the age density of the population at time t, in the sense that the total population at any time t in a specific age range $(\tilde{a}, \hat{a}) \subset (0, a_1)$ is

$$\int_{\tilde{a}}^{\hat{a}} p(a, t) da.$$

If the initial state ϕ is sufficiently smooth, then $p(a, t)$ satisfies the linear partial differential equation model (M.I.1):

$$\frac{\partial}{\partial t} p(a, t) + \underbrace{\frac{\partial}{\partial a} p(a, t)}_{aging} = \underbrace{-\mu(a) p(a, t)}_{mortality}, a \in (0, a_1), t > 0$$

$$p(0, t) = \underbrace{\int_0^{a_1} \beta(a) p(a, t) da}_{birth\ rate\ at\ time\ t}, t > 0$$

$$p(a, 0) = \phi(a), a \in (0, a_1), \phi \in Y$$

The mortality process is controlled by the age-dependent *mortality modulus* $\mu(a)$. The reproductive process is controlled by the age dependent *fertility modulus* $\beta(a)$. If the initial state $\phi \in Y$ is not sufficiently regular, then the formula $p(a, t) = (S(t)\phi)(a)$ is viewed as a generalized solution of (M.I.1). The advantage of the semigroup approach is that it enables description of the population processes as a dynamical system in the state space Y. Nonlinear versions of (M.I.1), as first investigated in [64], allow the mortality and fertility moduli to depend on the density $p(a,t)$ or a functional of the density, such as the total population $\int_0^{a_1} p(a,t)da$ at time t [43, 46, 122, 123, 127].

Size structured models have also been developed separately or in combination with age structured models [36, 44, 67, 93, 106, 116]. In these models size is viewed as a continuum variable s specific to individuals, such as mass, volume, length, maturity, bacterial or viral load, or other physiologic or demographic property. It is assumed that size increases in the same way for all individuals in the population, as controlled by a *growth modulus* $g(s)$. The interpretation of the growth modulus is that

$$\int_{\tilde{s}}^{\hat{s}} \frac{1}{g(s)} ds$$

is the time required for an individual to increase size from $\tilde{s}$ to $\hat{s}$, where $s_0 \leq \tilde{s} < \hat{s} \leq s_1$, with $s_0 \geq 0$ as the minimum size and $s_1 \leq \infty$ as the maximum size of individuals. If size structure is added to the age structured model (M.I.1), then the state space is $Y = L^1((0,a_1) \times (s_0,s_1); R)$, and the partial differential equation model for a nonlinear age-size structured population is (M.I.2):

$$\underbrace{\frac{\partial}{\partial t}p(a,s,t)}_{} + \underbrace{\frac{\partial}{\partial a}p(a,s,t)}_{aging} + \underbrace{\frac{\partial}{\partial s}(g(s)p(a,s,t))}_{growth}$$

$$= \underbrace{-\mu(a,s,p(a,s,t))\,p(a,s,t)}_{mortality}, \ a \in (0,a_1), s \in (s_0,s_1), t > 0$$

$$p(0,s,t) = \underbrace{\int_0^{a_1} \int_{s_0}^{s_1} \beta(a,\hat{s},s)p(a,\hat{s},t)d\hat{s}da}_{birth \ rate \ at \ time \ t}, s \in (s_0,s_1), t > 0$$

$$p(a,s,0) = \phi(a,s), a \in (0,a_1), s \in (s_0,s_1), \phi \in Y$$

where the total population at any time t in a specific age range $(\tilde{a},\hat{a}) \subset (0,a_1)$ and a specific size range $(\tilde{s},\hat{s}) \subset (s_0,s_1)$ is

$$\int_{\tilde{a}}^{\hat{a}} \int_{\tilde{s}}^{\hat{s}} p(a,s,t)dsda.$$

Many versions of the age-size structured model (M.I.2), both linear and nonlinear, have been investigated, and seminal treatments of such models are given by Metz and Diekmann [93] and Tucker and Zimmerman [116].

Our objective here is to extend age and size models to models incorporating spatial structure. Spatial structure in linear and nonlinear age or size structured models has also been investigated by many researchers, including [3–6, 13–15, 25, 31, 32, 37, 39–41, 54, 55, 58, 65, 70–72, 78–82, 87, 90, 101, 102, 108,

120, 121]. In these models individuals occupy position in a spatial environment $\Omega \subset R^n$, and spatial movement is typically controlled by diffusion or taxis processes. An example of a nonlinear model with age-size-spatial structure is model (M.I.3):

$$\frac{\partial}{\partial t}p(x,a,s,t) + \underbrace{\frac{\partial}{\partial a}p(x,a,s,t)}_{aging} + \underbrace{\frac{\partial}{\partial s}(g(s)p(x,a,s,t))}_{growth}$$

$$= \underbrace{\alpha(x)\triangle p(x,a,s,t)}_{diffusion} - \underbrace{\nabla \cdot (\chi(x)p(x,a,s,t))}_{taxis}$$

$$\underbrace{- \mu(x,a,s,p(x,a,s,t))}_{mortality}, x \in \Omega, a \in (0,a_1), s \in (s_0,s_1), t > 0$$

$$p(x,0,s,t) = \underbrace{\int_0^{a_1}\int_{s_0}^{s_1} \beta(x,a,\hat{s},s)p(x,a,\hat{s},t)d\hat{s}da}_{birth\ rate\ at\ time\ t}, x \in \Omega, s \in (s_0,s_1), t > 0$$

$$p(x,a,s,0) = \phi(x,a,x), x \in \Omega, a \in (0,a_1), s \in (s_0,s_1), \phi \in Y$$

where the total population at any time t in an age range $(\tilde{a},\hat{a}) \subset (0,a_1)$, a size range $(\tilde{s},\hat{s}) \subset (s_0,s_1)$, and a subset $\Omega' \subset \Omega$ is

$$\int_{\tilde{a}}^{\hat{a}}\int_{\tilde{s}}^{\hat{s}}\int_{\Omega'} p(x,a,s,t)\,dx\,ds\,da.$$

In model (M.I.3) the state space is $Y = L^1((0,a_1) \times (s_0,s_1);Z)$, where Z is a Banach space of functions defined on Ω. The mortality modulus μ in (M.I.3) is a function of the density $p(x,a,s,t)$, which allows the influence of crowding and the limitation of resources.

Our objective is to develop a semigroup approach to investigate linear and nonlinear versions of model (M.I.3). We consider first in Sect. 1.1 models with size and spatial structure, then in Sect. 1.2 models with age and spatial structure, and last in Sect. 1.3 models with age, size, and spatial structure. These models are mathematically complex, and their scientific applicability depends on extensive parametric input. Nevertheless, mathematical description of many biological processes requires elaborate mathematical formulation and intensive scientific validation. We discuss one such application in a model of tumor growth in Sect. 1.3. The growth of tumors in spatial environments is extremely complex, and there is an extensive literature of models, including [1, 2, 7, 9, 10, 18, 22, 27–29, 35, 38, 47, 48, 56, 57, 60, 83, 84, 95, 105, 107, 126]. A tumor grows in a characteristic way: by cell division as two daughter cells replace a dividing mother cell (a cell age dependent process), by individual cell growth and volume displacement (a cell size dependent process), and by occupation of available position in the tissue environment (a spatial dependent process). These processes, coupled to interaction with constituents in the spatial environment, transend molecular, cellular, and tissue scales, and are all important in the understanding of tumor pathology and therapy. Mathematical models such as (M.I.3), which incorporate age, size, and spatial structure, have the potential to advance this understanding.

1.2 Population Models Structured by Size and Spatial Position

Individuals are distinguished by a structure variable $s \in (s_0, s_1)$ corresponding to size, and a spatial position variable $x \in \Omega \subset R^n$. We seek a density function $p(x, s, t)$ that describes the distribution of population at time t with respect to size s and position x. The total population of individuals with size between $\hat{s}$ and $\tilde{s}$, $0 \le s_0 \le \hat{s} < \tilde{s} \le s_1 \le \infty$ and position $x \in \hat{\Omega} \subset \Omega$ at time t is

$$\int_{\hat{s}}^{\tilde{s}} \int_{\hat{\Omega}} p(x, s, t) dx ds.$$

If $s_1 < \infty$, individuals may attain size s_1, but they are no longer tracked in the model. We assume that all individuals increase their size s over time in the same way, as governed by a growth function $g(s)$ satisfying the following hypotheses:

(H.1.1) $s_1 < \infty$, $g : [s_0, s_1] \to [0, \infty)$ is continuous on $[s_0, s_1]$, positive on (s_0, s_1), and (a) $g(s_0) > 0$ and $g(s_1) > 0$, (b) $g(s_0) = 0$ and $g(s_1) > 0$, (c) $g(s_0) > 0$ and $g(s_1) = 0$, (d) $g(s_0) = 0$ and $g(s_1) = 0$.

(H.1.2) $s_1 = \infty$, $g : [s_0, \infty) \to [0, \infty)$ is uniformly continuous and bounded on $[s_0, \infty)$, positive on (s_0, ∞), and (a) $g(s_0) > 0$, (b) $g(s_0) = 0$.

The interpretation of the growth function $g(s)$ is as follows: $\int_{\hat{s}}^{\tilde{s}} \frac{1}{g(s)} ds$ is the time required for an individual to increase size from $\hat{s}$ to $\tilde{s}$. The cases in Hypotheses (H.1.1) and (H.1.2) distinguish the following possibilities:

For Hypothesis (H.1.1a): Individuals may grow from any size $\hat{s}$ greater than the minimum size s_0 to the maximum size s_1 in finite time bounded independently of $\hat{s}$ by $\int_{s_0}^{s_1} \frac{1}{g(s)} ds$;

For Hypothesis (H.1.1b): If $\int_{s_0}^{s_1} \frac{1}{g(s)} ds < \infty$, then individuals may grow from any size $\hat{s}$ greater than the minimum size s_0 to the maximum size s_1 in finite time bounded by this integral; if $\int_{s_0}^{s_1} \frac{1}{g(s)} ds = \infty$, then the time to grow from $\hat{s}$ to the maximum size s_1 is unbounded as $\hat{s} \downarrow s_0$.

For Hypothesis (H.1.1c): If $\int_{s_0}^{s_1} \frac{1}{g(s)} ds < \infty$, then individuals may grow from any size $\hat{s}$ greater than the minimum size s_0 to the maximum size s_1 in finite time bounded by this integral; if $\int_{s_0}^{s_1} \frac{1}{g(s)} ds = \infty$, then individuals approach, but never attain the maximum size s_1, as time goes to ∞.

The interpretations of Hypotheses (H.1.1d) and (H.1.2a), (H.1.2b) are similar.

It is assumed that individuals move in Ω as governed by a semigroup of bounded linear operators in a Banach space Z of functions defined on Ω, such as $L^1(\Omega; R)$ or $C(\Omega; R)$. General treatments of semigroups of linear or nonlinear operators in Banach spaces are given in [33,49,59,91,94,96,103,122]. We define

Definition 1.1. *A family of linear (or nonlinear) operators* $T(t), t \geq 0$ *in the Banach space* Z *is a strongly continuous semigroup iff* $T(0) = I$ *and (i)* $T(t)T(s)z = T(t+s)z$ *for all* $z \in Z$ *and* $t, s \geq 0$, *and (ii)* $T(t)z$ *is continuous in* t *for each* $z \in Z$. *The infinitesimal generator of* $T(t), t \geq 0$ *is the linear (or nonlinear) operator* A *defined as* $Az = \lim_{t \to +0} \frac{T(t)z - z}{t}$, *with domain* $D(A)$ *all* $z \in Z$ *for which this limit exists.*

If the operators $T(t), t \geq 0$ are bounded linear operators in Z, then $T(t), t \geq 0$ is a linear semigroup in Z. If the operators $T(t), t \geq 0$ are nonlinear operators in Z, then $T(t), t \geq 0$ is a nonlinear semigroup in Z. In a typical application to our problem, A is a diffusion operator or a chemotaxis operator or a combination of the two. We assume the following hypothesis:

(H.1.3) $T(t), t \geq 0$ is a strongly continuous semigroup of bounded linear operators in Z with infinitesimal generator A.

If $T(t), t \geq 0$ is a strongly continuous semigroup of bounded linear operators in Z, it is known [96] that there exist constants $M \geq 1$ and $\omega \in R$ such that

$$|T(t)| \leq Me^{\omega t}, \ t \geq 0. \tag{1.1}$$

The population density $p(x, s, t)$ may be viewed as a Z-valued function evaluated at $s \in (s_0, s_1)$: $p(x, s, t) = p(t)(s)(x)$, $x \in \Omega$, where $p(t) \in Y = L^1((s_0, s_1); Z)$ for $t \geq 0$. We first consider the case without mortality. If

$$\lim_{s \to s_0^+} \int_{s_0}^{s} \frac{1}{g(s)} ds = 0, \tag{1.2}$$

then Model (M.1.1) consists of the equations

$$\frac{\partial}{\partial t} p(s, t) + \frac{\partial}{\partial s}(g(s)\, p(s, t)) = Ap(s, t), s_0 < s < s_1, t > 0 \tag{1.3}$$

$$p(s, 0) = \phi(s), s_0 < s < s_1, \ \phi \in Y \tag{1.4}$$

$$p(s_0, t) = 0, t > 0 \tag{1.5}$$

If

$$\lim_{s \to s_0^+} \int_{s_0}^{s} \frac{1}{g(s)} ds = \infty, \tag{1.6}$$

then Model (M.1.2) consists only of the (1.3) and (1.4) (there is no boundary condition (1.5) in this case).

We will obtain generalized solutions of these model equations in the Banach space $Y = L^1((s_0, s_1); Z)$ by constructing a strongly continuous semigroup of bounded linear operators $S(t), t \geq 0$ in Y, where Y is the Banach space of Bochner integrable Z-valued functions on (s_0, s_1), as in [68], Sect. 3.7. The idea of the construction uses the method of characteristics to find curves in the (s, t) coordinated system such that $p(s, t)$ satisfies ordinary differential equations along these curves ([23], Sect. 3.4). Define

$$\sigma(v, s) = \int_s^v \frac{1}{g(\hat{v})} d\hat{v}, s, v \in (s_0, s_1) \tag{1.7}$$

and at the end-points s_0, s_1 if the integrals exist there. Then, for fixed $s \in (s_0, s_1), \sigma^{-1}(u, s)$ exists, under Hypothesis (H.1.1) or (H.1.2). Also, for fixed $t \geq 0$, define $\tau(v, t) = v - t, v \geq t$, and $\sigma^{-1}(u, s)$ and $\tau^{-1}(u, t) = u + t$ satisfy the initial value problems

$$\frac{d}{du} \sigma^{-1}(u, s) = g(\sigma^{-1}(u, s)), \ \sigma^{-1}(0, s) = s, s \in (s_0, s_1) \tag{1.8}$$

and

$$\frac{d}{du} \tau^{-1}(u, t) = 1, \ \tau^{-1}(0, t) = t, t \geq 0 \tag{1.9}$$

Example 1.1. Let $s_0 = 1, s_1 = 3$, and let $g(s) = \sqrt{s-1}\,(3-s), 1 \leq s \leq 3$. For this $g(s)$ Hypothesis (H.1.1d) holds. Observe that

$$\int_1^s \frac{1}{g(\hat{s})} d\hat{s} < \infty, 1 < s < 3, \ \int_s^3 \frac{1}{g(\hat{s})} d\hat{s} = \infty, 1 < s < 3.$$

For $v, s \in [1, 3)$,

$$\sigma(v, s) = \sqrt{2}\, ArcTanh \left[\frac{\sqrt{v-1}}{\sqrt{2}} \right] - \sqrt{2}\, ArcTanh \left[\frac{\sqrt{s-1}}{\sqrt{2}} \right],$$

and for $s \in [1, 3), \sigma(1, s) \leq u < \infty$,

$$\sigma^{-1}(u, s) = 1 + 2 \left(Tanh \left[\frac{1}{2} \left(\sqrt{2}u + 2ArcTanh \left[\frac{\sqrt{s-1}}{\sqrt{2}} \right] \right) \right] \right)^2$$

The functions $\sigma(v, s)$ and $\sigma^{-1}(u, s)$ in Example 1.1 are illustrated in Fig. 1.1.

Example 1.2. Let $s_0 = 0, s_1 = 1$, and let $g(s) = s, 0 \leq s \leq 1$. For this $g(s)$ Hypothesis (H.1.1b) holds. Observe that

$$\int_0^s \frac{1}{g(\hat{s})} d\hat{s} = \infty, 0 < s \leq 1, \ \int_s^1 \frac{1}{g(\hat{s})} d\hat{s} < \infty, 0 < s \leq 1.$$

For $v, s \in (0, 1), \sigma(v, s) = log(\frac{v}{s})$, and for $s \in (0, 1), -\infty \leq u < \sigma(1, s) = log(\frac{1}{s}), \sigma^{-1}(u, s) = s\, e^u$. The functions $\sigma(v, s)$ and $\sigma^{-1}(u, s)$ in Example 1.2 are illustrated in Fig. 1.2.

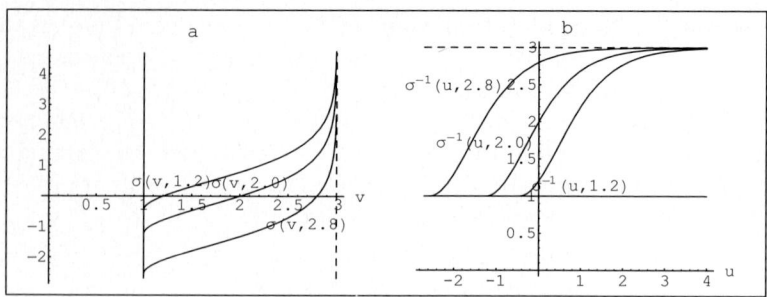

Fig. 1.1. For Example 1.1: (**a**) The graphs of $\sigma(v,s)$ for various values of $s \in (1,3)$. (**b**) The graphs of $\sigma^{-1}(u,s)$ for various values of $s \in (1,3)$

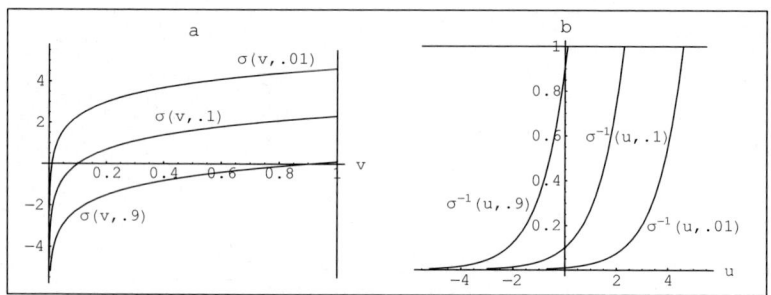

Fig. 1.2. For Example 1.2: (**a**) The graphs of $\sigma(v,s) = log(\frac{v}{s})$ for various values of $s \in (0,1)$. (**b**) The graphs of $\sigma^{-1}(u,s) = s\,e^u$ for various values of $s \in (0,1)$

Fix $s \in (s_0, s_1)$ and $t > 0$ and define $w(u) = p(\sigma^{-1}(u,s), \tau^{-1}(u,t))$. If $p(s,t)$ satisfies (1.3) and $g(s)$ is differentiable, then $w(u)$ satisfies

$$\frac{d}{du}w(u) = \frac{d}{du}p(\sigma^{-1}(u,s), \tau^{-1}(u,t)) \tag{1.10}$$

$$= \sigma^{-1\,\prime}(u,s)\frac{\partial}{\partial s}p(\sigma^{-1}(u,s), \tau^{-1}(u,t)) + \frac{\partial}{\partial t}p(\sigma^{-1}(u,s), \tau^{-1}(u,t))$$

$$= -g'(\sigma^{-1}(u,s))p(\sigma^{-1}(u,s), \tau^{-1}(u,t)) + Ap(\sigma^{-1}(u,s), \tau^{-1}(u,t))$$

$$= -g'(\sigma^{-1}(u,s))w(u) + Aw(u)$$

Define $\hat{w}(u) = w(u - t)$, and (1.10) yields

$$\frac{d}{du}\hat{w}(u) = w'(u - t) \tag{1.11}$$

$$= -g'(\sigma^{-1}(u - t, s))w(u - t) + Aw(u - t)$$

$$= -g'(\sigma^{-1}(u - t, s))\hat{w}(u) + A\hat{w}(u)$$

Integration of (1.11) from 0 to u yields

$$\hat{w}(u) = exp\left[-\int_0^u g'(\sigma^{-1}(\hat{u} - t, s))d\hat{u}\right]T(u)\hat{w}(0) \qquad (1.12)$$

$$= exp\left[-\int_{-t}^{u-t} g'(\sigma^{-1}(\tilde{u}, s))d\tilde{u}\right]T(u)\hat{w}(0)$$

Change the variable of integration in (1.12) to $v = \sigma^{-1}(\tilde{u}, s), dv/d\tilde{u} = g(v)$ to obtain

$$\hat{w}(u) = exp\left[-\int_{\sigma^{-1}(-t,s)}^{\sigma^{-1}(u-t,s)} \frac{g'(v)}{g(v)}dv\right]T(u)\hat{w}(0) \qquad (1.13)$$

$$= \frac{g(\sigma^{-1}(-t, s))}{g(\sigma^{-1}(u - t, s))}T(u)\hat{w}(0)$$

If $\sigma^{-1}(-t, s) > s_0 \Leftrightarrow t < \sigma(s, s_0) \Leftrightarrow \sigma^{-1}(t, s_0) < s$, set $u = t$ in (1.13) to obtain

$$p(s, t) = \hat{w}(t) = w(0) = \frac{g(\sigma^{-1}(-t, s))}{g(s)}T(t)w(-t) \qquad (1.14)$$

$$= \frac{g(\sigma^{-1}(-t, s))}{g(s)}T(t)p(\sigma^{-1}(-t, s), 0)$$

$$= \frac{g(\sigma^{-1}(-t, s))}{g(s)}T(t)\phi(\sigma^{-1}(-t, s)).$$

If $\sigma^{-1}(-t, s) < s_0 \Leftrightarrow t > \sigma(s, s_0) \Leftrightarrow \sigma^{-1}(t, s_0) > s$, set $p(s, t) = 0$. Note that if (1.6) holds, then $\sigma^{-1}(-t, s) > s_0 \Leftrightarrow \sigma^{-1}(t, s_0) < s$ for all $s \in (s_0, s_1), t > 0$.

For Example 1.1 the characteristic curves in the (s, t) coordinate system are illustrated in Fig. 1.3 and for Example 1.2 the characteristic curves in the (s, t) coordinate system are illustrated in Fig. 1.4.

From (1.14), we obtain the following formula for the generalized solutions of the size and spatial structured models (M.1.1) and (M.1.2): for $\phi \in Y, t \geq 0$,

$$p(s, t) = \begin{cases} \frac{g(\sigma^{-1}(-t,s))}{g(s)}T(t)\phi(\sigma^{-1}(-t, s)), & \sigma^{-1}(t, s_0) < s < s_1 \\ 0, & s_0 \leq s < \sigma^{-1}(t, s_0) \end{cases} \qquad (1.15)$$

We define a semigroup of linear operators $S(t), t \geq 0$ in Y by $(S(t)\phi)(s) = p(s, t), \phi \in Y, t \geq 0, s_0 < s < s_1$ where $p(s, t)$ is given by the formula in (1.15).

Theorem 1. *Let (H.1.1) or (H.1.2) and (H.1.3) hold. Let $\phi \in Y, t \geq 0$. If (1.2) holds, define for almost all s in (s_0, s_1)*

$$(S(t)\phi)(s) = \begin{cases} \frac{g(\sigma^{-1}(-t,s))}{g(s)}T(t)\phi(\sigma^{-1}(-t, s)), & \sigma^{-1}(t, s_0) < s < s_1 \\ 0, & s_0 \leq s < \sigma^{-1}(t, s_0) \end{cases} \qquad (1.16)$$

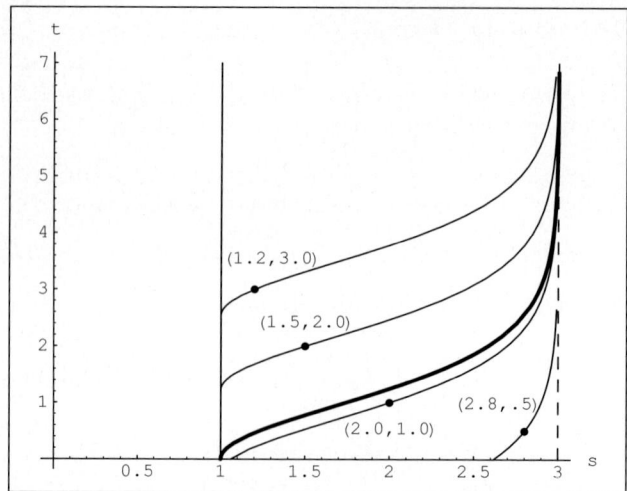

Fig. 1.3. The characteristic curves in the (s, t) coordinate system for Example 1.1 are $(\sigma^{-1}(u, s), u + t)$, where u is a parameter such that $-\sigma(s, s_0) = \sigma(s_0, s) \leq u < \sigma(s_1, s)$ if $\sigma(s, s_0) \leq t$, and $-t \leq u \leq \sigma(s_1, s)$ if $\sigma(s, s_0) > t$. The *dark curve* is $t = \sigma(s, s_0)$

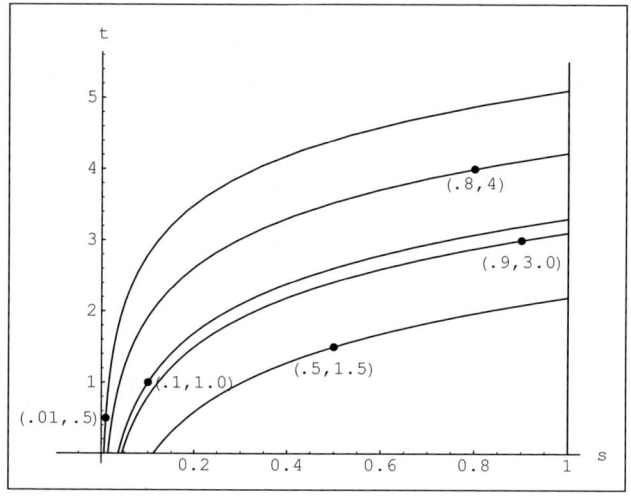

Fig. 1.4. The characteristic curves in the (s, t) coordinate system for Example 1.2 are $(\sigma^{-1}(u, s), u + t) = (s\, e^u, u + t)$, where u is a parameter such that $-t \leq u < \sigma(s_1, s) = -log(s)$ for all $s \in (0, 1], t > 0$

and if (1.6) holds, define

$$(S(t)\phi)(s) = \frac{g(\sigma^{-1}(-t, s))}{g(s)} T(t)\phi(\sigma^{-1}(-t, s)), \quad s_0 < s < s_1 \qquad (1.17)$$

$S(t), t \geq 0$ *is a strongly continuous semigroup of bounded linear operators in Y satisfying $\|S(t)\phi\|_Y \leq M\,e^{\omega t}\|\phi\|_Y$, $\phi \in Y, t \geq 0$. If (1.2) holds and $\sigma(s_0, s_1) < \infty$, then $S(t) \equiv 0$ for $t \geq \sigma(s_1, s_0)$. If Z is a Banach lattice and $T(t), t \geq 0$ is a positive semigroup in Z, then $S(t), t \geq 0$ is a positive semigroup in Y.*

Proof. First consider the case that (1.2) holds. We first show that $S(t)\phi \in Y$ for $t \geq 0, \phi \in Y$. For $\phi \in Y, t \geq 0$, (1.16) yields

$$\|S(t)\phi\|_Y = \int_{s_0}^{s_1} \|(S(t)\phi)(s)\|_Z ds \tag{1.18}$$

$$= \int_{\sigma^{-1}(t,s_0)}^{s_1} \frac{g(\sigma^{-1}(t,s))}{g(s)} \|T(t)\phi(\sigma^{-1}(t,s))\|_Z ds$$

Change the variable of integration in (1.18) $\hat{s} = \sigma^{-1}(-t, s) \Leftrightarrow \sigma(\hat{s}, s) = -t \Rightarrow \int_{\hat{s}}^{s} 1/g(v)dv = t \Rightarrow ds/d\hat{s} = g(s)/g(\hat{s})$, to obtain

$$\|S(t)\phi\|_Y \leq \int_{s_0}^{\sigma^{-1}(-t,s_1)} \|(T(t)\phi)(\hat{s})\|_Z d\hat{s} \tag{1.19}$$

$$\leq M\,e^{\omega t}\|\phi\|_Y.$$

From (1.19) we see that $S(t)\phi \in Y$, i.e. $\int_{s_0}^{s_1}(S(t)\phi)(s)ds$ exists in Z ([68], Theorem 3.7.4) and $|S(t)| \leq Me^{\omega t}, t \geq 0$. If $t \geq \sigma(s_1, s_0), S(t) \equiv 0$, since $\sigma^{-1}(\sigma(s_1, s_0), s_0) = s_1$ (all individuals present initially have attained the maximum size for $t > \sigma(s_1, s_0)$). $\square$

We next show that $S(t), t \geq 0$ satisfies Definition 1.1(i). Let $\phi \in Y, 0 < t_1 < t_2$, and $s_0 < s < s_1$.

Case 1. $\sigma^{-1}(-t_2, s) > s_0 \Leftrightarrow \sigma^{-1}(t_2, s_0) < s$. Observe that $\sigma^{-1}(-t_1, s) > \sigma^{-1}(-t_2, s) > s_0$, since $\sigma^{-1}(u, s)$ is increasing in u. Also, $\sigma^{-1}(u+v, s) = \sigma^{-1}(u, \sigma^{-1}(v, s))$, since if $c_1 = \sigma^{-1}(u+v, s) \Leftrightarrow u+v = \sigma(c_1, s), c_2 = \sigma^{-1}(v, s) \Leftrightarrow v = \sigma(c_2, s)$, and $c_3 = \sigma^{-1}(u, c_2) \Leftrightarrow u = \sigma(c_3, c_2)$, then

$$u + v = \int_{c_2}^{c_3} \frac{1}{g(\hat{v})} d\hat{v} + \int_{s}^{c_2} \frac{1}{g(\hat{v})} d\hat{v} = \int_{s}^{c_1} \frac{1}{g(\hat{v})} d\hat{v}$$

implies $c_1 = c_3$. Thus,

$$(S(t_1)S(t_2))\phi(s) = S(t_1) \frac{g(\sigma^{-1}(-t_2, s))}{g(s)} T(t_2)\phi(\sigma^{-1}(-t_2, s)) \tag{1.20}$$

$$= \frac{g(\sigma^{-1}(-t_2, s))}{g(s)} \frac{g(\sigma^{-1}(-t_1, \sigma^{-1}(-t_2, s)))}{g(\sigma^{-1}(-t_2, s))}$$

$$\times T(t_1)T(t_2)\phi(\sigma^{-1}(-t_1, \sigma^{-1}(-t_2, s)))$$

$$= \frac{g(\sigma^{-1}(-(t_1+t_2), s))}{g(s)} T(t_1 + t_2)\phi(\sigma^{-1}(-(t_1+t_2), s))$$

$$= (S(t_1 + t_2)\phi)(s)$$

Case 2. $\sigma^{-1}(-t_2, s) \le s_0 \Leftrightarrow \sigma^{-1}(t_2, s_0) \ge s$. A calculation similar to (1.20) shows that

$$(S(t_1)S(t_2)\phi)(s) = (S(t_1 + t_2)\phi)(s) \equiv 0.$$

We next show that $S(t), t \ge 0$ satisfies Definition 1.1(ii). Let Y_0 be a dense subset of Y consisting of continuous functions from $[s_0, s_1]$ to Z (if $s_1 = \infty$, then let these functions also have compact support in $[s_0, \infty)$) ([68], Sect. 3.7). Let $\phi \in Y_0$ and $t \ge 0$. From (1.16)

$$\|S(t)\phi - \phi\|_Y = \int_{s_0}^{s_1} \|(S(t)\phi)(s) - \phi(s)\|_Z ds \tag{1.21}$$

$$\le \int_{\sigma^{-1}(t, s_0)}^{s_1} \| \frac{g(\sigma^{-1}(-t, s))}{g(s)} \left(T(t)\phi(\sigma^{-1}(-t, s)) - \phi(\sigma^{-1}(-t, s)) \right) \|_Z ds$$

$$+ \int_{\sigma^{-1}(t, s_0)}^{s_1} \| \left(\frac{g(\sigma^{-1}(-t, s))}{g(s)} - 1 \right) \phi(\sigma^{-1}(-t, s)) \|_Z ds$$

$$+ \int_{\sigma^{-1}(t, s_0)}^{s_1} \| \phi(\sigma^{-1}(-t, s)) - \phi(s) \|_Z ds$$

$$+ \int_{s_0}^{\sigma^{-1}(t, s_0))} \| \phi(s) \|_Z ds$$

$$= I + II + III + IV$$

As above, change the integration variable $\hat{s} = \sigma^{-1}(-t, s) \Leftrightarrow s = \sigma^{-1}(t, \hat{s})$ in I and II to obtain

$$I = \int_{s_0}^{\sigma^{-1}(-t, s_1)} \frac{g(\hat{s})}{g(\sigma^{-1}(t, \hat{s}))} \|T(t)\phi(\hat{s}) - \phi(\hat{s})\|_Z d\hat{s},$$

$$II = \int_{s_0}^{\sigma^{-1}(-t, s_1)} \left| \frac{g(\hat{s})}{g(\sigma^{-1}(t, \hat{s}))} - 1 \right| \|\phi(\hat{s})\|_Z d\hat{s}$$

Then $I, II, III, IV \to 0$ as $t \to 0$, since $\sigma^{-1}(-t, \hat{s}) \to \hat{s}$ as $t \to 0$ uniformly for $\hat{s} \in supp(\phi) \subset (s_0, s_1)$, $T(t), t \ge 0$ is uniformly strongly continuous on compact subsets of Z, ϕ is continuous, and the range of ϕ on $supp(\phi)$ has compact closure in Z. Thus, $S(t), t \ge 0$ is strongly continuous in Y_0, and since (1.19) holds and Y_0 is dense in Y, $S(t), t \ge 0$ is strongly continuous in Y.

The positivity of $S(t), t \ge 0$ in Y follows immediately from (1.16), if $T(t), t \ge 0$ is a positive semigroup in the Banach lattice Z [11, 33, 49]. The case that (1.6) holds is similar. ∎

We give an example with size structure, but without spatial structure to illustrate Theorem 1.

Example 1.3. Let s_0, s_1, and $g(s)$ be defined as in Example 1.1. Let $Z = R$, let $Y = L^1((1, 3); R)$, let $T(t) = I, t \ge 0$, and let $\phi \in Y$ be defined as

$$\phi(s) = \begin{cases} 10(s - 1.1)^2(1.4 - s)^2 & 1.1 \le s \le 1.4 \\ (s - 2.0)^2(2.5 - s)^2 & 2.0 \le s \le 2.5 \\ 0 & \text{otherwise} \end{cases} \tag{1.22}$$

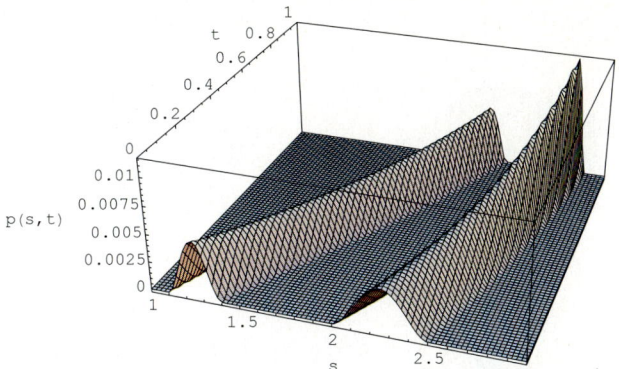

Fig. 1.5. The solution $p(s,t) = (S(t)\phi)(s)$ of the size structured model (without spatial structure) in Example 1.3 for the initial value $\phi(s)$ in (1.22). The largest individuals grow the slowest. The two peaks ultimately merge, and all individuals approach, but never attain, the maximum size 3. Since no individuals are lost or gained, the total population is conserved for time $t \geq 0$

The generalized solution $p(s,t) = (S(t)\phi)(s)$ obtained from (1.16) in Theorem 1 is graphed in Fig. 1.5.

An illustration of Theorem 1 with size structure and spatial structure is given in Example 1.4.

Example 1.4. Let s_0, s_1, and $g(s)$ be defined as in Example 1.3, and let ϕ be defined as in (1.22). Let $Z = L^1((x_0, x_1); R)$ with $x_0 = 0$ and $x_1 = 4$, let $Y = L^1((1,3); Z)$, and let $T(t), t \geq 0$, be the strongly continuous semigroup in Z with infinitesimal generator the diffusion operator A in Z with Neumann boundary conditions:

$$Az(x) = \alpha \frac{d^2 z}{dx^2}(x), \ x_0 < x < x_1, \tag{1.23}$$

$$D(A) = \left\{ z \in Z : \frac{d^2 z}{dx^2} \in Z, \frac{dz}{dx}(x_0) = \frac{dz}{dx}(x_1) = 0 \right\}$$

where $\alpha = .1$. For this semigroup the constants in (1.1) are $M = 1$ and $\omega = 0$. The initial value is $\Phi(s)(x) = \phi(s)\psi(x)$, where $\phi(s)$ is given by (1.22) and

$$\psi(x) = \begin{cases} 40000(x - 1.5)^2(2.5 - x)^2 & 1.5 \leq x \leq 2.5 \\ 0 & \text{otherwise} \end{cases} \tag{1.24}$$

The generalized solution $p(x, s, t) = (S(t)\Phi)(s)(x)$ obtained from (1.16) in Theorem 1 is graphed in Figs. 1.6 and 1.7.

The size structured models (M.1.1) and (M.1.2) can be extended to allow mortality, or more generally, linear or nonlinear gain or loss of individuals, by perturbation of the semigroup $S(t), t \geq 0$ in Theorem (1). The model equations have the following form (M.1.3):

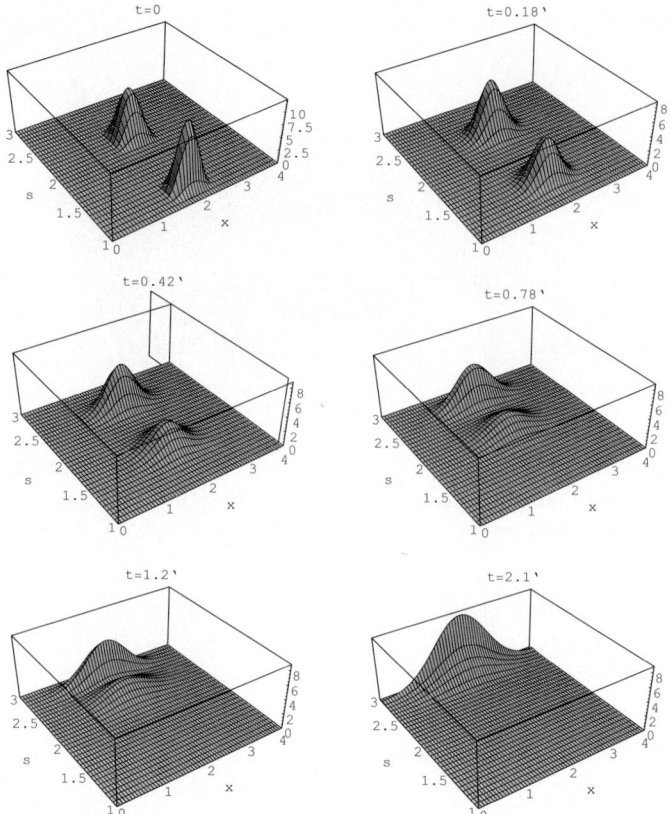

Fig. 1.6. The densities $p(x, s, t) = (S(t)\Phi)(s)(x)$ of the size and spatial structured model in Example 1.4 with initial value $\Phi(s)(x)$ for various times t. The largest individuals grow the slowest, and all individuals converge as $t \to \infty$ to the maximum size 3. The two initial concentrations are spread out in the spatial variable x by the diffusion process, and merged in the size variable as $s \to 3$ by the singularity of the grow rate $1/g(s)$ at $s = 3$. The total population is conserved both in the size and spatial variables

$$\frac{\partial}{\partial t} p(s, t) + \frac{\partial}{\partial s}(g(s)\, p(s, t)) = A p(s, t) + F(p(\cdot, t))(s), s_0 < s < s_1, t > 0 \quad (1.25)$$

$$p(s, 0) = \phi(s), s_0 < s < s_1 \quad (1.26)$$

$$p(s_0, t) = 0, t > 0. \quad (1.27)$$

Again, if (1.2) holds, then Model (M.1.3) consists of (1.25), (1.26), (1.27), and if (1.6) holds, then Model (M.1.4) consists of (1.25), (1.26) (there is no boundary condition in this case). Here F is a linear or nonlinear operator in Y satisfying the following global Lipschitz continuity hypothesis:

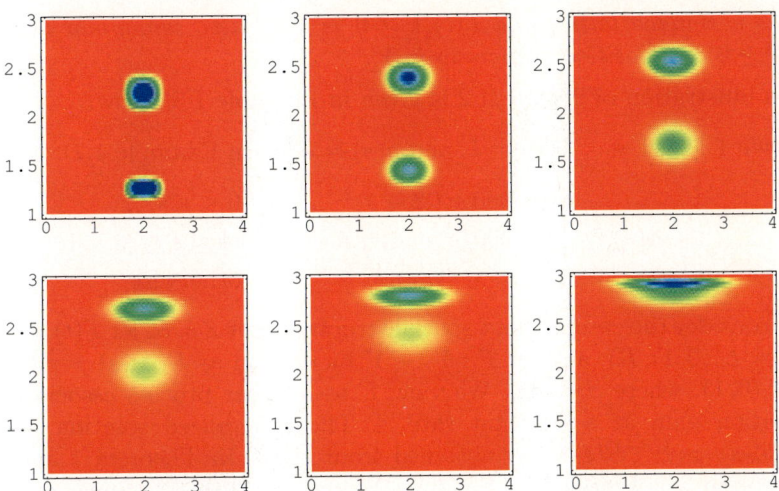

Fig. 1.7. The density plots in the (x, s) coordinate system of the distributions in Fig. 1.6

(H.1.4) $F : Y \to Y$ and there exits $L > 0$ such that $\|F(\phi_1) - F(\phi_2)\|_Y \leq L \|\phi_1 - \phi_2\|_Y$, for all $\phi_1, \phi_2 \in Y$.

Models (M.1.3) and (M.1.4) can be written abstractly as an ordinary differential equation in Y:

$$\frac{d}{dt}p(t) = \mathcal{A}p(t) + F(p(t)), t \geq 0, \ p(0) = \phi \in Y \qquad (1.28)$$

where $p : [0, \infty) \to Y$, $p(t)(s) = p(s, t)$, $\mathcal{A} : Y \to Y$,

$$(\mathcal{A}\phi)(s) = -\frac{d}{ds}(g(s)\phi(s)) + A\phi(s),$$

with $D(\mathcal{A})$ consisting of $\phi \in Y$ such that $-\frac{d}{ds}(g(s)\phi(s)) + A\phi(s) \in Y$, and, if (1.2) holds, $\phi(s_0) = 0$.

The proof of the following theorem is standard and may be found in [59, 91, 96] or [103]:

Theorem 2. *Let (H.1.1) or (H.1.2) and (H.1.3), (H.1.4) hold. Let $S(t), t \geq 0$ be the strongly continuous semigroup of bounded linear operators in $Y = L^1((s_0, s_1); Z)$ in Theorem 1 and let $\phi \in Y$. There is a unique solution $U(t)\phi, t \geq 0$ to the integral equation*

$$U(t)\phi = S(t)\phi + \int_0^t S(t - u)F(U(u)\phi)du, t \geq 0 \qquad (1.29)$$

Further, $U(t), t \geq 0$ is a strongly continuous semigroup of Lipschitz continuous linear or nonlinear operators in Y satisfying

$$\|U(t)(\phi_1) - U(t)(\phi_2)\|_Y \leq Me^{(ML+\omega)t} \|\phi_1 - \phi_2\|_Y, t \geq 0, \phi_1, \phi_2 \in Y \qquad (1.30)$$

If Z is a Banach lattice and $T(t), t \geq 0$ is a positive semigroup in Z, then $U(t), t \geq 0$ is a positive semigroup in Y.

An illustration of Theorem 2 is given in Example 1.5 below.

Example 1.5. Let $s_0 = 0$, $s_1 = 1$, and $g(s) = s$ (as in Example 1.2). Let

$$\phi(s) = \begin{cases} 10,000\, s^2(.2-s)^2 & 0 \leq s \leq .2 \\ 20,000\,(s-.4)^2(.6-s)^2 & .4 \leq s \leq .6 \\ 0 & \text{otherwise} \end{cases} \tag{1.31}$$

First, consider the case without spatial structure. Let $Z = R$, let $T(t) = I, t \geq 0$, let $Y = L^1(0,1); R)$, and let $F : Y \to Y$ be defined as $F(\phi)(s) = \beta\phi(s)$, $\phi \in Y$, $s \in (0,1)$, where $\beta = 2.0$. We view F as a linear process associated with population gain at a constant rate β. The generalized solution $p(s,t) = (U(t)\phi)(s) = e^{(\beta-1)t}\phi(e^{-t}s)$ obtained from (1.16) in Theorem 1 and (1.29) in Theorem 2 is graphed in Fig. 1.8. We remark that this linear semigroup $U(t), t \geq 0$ for this example is known to exhibit chaotic behavior for $\beta > 1$ [125]. We next consider the case with spatial structure. Let $Z = L^1((0,1); R)$, let $Y = L^1((0,1); Z)$, and let $T(t), t \geq 0$, be the strongly continuous linear semigroup in Z with infinitesimal generator the diffusion operator A in Z with Neumann boundary conditions as in (1.23), with $\alpha = .01$, $x_0 = 0$, $x_1 = 1.0$. Let the population source $F : Y = L^1((0,1); Z) \to Y$ be spatially dependent: for $\phi \in Y$, $s \in (s_0, s_1)$,

$$(F\phi(s))(x) = \begin{cases} 50(x-.5)(1.0-x)\phi(s) & 0.5 \leq x \leq 1.0 \\ 0 & 0.0 \leq x < 0.5 \end{cases} \tag{1.32}$$

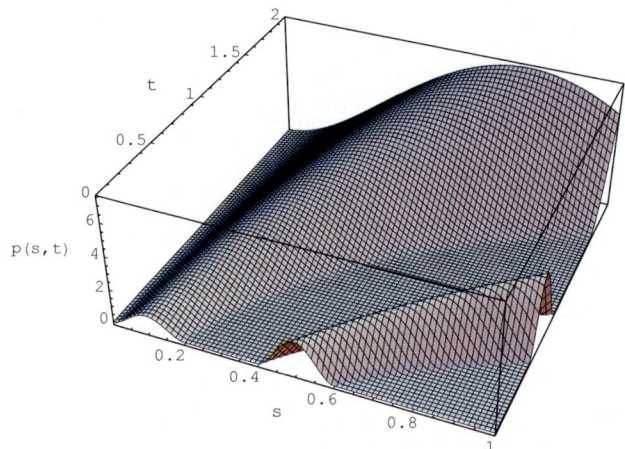

Fig. 1.8. The solution $p(s,t) = (U(t)\phi)(s)$ of the size structured model (without spatial structure) in Example 1.5 for the initial value $\phi(s)$ in (1.31). All individuals attain the maximum size 1 in finite time, and the time to reach size 1 from size s is $-log(s)$. Since there are individuals present initially with arbitrary small positive size s, the population is never extinguished, and grows without bound as $t \to \infty$

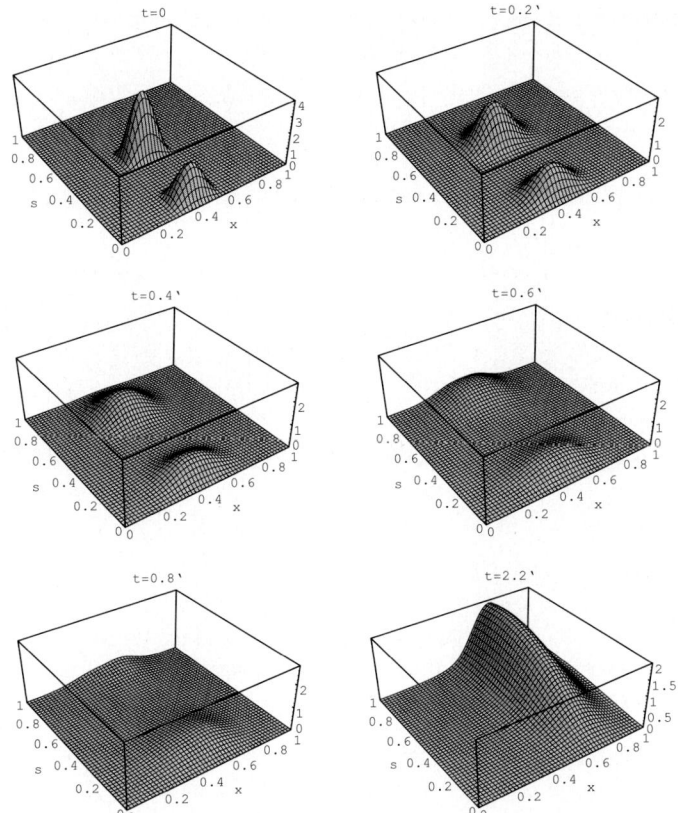

Fig. 1.9. The solution $p(s,t) = (U(t)\Phi)(s)$ of the size structured model (with spatial structure) in Example 1.5 for the initial value $\Phi(s)$ for various time values. The behavior with respect to the size distribution is similar to Fig. 1.8, with diffusion acting to spread the population in space. Since the source term $Fp(\cdot, t)(s)(x)$ is dependent on the spatial variable x through (1.32), there is a bias in the growth of the population over time in the right half of the spatial domain $(0, 1)$

Let $\Phi(s)(x) = \phi(s)\psi(x)$, where $\phi(s)$ is given by (1.31) and

$$\psi(x) = \begin{cases} 40{,}000(x - .3)^2(.6 - x)^2 & .3 \le x \le .6 \\ 0 & \text{otherwise} \end{cases} \qquad (1.33)$$

The generalized solution $p(s,t) = (U(t)\Phi)(s)$ for this case, obtained from (1.16) in Theorem 1 and (1.29) in Theorem 2, is graphed in Figs. 1.9 and 1.10.

An application of Theorem 2 to a linear model of a size structured proliferating cell population is given in Example 1.6.

Example 1.6. We first consider the case without diffusion. This case has been analyzed in [62] (see also [33, 44, 127]). Let $s_0 = 1.0, s_1 = 3.6, g(s) = 5.0 - s$, and let $Y = L^1(s_0, s_1); R)$. It is assumed that the minimum size of a mother

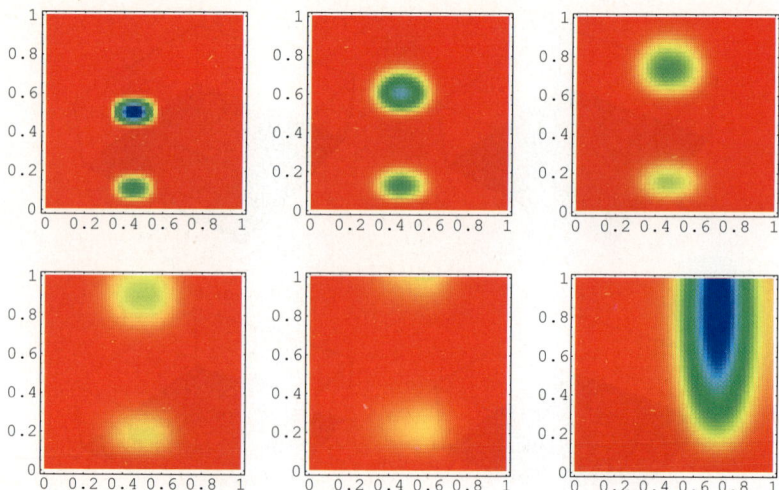

Fig. 1.10. The density plots in the (x, s) coordinate system of the distributions in Fig. 1.9

cell at division is $2s_0$ and the maximum is s_1. The condition $s_1/2 < 2s_0$ assures that every newborn cell must grow for some time before it can divide. The density $p(s,t)$ of the population with respect to size s satisfies the initial-boundary value problem

$$\frac{\partial}{\partial t}p(s,t) + \frac{\partial}{\partial s}(g(s)p(s,t)) = -(\beta(s) + \mu(s))p(s,t) + 4\beta(2s)p(2s,t) \quad (1.34)$$

$$p(s_0, t) = 0, t \geq 0 \quad (1.35)$$

$$p(s,0) = \phi(s), \phi \in Y, s_0 < s < s_1 \quad (1.36)$$

The factor 4 in (1.34) arises from the assumption that size is conserved during the division process, and each daughter cell inherits exactly one-half the size of the mother cell. The division modulus is

$$\beta(s) = \begin{cases} 5(s - 2.0)(3.6 - s) & 2.0 \leq s \leq 3.6 \\ 0 & \text{otherwise} \end{cases} \quad (1.37)$$

The mortality modulus is $\mu(s) \equiv .4$. The generalized solution of this model is obtained from Theorem 1, (1.16) with $T(t) = I, t \geq 0$, and Theorem 2 with $F : Y \to Y$ defined as $F(\phi)(s) = -(\beta(s) + \mu(s))\phi(s) + 4\beta(2s)\phi(2s), \phi \in Y, s \in (s_0, s_1)$. The generalized solution $p(s,t) = (S(t)\phi)(s)$ for

$$\phi(s) = \begin{cases} .5(s - 2.4)(2.8 - s) & 2.4 \leq s \leq 2.8 \\ 0 & \text{otherwise} \end{cases} \quad (1.38)$$

is graphed in Fig. 1.11. The model with spatial diffusion is obtained from Theorem 2 with $Z, T(t), t \geq 0, A, \alpha = .5$ as in Example 1.4, $Y = L^1$

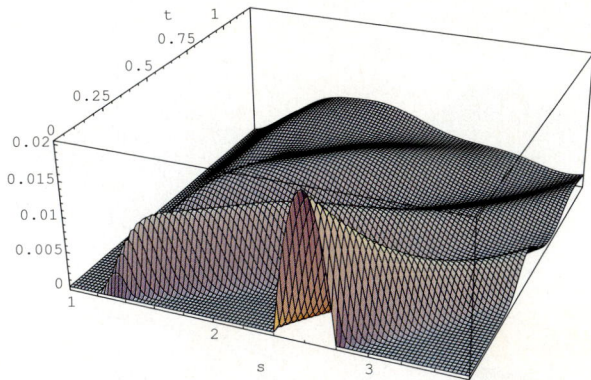

Fig. 1.11. The solution $p(s,t) = (S(t)\phi)(s)$ of the proliferating cell population model structured by cell size (without spatial structure) in Example 1.6 for the initial value $\phi(s)$ in (1.38). Since some cells present initially are large enough to divide at time $t = 0$, some newly divided daughter cells appear immediately. As the population grows exponentially in time, the initial size synchronization at time $t = 0$, with a concentration centered at 2.6, is gradually dispersed

$((s_0, s_1); Z)$, and $s_0, s_1, g, \beta, \mu, F$ as above. The solution $p(s,t) = (S(t)\Phi)(s)$, is graphed at various time points in Figs. 1.12 and 1.13 for the initial value $(\Phi(s))(x) = \phi(s)\psi(x)$, with ϕ as in (1.38) and ψ as follows:

$$\psi(x) = \begin{cases} 400(x - 1.5)(2.5 - x) & 1.5 \leq x \leq 2.5 \\ 0 & \text{otherwise} \end{cases} \tag{1.39}$$

The asymptotic behavior of the linear semigroup in Example 1.6 in the case without spatial structure has been analyzed in [44] and [62]. This behavior is known as *asynchronous* or *balanced exponential growth* [113, 114, 122, 124, 127].

Definition 1.2. *Let $T(t), t \geq 0$ be a strongly continuous semigroup of bounded linear operators in the Banach space Z with infinitesimal generator A. $T(t), t \geq 0$ has asynchronous exponential growth with intrinsic growth constant λ_0 if and only if there is a nonzero finite rank projection P_0 in Z (the spectral projection) such that*

$$\lim_{t \to \infty} e^{-\lambda_0 t} T(t) = P_0. \tag{1.40}$$

Asynchronous exponential growth means that the semigroup stabilizes as $t \to \infty$ to a finite-dimensional image of the state space of initial values, after multiplication by an exponential factor in time. If the state space Y is infinite dimensional, then any synchronization of information in the initial value is retained only in the finite dimensional space $P_0 Y$, and is thus asynchronized over time.

In [62] it is proved that if g is continuously differentiable and strictly positive on (s_0, s_1), $2g(s) \neq g(2s)$ for some $s \in (s_0, s_1)$ and $s_0 < s_1/2 < s_2$,

20   G.F. Webb

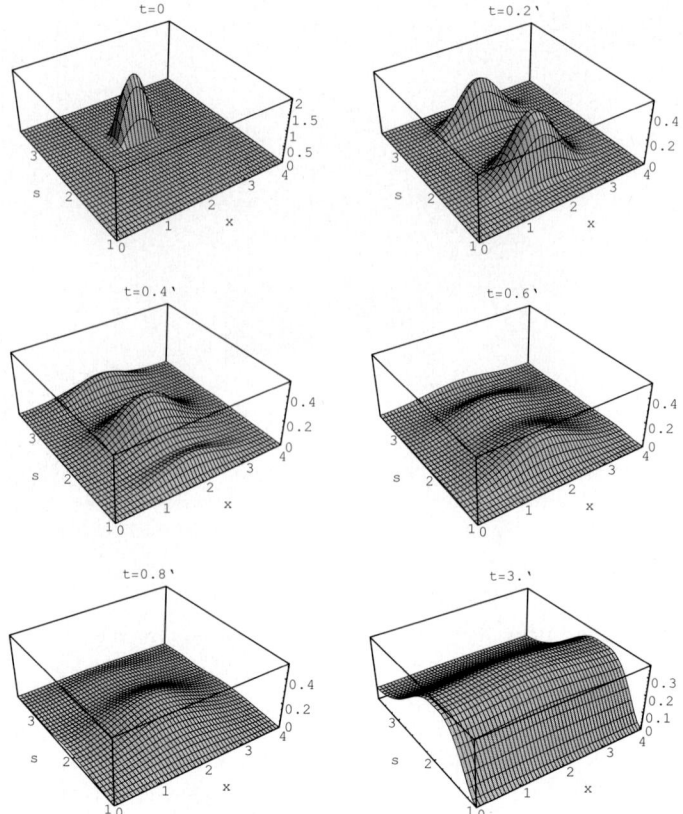

Fig. 1.12. The solution $p(s,t) = (U(t)\Phi)(s)$ of the size structured model (with spatial diffusion) in Example 1.6 for the initial value $\Phi(s)$ for various time values. In the last plot for time $t = 3.0$, the size and spatial structure have almost completely dispersed from the initial value, and the size structure is essentially uniform in x

where s_2 is the minimum age of division, then the linear semigroup $S(t), t \geq 0$ (without spatial dependence) in Example 1.6 has asynchronous exponential growth with P_0 in (1.40) having rank 1. If $2g(s) \equiv g(2s)$ for $s \in (s_0, s_1)$, then this semigroup has *periodic exponential growth* [45, 62, 127].

For the semigroup $S(t), t \geq 0$ in Theorem 1, we can prove

Theorem 3. *Let (H.1.1) or (H.1.2) and (H.1.3) hold. Let $S(t), t \geq 0$ be the linear semigroup in Theorem 1 with $T(t), t \geq 0$ as in (H.1.3). Let $S_1(t), t \geq 0$ be the linear semigroup in Theorem 1 obtained with $T(t) \equiv I$ in Z as in (H.1.3). If $S_1(t), t \geq 0$ has asynchronous exponential growth in $Y = L^1((s_0, s_1); Z)$ with intrinsic growth constant λ_1 and with spectral projection P_1, and $T(t), t \geq 0$ has asynchronous exponential growth in Z with intrinsic growth constant λ_0 and with spectral projection P_0, then $S(t), t \geq 0$ has*

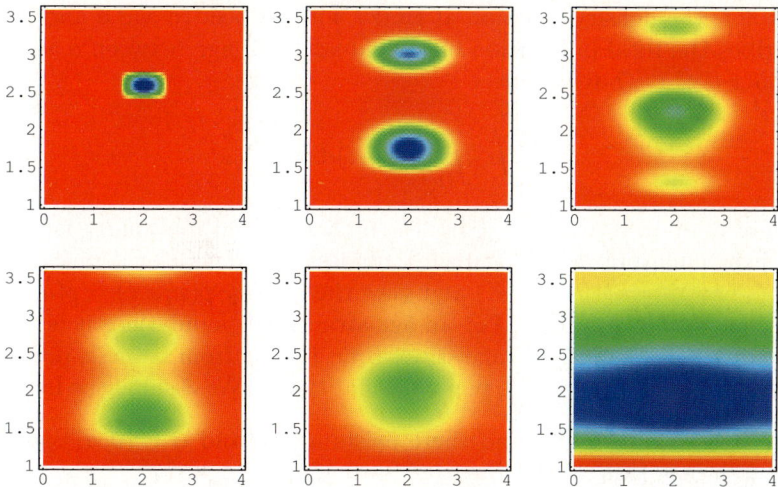

Fig. 1.13. The density plots in the (x, s) coordinate system of the distributions in Fig. 1.12

asynchronous exponential growth in Y with intrinsic growth constant $\lambda_0 + \lambda_1$ and with spectral projection $(P\phi)(s) = P_0(P_1\phi(s)), \phi \in Y, s \in (s_0, s_1)$.

Proof. From (1.16) and (1.17), $(S(t)\phi)(s) = T(t)(S_1(t)\phi(s))$, which means that

$$\lim_{t \to \infty} e^{-(\lambda_0 + \lambda_1)t} (S(t)\phi)(s) = \lim_{t \to \infty} e^{-\lambda_0 t} T(t) e^{-\lambda_1 t}(S_1(t)\phi)(s)$$
$$= P_0(P_1(\phi(s))), \ \phi \in Y, \ s \in (s_0, s_1).$$

Also, P is linear, bounded, and a projection in Y, since

$$P^2\phi = P(\lim_{t \to \infty} e^{-(\lambda_0 + \lambda_1)t} S_1(t)\phi)$$
$$= \lim_{\hat{t} \to \infty} e^{-(\lambda_0 + \lambda_1)\hat{t}} S_1(\hat{t}) \left(\lim_{t \to \infty} e^{-(\lambda_0 + \lambda_1)t} S_1(t)\phi \right)$$
$$= \lim_{t + \hat{t} \to \infty} e^{-(\lambda_0 + \lambda_1)(t + \hat{t})} S_1(t + \hat{t})\phi$$
$$= P\phi.$$

Further, P has finite rank, if P_0 and P_1 have finite rank.   □

1.3 Population Models Structured by Age and Spatial Position

Individuals are distinguished by age $a \in (0, a_1)$, where $a_1 \leq \infty$ is the maximum age (if $a_1 < \infty$, individuals may attain age greater than a_1, but they are no longer tracked in the model), and spatial position variable $x \in \Omega \subset R^n$. As in Sect. 1.1, the density function at time t with respect to age a and position x describes the total population of individuals with age between $\hat{a}$ and $\tilde{a}$ and position $x \in \hat{\Omega} \subset \Omega$ at time t:

$$\int_{\hat{a}}^{\tilde{a}} \int_{\hat{\Omega}} p(x, a, t) dx da.$$

We again assume Hypothesis (H.1.3), so that spatial movement is governed by a linear semigroup $T(t), t \geq 0$ with infinitesimal generator A in the Banach space Z. Let $Y = L^1((0, a_1); Z)$ and view the density $p(x, a, t) = p(a, t)(x)$ as a function from $[0, a_1)$ to Z. For an age-structured population, it is assumed that individuals born (with age 0) at time t arise from a birth process dependent on the existing fertile population at time t. We assume the hypothesis

(H.2.1) $\beta \in C([0, a_1); R^+)$ and there exists $\bar{a} > 0$ and $\bar{\beta} \in (0, a_1)$ such that $\beta(a) \equiv 0$ for $a \in [\bar{a}, a_1)$ and $\beta(a) \leq \bar{\beta}$ for $0 \leq a \leq \bar{a}$.

We again first consider the case without mortality. The model equations are (M.2.1):

$$\frac{\partial}{\partial t} p(a, t) + \frac{\partial}{\partial a} p(a, t)) = A p(a, t), a \in (0, a_1), t > 0 \tag{1.41}$$

$$p(a, 0) = \phi(a), a \in (0, a_1), \phi \in Y \tag{1.42}$$

$$p(0, t) = \int_0^{a_1} \beta(a) p(a, t) da, t > 0. \tag{1.43}$$

The analysis of Sect. 1.1 applies to model (M.2.1). Define $w(u) = p(a + u, t + u), u \leq 0$. Then, $w(0) = p(a, t)$ and

$$\frac{d}{du} w(u) = \frac{\partial}{\partial t} p(a + u, t + u) + \frac{\partial}{\partial a} p(a + u, t + u) \tag{1.44}$$

$$= A p(a + u, t + u)$$

$$= A w(u)$$

As in Sect. 1.1, if $a > t$, then

$$p(a, t) = w(0) = T(t) w(-t) = T(t) p(a - t, 0) = T(t) \phi(a - t) \tag{1.45}$$

and, if $0 < a < t$, then

$$p(a, t) = w(0) = T(a) w(-a) = T(a) p(0, t - a) \tag{1.46}$$

Thus,

$$p(a, t) = \begin{cases} T(t) \phi(a - t), & t < a < a_1 \\ T(a) p(0, t - a), & 0 < a < t \end{cases} \tag{1.47}$$

To obtain $p(a,t)$ for $a < t$, we must solve the boundary condition (1.43) using (1.47):

$$p(0,t) = \int_0^{a_1} \beta(a)p(a,t)da \tag{1.48}$$

$$= \int_0^t \beta(a)T(a)p(0,t-a)da + \int_t^{a_1} \beta(a)T(t)\phi(a-t)da$$

(where the last integral in (1.48) is 0 if $t > a_1$). Set $b_\phi(t) = p(0,t), t \geq 0$, and (1.48) is a linear Volterra integral equation for b_ϕ in Z:

$$b_\phi(t) = \int_0^t \beta(a)T(a)b_\phi(t-a)da + T(t)\int_0^{a_1-t} \beta(a+t)\phi(a)da \tag{1.49}$$

A unique solution b_ϕ of (1.49), continuous from $[0,\infty)$ to Z, is obtained by standard arguments [63]. To obtain generalized solutions of model (M.2.1), we define a semigroup $S(t), t \geq 0$ in Y as follows: for $\phi \in Y, t \geq 0$, almost all $a \in (0, a_1)$,

$$(S(t)\phi)(a) = \begin{cases} T(t)\phi(a-t), & t < a < a_1 \\ T(a)b_\phi(t-a), & 0 < a < t \end{cases} \tag{1.50}$$

Theorem 4. *Let (H.1.3) and (H.2.1) hold. $S(t), t \geq 0$ is a strongly continuous semigroup of bounded linear operators in Y satisfying $\|S(t)\phi\|_Y \leq M$ $e^{(\bar{\beta}M+\omega)t}\|\phi\|_Y$, $\phi \in Y, t \geq 0$. If Z is a Banach lattice and $T(t), t \geq 0$ is a positive semigroup in Z, then $S(t), t \geq 0$ is a positive semigroup in Y.*

Proof. We first observe that for $\phi \in Y, t \geq 0$,

$$\|b_\phi(t)\|_Z \leq \int_0^t \bar{\beta}Me^{\omega(t-a)}\|b_\phi(a)\|_Z da + Me^{\omega t}\bar{\beta}\int_0^{a_1}\|\phi(a)\|_Z da$$

which implies

$$e^{-\omega t}\|b_\phi(t)\|_Z \leq \bar{\beta}M\int_0^t e^{-\omega a}\|b_\phi(a)\|_Z da + M\bar{\beta}\int_0^{a_1}\|\phi(a)\|_Z da$$

By Gronwall's inequality

$$\|b_\phi(t)\|_Z \leq \bar{\beta}Me^{(\bar{\beta}M+\omega)t}\|\phi\|_Y \tag{1.51}$$

Thus, (1.51) implies

$$\|S(t)\phi\|_Y = \int_0^{a_1}\|(S(t)\phi)(a)\|_Z da$$

$$= \int_0^t \|T(a)b_\phi(t-a)\|_Z da + \int_t^{a_1}\|T(t)\phi(a-t)\|_Z da$$

$$\leq \int_0^t Me^{\omega a}M\bar{\beta}e^{(\bar{\beta}M+\omega)(t-a)}\|\phi\|_Y da + Me^{\omega t}\|\phi\|_Y$$

$$= Me^{(\bar{\beta}M+\omega)t}\|\phi\|_Y$$

To prove the semigroup property (i) in Definition 1.1, we next prove that for $\phi \in Y$,

$$b_{S(t_1)\phi}(t) = b_\phi(t + t_1), t_1, t > 0 \qquad (1.52)$$

Observe from (1.49) and (1.50) that

$$b_{S(t_1)\phi}(t) = \int_0^t \beta(t-a)T(t-a)b_{S(t_1)\phi}(a)da \qquad (1.53)$$

$$+ \int_0^{a_1-t} \beta(a+t)T(t)(S(t_1)\phi)(a)da$$

Also,

$$b_\phi(t + t_1) = \int_0^{t_1} \beta(t + t_1 - a)T(t + t_1 - a)b_\phi(a)da \qquad (1.54)$$

$$+ \int_{t_1}^{t+t_1} \beta(t + t_1 - a)T(t + t_1 - a)b_\phi(a)da$$

$$+ \int_0^{a_1-t-t_1} \beta(t + t_1 + a)T(t + t_1)\phi(a)da$$

$$= \int_0^{t_1} \beta(a+t)T(a+t)b_\phi(t_1 - a)da$$

$$+ \int_0^t \beta(t-a)T(t-a)b_\phi(a + t_1)da$$

$$+ \int_{t_1}^{a_1-t} \beta(a+t)T(t)T(t_1)\phi(a - t_1)da$$

$$= \int_0^t \beta(t-a)T(t-a)b_\phi(a + t_1)da$$

$$+ T(t)\int_0^{a_1-t} \beta(a+t)(S(t_1)\phi)(a)da$$

By the uniqueness of solutions to (1.49), we then obtain (1.52), which implies $(S(t)S(t_1)\phi)(0) = (S(t + t_1)\phi)(0), t, t_1 \geq 0$.

For $\phi \in Y, 0 \leq t < t_1$, and $t + t_1 < a$,

$$(S(t)S(t_1)\phi)(a) = T(t)(S(t_1)\phi)(a - t) \qquad (1.55)$$

$$= T(t)T(t_1)\phi(a - t - t_1)$$

$$= T(t + t_1)\phi(a - (t + t_1))$$

$$= (S(t + t_1)\phi)(a)$$

For $t < a < t + t_1$,

$$(S(t)S(t_1)\phi)(a) = T(t)(S(t_1)\phi)(a - t) \qquad (1.56)$$

$$= T(t)T(a - t)(S(t_1 - (a - t))\phi)(0)$$

$$= T(a)(S(t + t_1 - a)\phi)(0)$$

$$= (S(t + t_1)\phi)(a)$$

For $a < t$, by (1.52)

$$(S(t)S(t_1)\phi)(a) = T(a)(S(t-a)(S(t_1)\phi)(0) \tag{1.57}$$
$$= T(a)(S(t+t_1-a)\phi)(0)$$
$$= (S(t+t_1)\phi)(a)$$

Thus, (1.55)–(1.57) imply $S(t+t_1) = S(t)S(t_1)$ for $t, t_1 \geq 0$.

To prove the strong continuity property (ii) in Definition 1.1, we let Y_0 be a dense set of continuous functions in Y with compact support in $(0, a_1)$. Then,

$$\|S(t)\phi - \phi\|_Y = \int_0^{a_1} \|(S(t)\phi)(a) - \phi(a)\|_Z da \tag{1.58}$$

$$= \int_0^t \|T(a)(S(t-a)\phi)(0) - \phi(a)\|_Z da$$

$$+ \int_t^{a_1} \|T(t)\phi(a-t) - \phi(a)\|_Z da$$

The last term in (1.58) $\to 0$ as $t \to 0$ as in the proof of Theorem 1. The next to last term in (1.58) $\to 0$ as $t \to 0$, since $\|T(a)(S(t-a)\phi)(0) - \phi(a)\|_Z$ is integrable in a.

The positivity of $S(t), t \geq 0$ in Y follows immediately from (1.49) and (1.50), if $T(t), t \geq 0$ is a positive semigroup in the Banach lattice Z, since $\beta(a) \geq 0$ for $a \in (0, a_1)$. ∎

The age structured model (M.2.1) can be extended to allow nonlinear gain or loss of individuals by perturbation of the semigroup $S(t), t \geq 0$ in Theorem 4. The model equations have the following form (M.2.2):

$$\frac{\partial}{\partial t}p(a,t) + \frac{\partial}{\partial a}p(a,t) = Ap(a,t) + F(p(\cdot,t))(a), a \in (0, a_1), t > 0 \tag{1.59}$$

$$p(a,0) = \phi(a), a \in (0, a_1), \phi \in Y \tag{1.60}$$

$$p(0,t) = \int_0^{a_1} \beta(a)p(a,t)da, t > 0 \tag{1.61}$$

where F satisfies hypothesis (H.1.4). Model (M.2.2) can be written abstractly as an ordinary differential equation in Y:

$$\frac{d}{dt}p(t) = \mathcal{A}p(t) + F(p(t)), t \geq 0, p(0) = \phi \in Y \tag{1.62}$$

where $p : [0, \infty) \to Y$, $p(t)(a) = p(a,t)$, $\mathcal{A} : Y \to Y$,

$$(\mathcal{A}\phi)(a) = -\frac{d}{da}\phi(a) + A\phi(a),$$

with $D(\mathcal{A})$ consisting of $\phi \in Y$ such that $-\frac{d}{da}\phi(a) + A\phi(a) \in Y$, and $\phi(0) = \int_0^{a_1} \beta(a)\phi(a)da$.

The following theorem is analogous to Theorem 2:

Theorem 5. *Let (H.1.3), (H.1.4), (H.2.1) hold. Let $S(t), t \geq 0$ be the strongly continuous semigroup of bounded linear operators in $Y = L^1((0, \infty); Z)$ in Theorem 4 and let $\phi \in Y$. There is a unique solution $U(t)\phi, t \geq 0$ to the integral equation (1.29). Further, $U(t), t \geq 0$ is a strongly continuous semigroup of Lipschitz continuous nonlinear operators in Y satisfying*

$$\|U(t)(\phi_1) - U(t)(\phi_2)\|_Y \leq Me^{(M(\bar{\beta}+L)+\omega)t} \|\phi_1 - \phi_2\|_Y, \, t \geq 0, \phi_1, \phi_2 \in Y \tag{1.63}$$

If Z is a Banach lattice and $T(t), t \geq 0$ is a positive semigroup in Z, then $U(t), t \geq 0$ is a positive semigroup in Y.

We give a linear example with age structure, with mortality, but without spatial structure, to illustrate Theorem 5.

Example 2.1. Let $Z = R$, $Y = L^1((0, a_1); R)$, $T(t) = I, t \geq 0$, let $\phi \in Y$ be defined as

$$\phi(a) = \begin{cases} (a - .4)(1.4 - a) & .4 \leq a \leq 1.4 \\ 0 & \text{otherwise} \end{cases} \tag{1.64}$$

and let the fertility modulus $\beta(a)$ be defined as

$$\beta(a) = \begin{cases} 6.0(a - 2.0)(4.0 - a)e^{(-3.0(a-2.0))} & 2.0 \leq a \leq 4.0 \\ 0 & \text{otherwise} \end{cases} \tag{1.65}$$

Let the mortality of individuals be given by $\mu(a) = .05a$, (*i.e.*, the probability that an individual survives to age a_1 from age a_2 is $e^{-\mu(a_2-a_1)}$). Let $F : Y \to Y$ be defined by $F(\phi)(a) = -\mu(a)\phi(a), a \in (0, a_1), \phi \in Y$. The generalized solution $p(a, t) = (U(t)\phi)(a)$ to the model (M.2.2) obtained from $S(t), t \geq 0$ in Theorem 4 and $U(t), t \geq 0$ in Theorem 5, is graphed in Fig. 1.14.

For Example 2.1 (without spatial structure) it is known that the linear semigroup $U(t), t \geq 0$ of this linear age structured model exhibits asynchronous exponential growth under the hypothesis $\beta(a), \mu(a) \in L^\infty_+((0, a_1); R)$ with $\beta(a)$ not identically 0 on $(0, a_1)$ [33, 122, 127]. The intrinsic growth constant λ_0 in Definition 1.2 is the unique real-valued solution of the characteristic equation

$$1 = 2 \int_0^{a_1} e^{-\lambda_0 a} \pi(a, 0)\beta(a)da \tag{1.66}$$

where

$$\pi(a, b) = exp\left(-\int_b^a (\beta(u) + \mu(u))du\right).$$

The projection P_0 in Definition 1.2 has 1-dimensional range and is given by the formula

$$P_0(\phi)(a) = \frac{e^{-\lambda_0 a}\pi(a, 0) \int_0^{a_1} \beta(b)e^{-\lambda_0 b} \left(\int_0^b e^{\lambda_0 u}\pi(b, u)\phi(u)du\right) db}{\int_0^{a_1} \beta(b)be^{-\lambda_0 b}\pi(b, 0)db} \tag{1.67}$$

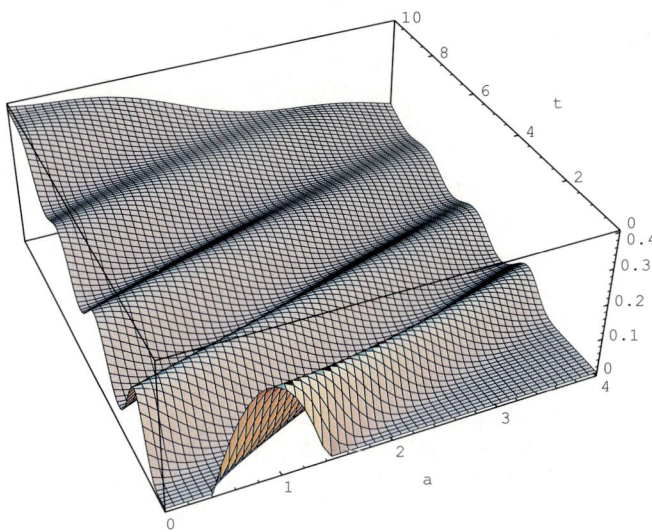

Fig. 1.14. The solution $p(a,t) = (U(t)\phi)(a)$ of the linear age structured model (without spatial structure) in Example 2.1 for the initial value $\phi(a)$ in (1.64). Since the initial population all has age $a \le 1.4$, and the minimum reproductive age is $a = 2.0$, no individuals are born until time $t = .6$. The initial sychronization of the population, centered at $a = .9$, is dispersed over time. The population grows exponentially to ∞, and exhibits asynchronous exponential growth with intrinsic growth constant $\lambda_0 = .0072$ satisfying (1.66), and spectral projection P_0 as in (1.67) with rank 1

Asynchronous exponential growth means that the proportion of the population in any age range $(\hat{a}, \tilde{a})$ stabilizes as $t \to \infty$ in the sense that

$$lim_{t \to \infty} \frac{\int_{\hat{a}}^{\tilde{a}} p(a,t)da}{\int_0^{a_1} p(a,t)da} = \frac{\int_{\hat{a}}^{\tilde{a}} (P\phi)(a)da}{\int_0^{a_1} (P\phi)(a)da}.$$

Since the projection P_0 above has rank 1, the limit above is independent of the initial age distribution $\phi(a)$. This property is observed in proliferating cell cultures, which lose synchrony of the initial age distribution after a few generations. The loss of the initial synchronization results from the hypothesis that not every cell divides at exactly the same age, and there is a dispersion of the ages of division controlled by the division modulus $\beta(a)$.

An illustration of Theorem 5 with spatial structure is given in Example 2.2. The assumptions on the age structure are the same as in Example 2.1, except that the mortality also depends on the spatial variable x.

Example 2.2. Let $Z = L^1((0,4); R)$, $Y = L^1((0, a_1); Z)$, $a_1 = 4$, let $T(t), t \ge 0$, be the strongly continuous linear semigroup in Z with infinitesimal generator the diffusion operator A in Z with Neumann boundary conditions as in (1.23), with $\alpha = .2$, $x_0 = 0.0$, $x_1 = 4.0$. Let the fertility modulus $\beta(a)$ be as in (1.65), and let $S(t), t \geq 0$ be the linear semigroup in Y as in Theorem 4. Let the mortality modulus be defined as follows: for $a \in (0, a_1)$,

$$\mu(x, a) = \begin{cases} .05a + 2.0(x - 2.5)(3.5 - x) & 2.5 \leq x \leq 3.5 \\ .05a & \text{otherwise} \end{cases} \tag{1.68}$$

so that individuals located in the right side of the spatial domain $(0, 4)$ have higher mortality. Let $F : Y \to Y$ be defined as $F\phi(a)(x) = \mu(x)\phi(a)$, $\phi \in Y, a \in (0, a_1), x \in (0, 4)$. Let $\Phi(a)(x) = \phi(a)\psi(x)$, where $\phi(a)$ is given by (1.64) and

$$\psi(x) = \begin{cases} 4(x - 1.0)^2(3.0 - x)^2 & 1.0 \leq x \leq 3.0 \\ 0 & \text{otherwise} \end{cases} \tag{1.69}$$

The generalized solution $p(a, t) = (U(t)\Phi)(a)$ to the model (M.2.2) obtained from $S(t), t \geq 0$ in Theorem 4 and $U(t), t \geq 0$ in Theorem 5, is graphed in Figs. 1.15 and 1.16.

A nonlinear illustration of Theorem 5 (without spatial structure) is given in Example 2.3. The assumptions are the same as in Example 2.1, except that the mortality depends nonlinearly on the density.

Example 2.3. Let $Z, Y, T(t), t \geq 0, S(t), t \geq 0$, let

$$\beta(a) = \begin{cases} 10.0(a - 2.0)(4.0 - a)e^{(-2.0(a-2.0))} & 2.0 \leq a \leq 4.0 \\ 0 & \text{otherwise} \end{cases} \tag{1.70}$$

let $\phi \in Y$ be defined as

$$\phi(a) = \begin{cases} 5.0 \, a \, (1.0 - a) & 0.0 \leq a \leq 1.0 \\ 0 & \text{otherwise} \end{cases} \tag{1.71}$$

and let $F : Y \to Y$ be defined as

$$F(\phi)(a) = -\left(\mu(a) + \gamma \int_0^{a_1} \phi(\hat{a})d\hat{a} \right) \phi(a), \ \phi \in Y, a \in (0, a_1),$$

where $\mu(a) = .1 \, a$ and $\gamma = .1$. It is known that for this nonlinear age structured model, where the mortality is influenced by crowding, nonlocally through the total population density, the solutions converge to a unique globally attracting equilibrium $\phi_0(a)$ ([46, 123]). This equilibrium solution has the form

$$\phi_0(a) = \frac{\lambda_0 \, P_0(a)}{\gamma \int_0^{a_1}(P_0\phi(a))da}, a \in (0, a_1), \tag{1.72}$$

where $\lambda_0 = .139$ is the solution of the characteristic equation (1.66), P_0 is the rank 1 projection in Y given by (1.67), with $\beta(a)$ as in (1.70) and $\mu(a) = .1 \, a$ as above. The generalized solution $p(a, t) = (U(t)\phi)(a)$ to the model (M.2.2)

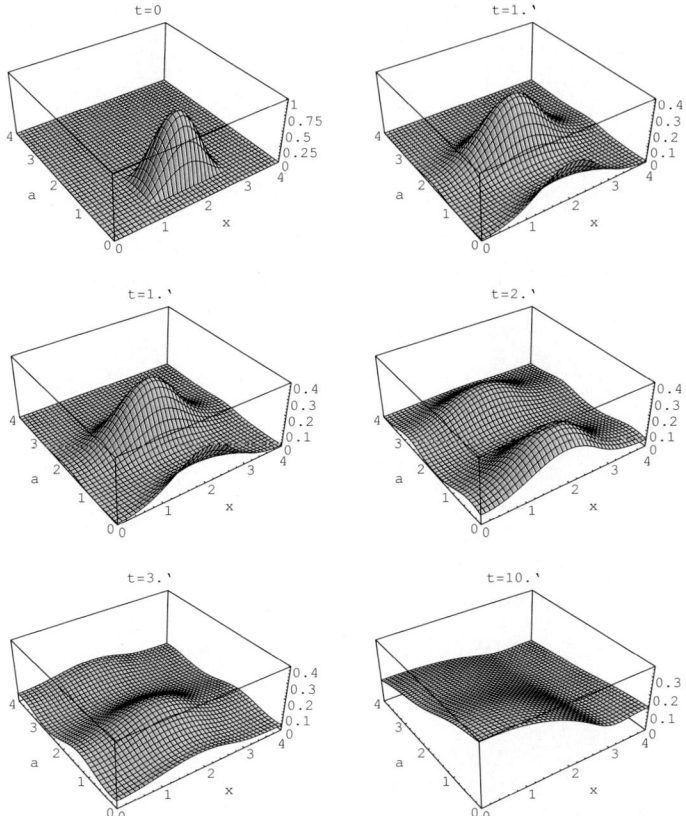

Fig. 1.15. The solution $p(a,t) = (U(t)\varPhi)(a)$ of the linear age structured model with spatial structure in Example 2.2. The initial structure of the population is asynchronized in age as in Example 2.1, but also in space, as the population grows exponentially over time. The spatial distribution shows higher mortality over time for individuals located in the right side of the spatial domain $\varOmega = (0,4)$

obtained from $S(t), t \geq 0$ in Theorem 4 and $U(t), t \geq 0$ in Theorem 5, is graphed in Fig. 1.17.

A nonlinear illustration of Theorem 5 with spatial structure is given in Example 2.4.

Example 2.4. Let $Z = L^1((x_0, x_1); R)$, $Y = L^1((0, a_1); Z)$, $a_1 = 4$, let $T(t), t \geq 0$, be the strongly continuous linear semigroup in Z with infinitesimal generator the diffusion operator A in Z with Neumann boundary conditions as in (1.23), with $\alpha = .2$, $x_0 = 0.0$, $x_1 = 4.0$. Let the fertility modulus $\beta(a)$ be as in (1.70), let $F : Y \to Y$ be defined as

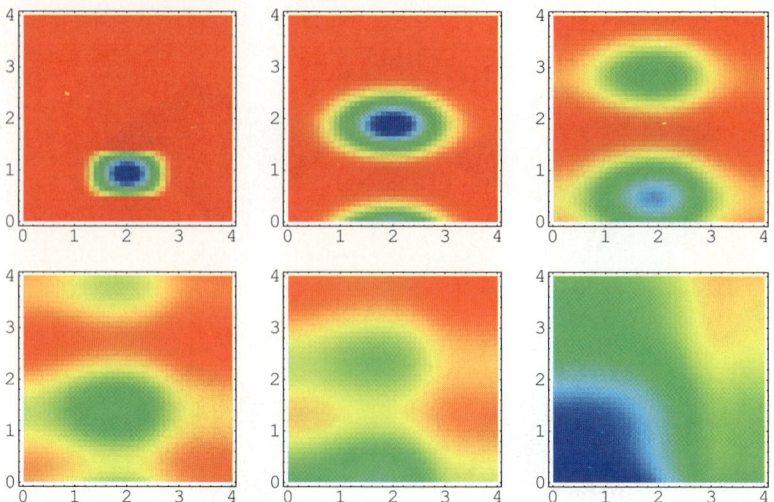

Fig. 1.16. The density plots in the (x, a) coordinate system of the distributions in Fig. 1.15

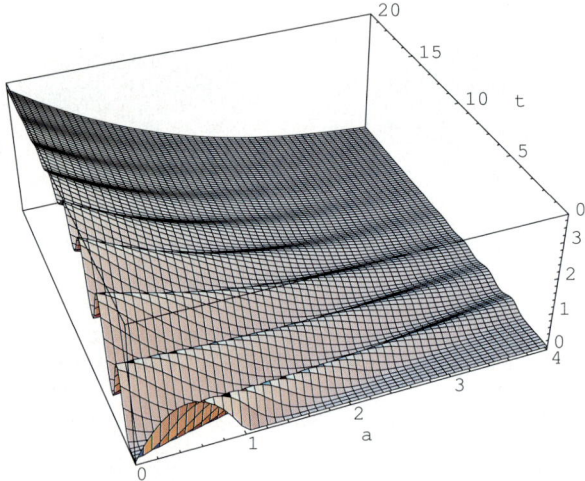

Fig. 1.17. The solution $p(a, t) = (U(t)\phi)(a)$ of the nonlinear age structured model (without spatial structure) in Example 2.3. The initial structure of the population is asynchronized in age and converges over time to a globally attracting equilibrium $\phi_0(a)$ given by (1.72)

$$F(\phi)(a)(x) = -\left(\xi(x) + \mu(a) + \gamma \int_0^{a_1} \phi(\hat{a})d\hat{a}\right)\phi(a),$$

$$\phi \in Y, x \in (x_0, x_1), \ a \in (0, a_1),$$

where $\xi(z) = .2\,(4.0 - x)$, $\mu(a) = .1\,a$, and $\gamma = .1$. Let $\Phi(a)(x) = \phi(a)\psi(x)$, where $\phi(a)$ is given by (1.71) and

$$\psi(x) = \begin{cases} 10.0(x - .5)^2(1.5 - x)^2 & .5 \leq x \leq 1.5 \\ 5.0(x - 2.5)^2(3.5 - x)^2 & 2.5 \leq x \leq 3.5 \\ 0 & \text{otherwise} \end{cases} \qquad (1.73)$$

Let $S(t), t \geq 0$ be the linear semigroup in Y as in Theorem 4. The generalized solution $p(a,t) = (U(t)\Phi)(a)$ to the model (M.2.2) obtained from $S(t), t \geq 0$ in Theorem 4 and $U(t), t \geq 0$ in Theorem 5, is graphed in Figs. 1.18 and 1.19.

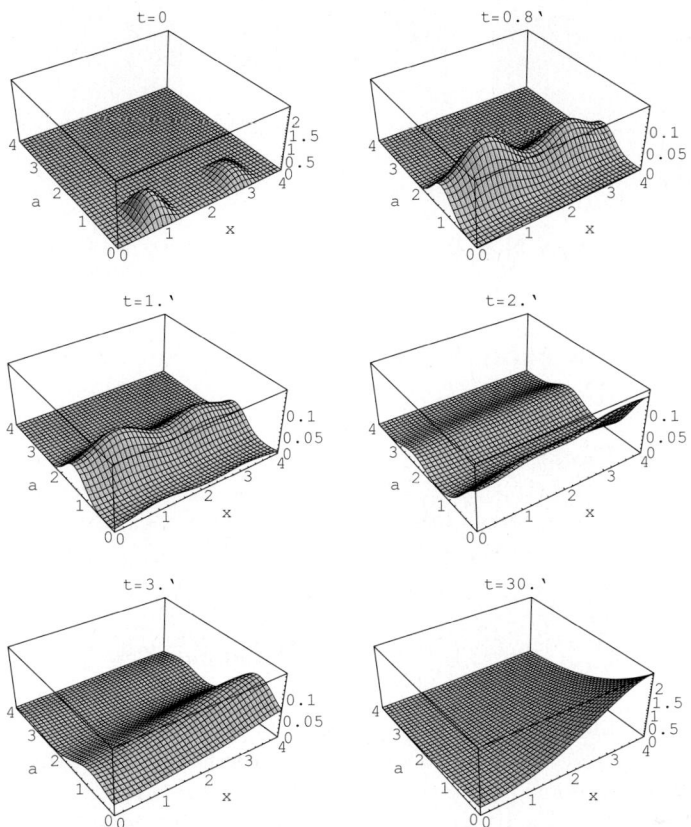

Fig. 1.18. The solution $p(a,t) = (U(t)\Phi)(a)$ of the nonlinear age and spatial structured model in Example 2.4. The initial structure of the population, with two concentrations of juveniles in the spatial domain $\Omega = (0,4)$, is asynchronized in age and space, and converges to a globally attracting equilibrium. The global attractor has age structure with decreasing density as individuals age, and spatial structure with higher density in the right side of Ω

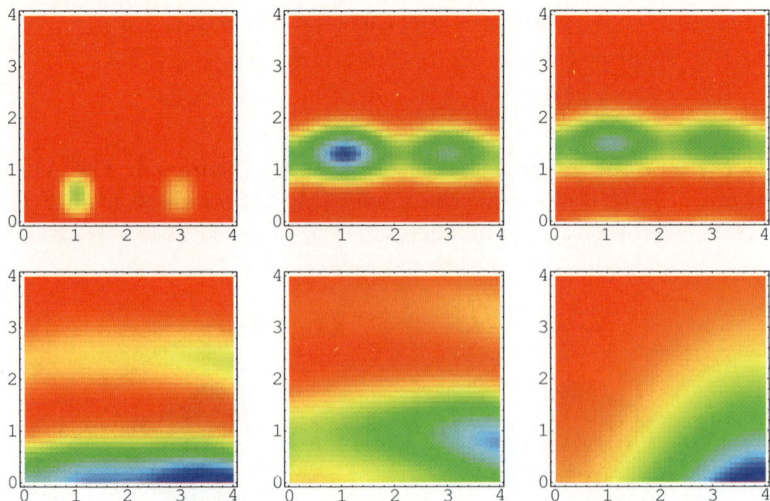

Fig. 1.19. The density plots in the (x, a) coordinate system of the distributions in Fig. 1.18

1.4 Population Models Structured by Age, Size, and Spatial Position

The analysis of Sects. 1.1 and 1.2 can be combined to treat age-size-space structured models. As in Sects. 1.1 and 1.2, the density function at time t, with respect to age a, size s, and position x, describes the total population of individuals with age between $\hat{a}$ and $\tilde{a}$, size between $\hat{s}$ and $\tilde{s}$ and position $x \in \hat{\Omega} \subset \Omega$ at time t:

$$\int_{\hat{a}}^{\tilde{a}} \int_{\hat{s}}^{\tilde{s}} \int_{\hat{\Omega}} p(x, a, s, t) dx ds da.$$

We again assume Hypothesis (H.1.3), so that spatial movement is governed by a linear semigroup $T(t), t \geq 0$ with infinitesimal generator A in the Banach space Z. We first consider the case that the density $p(a, s, t)$ is in the state space $Y = L^1((0, a_1) \times (0, s_2); Z)$, where $a_1 \leq \infty$ is the maximum age and $s_2 \leq \infty$ is the maximum size. Further, we assume that the growth modulus $g(s) = 1$ for $s \in [0, s_2)$. The equations for the model (M.3.1) are:

$$\frac{\partial}{\partial t} p(a, s, t) + \frac{\partial}{\partial a} p(a, s, t) + \frac{\partial}{\partial s} p(a, s, t) \tag{1.74}$$
$$= Ap(a, s, t), a \in (0, a_1), s \in (0, s_2), t > 0$$

$$p(a, s, 0) = \phi(a, s), a \in (0, a_1), s \in (0, s_2), \phi \in Y \tag{1.75}$$

$$p(0, s, t) = \int_0^{a_1} \int_0^{s_2} \beta(a, \hat{s}, s) p(a, \hat{s}, t) d\hat{s} \, da, s \in (0, s_2), t > 0 \tag{1.76}$$

$$p(a, 0, t) = 0, a \in (0, a_1), t > 0 \qquad (1.77)$$

where $\beta(a, \hat{s}, s)$ satisfies the following hypothesis:

(H.3.1) $\beta : [0, a_1) \times [0, s_2) \times [0, s_2) \rightarrow [0, \infty)$, β is continuous, and there exists $\bar{a} \in (0, a_1), \bar{s} \in (0, s_2)$ such that $\beta(a, \hat{s}, s) = 0$ for $\bar{a} < a < a_1$, and $\bar{s} < \hat{s}, s < s_2$.

As in Sects. 1.1 and 1.2, the method of characteristics yields the following formula for the density $p(a, s, t)$:

$$p(a, s, t) = \begin{cases} T(t)\phi(a - t, s - t), & 0 < t < a, 0 < t < s \\ T(a)p(0, s - a, t - a), & 0 < a < t, 0 < a < s \\ 0 & a > t, s < t \text{ or } a < t, a > s \end{cases} \qquad (1.78)$$

As in Sect. 1.2, we use (1.76) and (1.78) to obtain an equation for the birth-rate function $b_\phi(s, t) := p(0, s, t)$, where $b_\phi : (0, s_2) \times (0, \infty) \rightarrow Z$ satisfies the following integral equation:

$$b_\phi(s, t) = \int_0^t \int_a^{s_2} \beta(a, \hat{s}, s)T(a)p(0, \hat{s} - a, t - a)d\hat{s}da \qquad (1.79)$$

$$+ T(t) \int_t^{a_1} \int_t^{s_2} \beta(a, \hat{s}, s)\phi(a - t, \hat{s} - t)d\hat{s}da$$

$$= \int_0^t \int_a^{s_2} \beta(a, \hat{s}, s)T(a)b_\phi(\hat{s} - a, t - a)d\hat{s}da$$

$$+ T(t) \int_0^{a_1 - t} \int_0^{s_2 - t} \beta(a + t, \hat{s} + t, s)\phi(a, \hat{s})d\hat{s}da$$

(where the last integral in (1.79) is 0 if $t > a_1$ or $t > s_2$).

Let $V = \{f \in C([0, s_2); Z) : \lim_{s \rightarrow s_2} f(s) = 0\}$, let $W = C([0, t_1]; V)$, where $t_1 > 0$. Let

$$c_\phi(s, t) = \begin{cases} T(t) \int_0^{a_1 - t} \int_0^{s_2 - t} \beta(a + t, \hat{s} + t, s)\phi(a, \hat{s})d\hat{s}da, t < a_1 \text{ and } t < s_2 \\ 0, \text{ otherwise} \end{cases}$$

Let $C_\phi(t)(s) = c_\phi(s, t)$, $0 \le t \le t_1, s \in (0, s_2)$, and then $C_\phi \in W$. Define $K : [0, \infty) \rightarrow B(V)$ (the space of bounded linear operators in V) by

$$(K(a)f)(s) = T(a) \int_a^{s_2} \beta(a, \hat{s}, s)f(\hat{s} - a)d\hat{s}, \ f \in V, a \in (0, a_1).$$

Then, $K(a)$ is well-defined, since $T(t), t \ge 0$ is uniformly strongly continuous on compact sets of Z, and β and f are continuous. Equation (1.79) may now be written as an abstract linear Volterra integral equation in V:

$$B_\phi(t) = \int_0^t K(a)B_\phi(t - a)da + C_\phi(t), t \ge 0 \qquad (1.80)$$

where $B_\phi \in W$ and $B_\phi(t)(s) = b_\phi(s,t)$. The existence of a unique solution to (1.80), and thus to (1.79), is obtained by standard arguments ([63]). We thus define the family of linear operators $S(t), t \geq 0$ in Y by the following formula:

$$(S(t)\phi)(a,s) = \begin{cases} T(t)\phi(a-t, s-t), & 0 < t < a, 0 < t < s \\ T(a)b_\phi(s-a, t-a), & 0 < a < t, 0 < a < s \\ 0 & a > t, s < t \text{ or } a < t, a > s \end{cases} \qquad (1.81)$$

The proof of the following theorem is very similar to the proof of Theorem 4:

Theorem 6. *Let (H.1.3) and (H.3.1) hold. $S(t), t \geq 0$ defined in (1.81), is a strongly continuous semigroup of bounded linear operators in $Y = L^1((0, a_1) \times (0, s_2); Z)$ satisfying $\|S(t)\phi\|_Y \leq M e^{(\bar\beta M + \omega)t} \|\phi\|_Y$, $\phi \in Y, t \geq 0$. If Z is a Banach lattice and $T(t), t \geq 0$ is a positive semigroup in Z, then $S(t), t \geq 0$ is a positive semigroup in Y.*

As in Sects. 1.1 and 1.2, the model (M.3.1) can be extended to a nonlinear model allowing population loss or gain (M.3.2):

$$\frac{\partial}{\partial t}p(a,s,t) + \frac{\partial}{\partial a}p(a,s,t) + \frac{\partial}{\partial s}p(a,s,t) \qquad (1.82)$$
$$= Ap(a,s,t) + F(p(\cdot,\cdot,t))(a,s), a \in (0, a_1), s \in (0, s_2), t > 0$$

$$p(a,s,0) = \phi(a,s), a \in (0, a_1), s \in (0, s_2), \phi \in Y \qquad (1.83)$$

$$p(0,s,t) = \int_0^{a_1} \int_0^{s_2} \beta(a, \hat{s}, s)p(a, \hat{s}, t)\, d\hat{s}\, da, s \in (0, s_2), t > 0 \qquad (1.84)$$

$$p(a,0,t) = 0, a \in (0, a_1), t > 0 \qquad (1.85)$$

The following theorem is analogous to Theorem 5:

Theorem 7. *Let (H.1.3), (H.1.4), (H.3.1) hold. Let $S(t), t \geq 0$ be the strongly continuous semigroup of bounded linear operators in $Y = L^1((0, a_1) \times (0, s_2); Z)$ in Theorem 6 and let $\phi \in Y$. There is a unique solution $U(t)\phi, t \geq 0$ to the integral equation (1.29). Further, $U(t), t \geq 0$ is a strongly continuous semigroup of Lipschitz continuous nonlinear operators in Y satisfying (1.63). If Z is a Banach lattice and $T(t), t \geq 0$ is a positive semigroup in Z, then $U(t), t \geq 0$ is a positive semigroup in Y.*

The models (M.3.1) and (M.3.2) can be used to treat the following model allowing variable growth rates in size (M.3.3):

$$\frac{\partial}{\partial t}\hat{p}(a,s,t) + \frac{\partial}{\partial a}\hat{p}(a,s,t) + \frac{\partial}{\partial s}(g(s)\hat{p}(a,s,t)) \qquad (1.86)$$
$$= A\hat{p}(a,s,t) + \hat{F}(\hat{p}(\cdot,\cdot,t))(a,s), a \in (0, a_1), s \in (s_0, s_1), t > 0$$

$$\hat{p}(a,s,0) = \hat{\phi}(a,s), a \in (0, a_1), s \in (s_0, s_1), \hat{\phi} \in \hat{Y} \qquad (1.87)$$

$$\hat{p}(0, s, t) = \int_0^{a_1} \int_{s_0}^{s_1} \hat{\beta}(a, \tilde{s}, s)\hat{p}(a, \tilde{s}, t)\, d\tilde{s}\, da, s \in (s_0, s_1), t > 0 \qquad (1.88)$$

$$\hat{p}(a, 0, t) = 0, a \in (0, a_1), t > 0 \qquad (1.89)$$

The model (M.3.3) can be transformed to model (M.3.2) under the following hypothesis:

(H.3.2) $g : [s_0, s_1) \to [0, \infty), 0 \leq s_0 < s_1 \leq \infty$, g is positive on $[s_0, s_1)$, g is continuously differentiable on $[s_0, s_1)$, g' is bounded on $[s_0, s_1)$, and

$$s_2 := \lim_{s \to s_1} \int_{s_0}^s \frac{1}{g(\hat{s})} d\hat{s} \leq \infty.$$

Let $\sigma(s, s_0) = \int_{s_0}^s \frac{1}{g(v)} dv, s \in (s_0, s_1)$ as in (1.7). Let Z be a Banach space, let $Y = L^1((0, a_1) \times (0, s_2); Z)$ and let $\hat{Y} = L^1((0, a_1) \times (s_0, s_1); Z)$. Define $J : \hat{Y} \to Y$ and $J^{-1} : Y \to \hat{Y}$ by

$$(J\hat{\phi})(a, s) = \hat{\phi}(a, \sigma^{-1}(s, s_0)), \hat{\phi} \in \hat{Y}, a \in (0, a_1), s \in (0, s_2)$$
$$(J^{-1}\phi)(a, s) = \phi(a, \sigma(s, s_0)), \phi \in Y, a \in (0, a_1), s \in (s_0, s_1).$$

Model (M.3.3) can be written abstractly as an ordinary differential equation in $\hat{Y}$:

$$\frac{d}{dt}\hat{p}(t) = \hat{A}\hat{p}(t) + \hat{F}(\hat{p}(t)), t \geq 0, \hat{p}(0) = \hat{\phi} \in \hat{Y} \qquad (1.90)$$

where $\hat{p} : [0, \infty) \to \hat{Y}, \hat{p}(t)(a, s) = \hat{p}(a, s, t), \hat{A} : \hat{Y} \to \hat{Y}$,

$$(\hat{A}\hat{\phi})(a, s) = -\frac{\partial}{\partial a}\hat{\phi}(a, s) - \frac{\partial}{\partial s}(g(s)\hat{\phi}(a, s)) + A\hat{\phi}(a, s),$$

with $D(\hat{A})$ consisting of $\hat{\phi} \in \hat{Y}$ such that $-\frac{\partial}{\partial a}\hat{\phi}(a, s) - \frac{\partial}{\partial s}(g(s)\hat{\phi}(a, s)) + A\hat{\phi}(a, s) \in \hat{Y}$, and

$$\hat{\phi}(0, s) = \int_0^{a_1} \int_{s_0}^{s_1} \hat{\beta}(a, \hat{s}, s)\hat{\phi}(a, \hat{s})d\hat{s}da, s \in (s_0, s_1), \hat{\phi}(a, s_0) = 0, a \in (0, a_1).$$

The transformation of (M.3.3) to (M.3.2) is accomplished through the formula $\hat{p}(a, s, t) = p(a, \sigma(s, s_0), t), a \in (0, a_1), s \in (s_0, s_1), t > 0$. We first consider the case $\hat{F} \equiv 0$. Let $\hat{\beta}$ satisfy

(H.3.3) $\hat{\beta} : [0, a_1) \times (s_0, s_1) \times (s_0, s_1) \to [0, \infty)$, β is continuous, and there exists $\hat{a} \in (0, a_1)$, $\hat{s} \in (s_0, s_1)$ such that $\hat{\beta}(a, \tilde{s}, s) = 0$ for $\hat{a} < a < a_1$, or $\hat{s} < s, \tilde{s} < s_1$.

Theorem 8. *Let (H.1.3), (H.3.2), and (H.3.3) hold. Let $\beta : (0, a_1) \times (0, s_2) \times (0, s_2) \to [0, \infty)$ be defined for $a \in (0, a_1)$ and $0 < \tilde{s}, s < s_2$, by*

$$\beta(a, \tilde{s}, s) = g(\sigma^{-1}(\tilde{s}, s_0))\, \hat{\beta}(a, \sigma^{-1}(\tilde{s}, s_0), \sigma^{-1}(s, s_0)) \qquad (1.91)$$

and let $F : Y \to Y$ be defined by $F\phi(a,s) = -g'(\sigma^{-1}(s,s_0))\phi(a,s), \phi \in Y$, $a \in (0,a_1), s \in (0,s_2)$. Let $U(t), t \geq 0$ be the strongly continuous semigroup of bounded linear operators in Y as in Theorem 7 for this β and F. Then, $\hat{S}(t), t \geq 0$ is a strongly continuous semigroup of bounded linear operators in $\hat{Y}$, where $\hat{S}(t) = J^{-1}U(t)J, t \geq 0$. Let $\hat{\phi} \in \hat{Y}$ and let $p(a,s,t) = (U(t)J\hat{Y})(a,s), a \in (0,a_1), s \in (0,s_2), t > 0$. If $p(a,s,t)$ is a classical solution of (M.3.2), then $\hat{p}(a,s,t) = p(a, \sigma(s,s_0),t), a \in (0,a_1), s \in (s_0,s_1), t > 0$, is a classical solution of (M.3.3) with $\hat{F} \equiv 0$.

Proof. That $\hat{S}(t), t \geq 0$ is a strongly continuous semigroup of bounded linear operators in $\hat{Y}$, is immediate. For $0 < a_1, s_0 < s < s_1, t > 0$,

$$\frac{\partial}{\partial t}\hat{p}(a,s,t) + \frac{\partial}{\partial a}\hat{p}(a,s,t) + \frac{\partial}{\partial s}(g(s)\hat{p}(a,s,t))$$

$$= \frac{\partial}{\partial t}p(a,\sigma(s,s_0),t) + \frac{\partial}{\partial a}p(a,\sigma(s,s_0),t)$$

$$+ g(s)\frac{\partial}{\partial s}p(a,\sigma(s,s_0),t)\frac{d}{ds}\sigma(s,s_0) + g'(s)p(a,\sigma(s,s_0),t)$$

$$= A\,p(a,\sigma(s,s_0),t) + F(p(a,\sigma(s,s_0),t) + g'(s)p(a,\sigma(s,s_0),t)$$

$$= A\,\hat{p}(a,s,t)$$

For $s \in (s_0,s_1), t > 0$, the definition of β in (1.91), and a change of integration variable $s' = \sigma^{-1}(\tilde{s},s_0) \Leftrightarrow \sigma(s',s_0) = \tilde{s}, d\tilde{s}/ds' = 1/g(s')$ yields

$$\hat{p}(0,s,t) = p(0,\sigma(s,s_0),t)$$

$$= \int_0^{a_1}\int_0^{s_2}\beta(a,\tilde{s},\sigma(s,s_0))p(a,\tilde{s},t)\,d\tilde{s}da$$

$$= \int_0^{a_1}\int_0^{s_2}g(\sigma^{-1}(\tilde{s},s_0))\hat{\beta}(a,\sigma^{-1}(\tilde{s},s_0),s)p(a,\tilde{s},t)\,d\tilde{s}da$$

$$= \int_0^{a_1}\int_{s_0}^{s_1}g(s')\hat{\beta}(a,s',s)p(a,\sigma(s',s_0),t)\frac{ds'}{g(s')}da$$

$$= \int_0^{a_1}\int_{s_0}^{s_1}\hat{\beta}(a,s',s)\hat{p}(a,s',t)ds'da$$

Further, $\hat{p}(a,s_0,t) = p(s,\sigma(s_0 s_0),t) = p(a,0,t) = 0, a \in (0,a_1), t > 0$ and $\hat{p}(a,s,0) = p(a,\sigma(s,s_0),0) = J\hat{\phi}(a,\sigma(s,s_0)) = \hat{\phi}(a,s), a \in (0,a_1), s \in (s_0,s_1)$. Thus, $\hat{p}(a,s,t)$ is a classical solution of (M.3.3) with $\hat{F} \equiv 0$. ∎

The generalized solutions of the model (M.3.3) for the case $\hat{F} \neq 0$ are obtained from the following theorem:

Theorem 9. *Let (H.1.3), (H.3.1) hold, let* $\hat{Y} = L^1((0, a_1) \times (s_0, s_1); Z)$ *and let* $\hat{F}$ *satisfy (H.1.4) in* $\hat{Y}$. *Let* $\hat{S}(t), t \geq 0$ *be the strongly continuous semigroup of bounded linear operators in* $\hat{Y}$ *as in Theorem 8 and let* $\hat{\phi} \in \hat{Y}$. *There is a unique solution* $\hat{U}(t)\hat{\phi}, t \geq 0$ *to the integral equation*

$$\hat{U}(t)\hat{\phi} = \hat{S}(t)\hat{\phi} + \int_0^t \hat{S}(t - u)\hat{F}(\hat{U}(u)\hat{\phi}) \, du, t \geq 0 \qquad (1.92)$$

Further, $\hat{U}(t), t \geq 0$ *is a strongly continuous semigroup of Lipschitz continuous nonlinear operators in* Y *satisfying*

$$\|\hat{U}(t)(\hat{\phi}_1) - \hat{U}(t)(\hat{\phi}_2)\|_{\hat{Y}} \leq Me^{(M(\bar{\beta}+L)+\omega)t} \|\hat{\phi}_1 - \hat{\phi}_2\|_{\hat{Y}}, t \geq 0, \hat{\phi}_1, \hat{\phi}_2 \in \hat{Y}$$
$$(1.93)$$

If Z *is a Banach lattice and* $T(t), t \geq 0$ *is a positive semigroup in* Z, *then* $\hat{U}(t), t \geq 0$ *is a positive semigroup in* $\hat{Y}$.

We give an illustration of age, size, and spatial structure in a model of tumor growth. The model is derived from the hybrid discrete-continuous tumor invasion model of Anderson [8]. The model we present is continuous in all variables, and the individual processes of cells are modeled according to cell age and cell size. As in [8], the model is based on the population densities of proliferating and quiescent tumor cells, the density of surrounding tissue macromolecules, the concentration of matrix degradative enzyme, and the concentration of oxygen. The tumor is contained in a region of tissue Ω in 1, 2 or 3 dimensions. The proliferating and quiescent cells are distinguished by type $i = 1, 2, \ldots, n$ corresponding to a sequence of mutations, by position $x \in \Omega$, by age $a \geq 0$ (newly divided cells have age 0), and by cell size s between s_0 and s_1. The cell types correspond to increasingly aggressive phenotyes, through a progression of mutations. Higher indexed phenotypes may have higher mutation rates, shorter cycle times, lower death rates, decreased transition from proliferation to quiescence, increased recruitment from quiescence to proliferation, and reduced cell–cell adhesion. Age for proliferating cells corresponds to position in the cell cycle. Age for quiescent cells corresponds to arrested position in the cell cycle (the age of a quiescent cell is fixed at the age it had when it transitioned from proliferation to quiescence, and if a quiescent cell transitions back to quiescence, then it aging resumes). Size can be interpreted appropriately as mass, diameter, volume, DNA content, or any other quantifiable property of individual cells that is conserved with cell division.

The dependent variables of the model are as follows:

$p_i(x, a, s, t)$ = proliferating tumor cells of type i in the tumor at position x, age a, and size s at time t, where $i = 1, 2, 3, \ldots, n$ corresponds to a linear sequence of mutated phenotypes of increasing aggressiveness;

$q_i(x, a, s, t)$ = quiescent tumor cells of type i in the tumor at position x, arrested age a, and size s at time t;

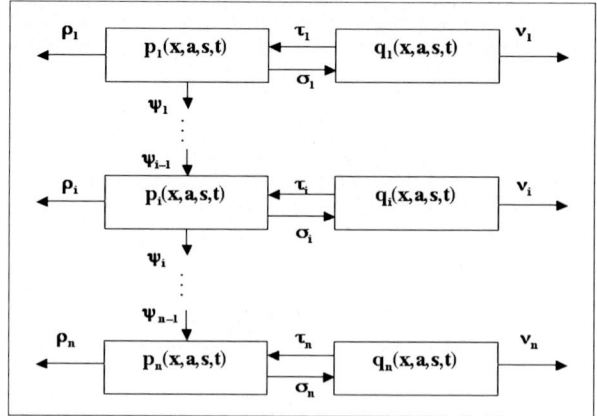

Fig. 1.20. The proliferating cell compartments $p_i(x, a, s, t)$ and quiescent cell compartments $q_i(x, a, s, t)$, $i = 1, 2, \ldots n$ of increasingly aggressive mutated cell types. Transition from proliferation to quiescence occurs at rate σ_i, transition from quiescence to proliferation occurs at rate τ_i, proliferating cells die at rate ρ_i, and quiescent cells die at rate ν_i. The fraction of proliferating cells of type p_i that mutate to type p_{i+1} during mitosis is $\psi_i, i = 1, 2, \ldots, n - 1$

$m(x, t) = $ matrix degradative enzyme (MDE) concentration at position x at time t;

$f(x, t) = $ surrounding tissue macromolecule (MM) concentration at position x at time t;

$c(x, t) = $ oxygen concentration at position x at time t.

The proliferating and quiescent phenotype cell compartments $p_i(x, a, s, t)$, and $q_i(x, a, s, t)$, $i = 1, 2, \ldots, n$, are illustrated in Fig. 1.20.

Let

$$p(x, t) = \sum_{i=1}^{n} \int_{0}^{a_1} \int_{s_0}^{s_1} p_i(x, a, s, t)\,ds\,da$$

be the total population densities in x of proliferating cells of all types at time t, let

$$q(x, t) = \sum_{i=1}^{n} \int_{0}^{a_1} \int_{s_0}^{s_1} q_i(x, a, s, t)\,ds\,da$$

be the total population densities in x of quiescent cells of all types at time t, and let $n(x, t) = p(x, t) + q(x, t)$.

The model equations are as follows (M.3.4):

$$\frac{\partial}{\partial t} f(x, t) = - \underbrace{\delta(x)m(x, t)f(x, t)}_{degradation} \qquad (1.94)$$

$$\frac{\partial}{\partial t}m(x,t) = \underbrace{\nabla \cdot (\xi(x) \nabla m(x,t))}_{diffusion} + \underbrace{\mu(x)p(x,t)}_{production} - \underbrace{\lambda(x)m(x,t)}_{decay} \qquad (1.95)$$

$$\frac{\partial}{\partial t}c(x,t) = \underbrace{\nabla \cdot (\eta(x) \nabla c(x,t))}_{diffusion} + \underbrace{\beta(x)f(x,t)}_{production} \qquad (1.96)$$

$$- \underbrace{\omega(x,p(x,t))c(x,t)}_{uptake} - \underbrace{\epsilon(x)c(x,t)}_{decay}$$

$$\frac{\partial}{\partial t}p_i(x,a,s,t) + \underbrace{\frac{\partial}{\partial a}p_i(x,a,s,t)}_{cell\ aging} + \underbrace{\frac{\partial}{\partial s}(g_i(s)p_i(x,a,s,t))}_{cell\ growth} \qquad (1.97)$$

$$= \underbrace{\nabla \cdot (\alpha_i(x) \nabla p_i(x,a,s,t))}_{diffusion} - \underbrace{\chi_i \nabla \cdot (p_i(x,a,s,t) \nabla f(x,t))}_{haptotaxis}$$

$$- \underbrace{\rho_i(x,a,s,c(x,t),n(x,t))p_i(x,a,s,t)}_{cell\ death\ from\ insufficient\ oxygen}$$

$$- \underbrace{\theta_i(x,a,s,c(x,t),n(x,t))p_i(x,a,s,t)}_{loss\ of\ mother\ cells\ from\ division}$$

$$- \underbrace{\sigma_i(x,a,s,c(x,t),n(x,t))p_i(x,a,s,t)}_{exit\ to\ quiescence}$$

$$+ \underbrace{\tau_i(x,a,s,c(x,t),n(x,t))q_i(x,a,s,t)}_{entry\ from\ quiescence}$$

$$\underbrace{p_i(x,0,s,t)}_{newborn\ type\ i\ cells} \qquad (1.98)$$

$$= 2(1-\psi_i) \underbrace{\int_0^{a_1} \int_{s_0}^{s_1} \kappa_i(\hat{s},s)\theta_i(x,a,\hat{s},c(x,t),n(x,t))p_i(x,a,s,t)d\hat{s}da}_{division\ rate\ of\ type\ i\ cells}$$

$$+ 2\psi_{i-1} \underbrace{\int_0^{a_1} \int_{s_0}^{s_1} \kappa_i(\hat{s},s)\theta_{i-1}(x,a,\hat{s},c(x,t),n(x,t))p_{i-1}(x,a,s,t)d\hat{s}da}_{division\ rate\ of\ type\ i-1\ cells}$$

$$\frac{\partial}{\partial t}q_i(x,a,s,t) = \underbrace{\nabla \cdot (\gamma_i(x) \nabla q_i(x,a,s,t))}_{diffusion} \qquad (1.99)$$

$$- \underbrace{\nu_i(x,a,s,c(x,t),n(x,t))q_i(x,a,s,t)}_{cell\ death\ from\ insufficient\ oxygen}$$

$$+ \underbrace{\sigma_i(x,a,s,c(x,t),n(x,t))p_i(x,a,s,t)}_{entry\ from\ proliferation}$$

$$- \underbrace{\tau_i(x,a,s,c(x,t),n(x,t))q_i(x,a,s,t)}_{exit\ to\ proliferation}$$

$$p_i(x, a, s_0, t) = q_i(x, a, s_0, t) = 0 \qquad (1.100)$$

These equations are combined with initial conditions and no-flux boundary conditions on the boundary $\partial\Omega$ of Ω.

The diffusion terms in equations (1.97) and (1.99) account for cell–cell adhesion and cell movement due to random motility. The diffusion coefficients α_i and γ_i can be allowed to depend on decreased cell–cell adhesion and increased motility for progressively more aggressive phenotypes $i = 1, 2, \ldots, n$. The proliferation-quiescence transition rates σ_i, quiescence-proliferation transition rates τ_i, proliferating cell death rates ρ_i, and quiescent cell death rates ν_i depend on the supply of oxygen $c(x, t)$ and the population density $n(x, t)$. In equations (1.97),(1.98), $\theta_i(x, a, \hat{s}, c(x, t), n(x, t))$ is the rate at which cells of type i, age a, and size $\hat{s}$ divide at position x per unit time, where it is assumed that a mother cell divides into two daughter cells of unequal size. The factor ψ_i is the fraction of daughter cells of dividing mother cells of type i with type $i+1$ mutation (we set $\psi_0 = 0$). The birth rate $\kappa_i(\hat{s}, s)\theta_i(x, a, \hat{s}, c(x, t), n(x, t))$ of daughter cells of size s from mother cells of size $\hat{s}$ depends on the supply of oxygen $c(x, t)$, and also the availability of space, through the density of cells of all phenotypes $n(x, t)$. The kernel $\kappa_i(\hat{s}, s)$ is the probability that a daughter cell of size s results from the division of a mother cell of size $\hat{s}$. This kernel satisfies the symmetry condition $\kappa_i(\hat{s}, s) = \kappa_i(\hat{s}, \hat{s} - s)$. If it is assumed that all dividing cells divide into 2 daughter cells of equal size, then $\kappa_i(\hat{s}, s)$ is the delta function $\kappa_i(\hat{s}, s) = \delta(s - \frac{\hat{s}}{2})$, and (1.98) is replaced by

$$\underbrace{p_i(x, 0, s, t)}_{newborn\ type\ i\ cells} \qquad (1.101)$$

$$= 4(1 - \psi_i) \underbrace{\int_0^\infty \theta_i(x, a, 2s, c(x, t), n(x, t))p_i(x, a, 2s, t)da}_{division\ rate\ of\ type\ i\ cells}$$

$$+ 4\psi_{i-1} \underbrace{\int_0^\infty \theta_{i-1}(x, a, 2s, c(x, t), n(x, t))p_{i-1}(x, a, 2s, t)da}_{division\ rate\ of\ type\ i-1\ cells}$$

The movement of cells results not only from random motion due to cell motility, but also from directed motion due to haptotaxis. The haptotatic motion is induced by the matrix degradation enzyme MDE, which is produced by the proliferating cells, diffuses in the tumor environment, and provides oxygen and available space by degradation of the environmental tissue macromolecules MM. In this model the properties of individual cells are viewed as dynamic rates, based on cell age, size, and position, rather than as probabilities of movement on a discrete spatial lattice, as in [8].

Example 3.1. Consider model (M.3.4) in a special case, with homogeneous matrix degradation enzyme ($m(x, t) \equiv 1$, $\xi(x) \equiv 0$, $\mu(x) \equiv 0$, $\lambda(x) \equiv 0$), without haptotaxis ($\chi_i = 0$), with only one phenotype class ($n = 1$), with constant diffusion coefficients ($\eta(x) \equiv \eta, \alpha(x) \equiv \alpha, \gamma(x) \equiv \gamma$), and linear birth rate ($\theta_i(x, a, s)$, $i = 1, 2, \ldots, n$ independent of $c(x, t)$ and $n(x, t)$ in (1.98) and (1.101)). The model equations in this special case are (M.3.5):

$$\frac{\partial}{\partial t} f(x, t) = -\delta(x) f(x, t) \tag{1.102}$$

$$\frac{\partial}{\partial t} c(x, t) = \eta \triangle c(x, t) + \beta(x) f(x, t) - \omega(x, p(x, t)) c(x, t) \tag{1.103}$$

$$\frac{\partial}{\partial t} p(x, a, s, t) + \frac{\partial}{\partial a} p(x, a, s, t) + \frac{\partial}{\partial s} (g(s) p(x, a, s, t)) \tag{1.104}$$

$$= \alpha \triangle p(x, a, s, t) - \rho(x, a, s, c(x, t), n(x, t)) p(x, a, s, t)$$

$$- \theta(x, a, s) \, p(x, a, s, t) - \sigma(x, a, s, c(x, t), n(x, t)) p(x, a, s, t)$$

$$+ \tau(x, a, s, c(x, t), n(x, t)) q(x, a, s, t)$$

$$p(x, 0, s, t) = 2 \int_0^{a_1} \int_{s_0}^{s_1} \kappa(\hat{s}, s) \theta(x, a, \hat{s}) p(x, a, s, t) d\hat{s} da \tag{1.105}$$

$$\frac{\partial}{\partial t} q(x, a, s, t) = \gamma \triangle q(x, a, s, t) - \nu(x, a, s, c(x, t), n(x, t)) q(x, a, s, t) \tag{1.106}$$
$$+ \sigma(x, a, s, c(x, t), n(x, t)) p(x, a, s, t) - \tau(x, a, s, c(x, t), n(x, t)) q(x, a, s, t)$$

$$p(x, a, s_0, t) = q(x, a, s_0, t) = 0 \tag{1.107}$$

Let $Z, T(t), t \geq 0$, and A be as in Example 1.4, and let $\hat{Y}$ be as in Theorem 9. The system of equations (1.102)-(1.107) can be written abstractly in the Banach space $\mathcal{X} = Z \times Z \times \hat{Y} \times \hat{Y}$ as

$$\frac{d}{dt} \begin{bmatrix} f(t) \\ c(t) \\ p(t) \\ q(t) \end{bmatrix} = A \begin{bmatrix} f(t) \\ c(t) \\ p(t) \\ q(t) \end{bmatrix} + \mathcal{F} \left(\begin{bmatrix} f(t) \\ c(t) \\ p(t) \\ q(t) \end{bmatrix} \right), \quad \begin{bmatrix} f(0) \\ c(0) \\ p(0) \\ q(0) \end{bmatrix} = \begin{bmatrix} f_0 \\ c_0 \\ p_0 \\ q_0 \end{bmatrix}$$

where $A : \mathcal{X} \to \mathcal{X}$,

$$A \begin{bmatrix} f \\ c \\ p \\ q \end{bmatrix} = \begin{bmatrix} -\delta f \\ \eta \triangle c \\ -\frac{\partial}{\partial a} p - \frac{\partial}{\partial s} (g(s) p) + \alpha \triangle - \theta p \\ \gamma \triangle q \end{bmatrix},$$

$$D(A) = \{ [f, c, p, q]^T \in \mathcal{X} : c, p, q \in D(A),$$
$$p(0, s) = \int_0^{a_1} \int_{s_0}^{s_1} \kappa(\hat{s}, s) \theta(a, \hat{s}) p(a, \hat{s}) d\hat{s} da, p(a, s_0) = q(a, s_0) = 0 \}$$

and $\mathcal{F} : \mathcal{X} \to \mathcal{X}$, $D(\mathcal{F}) = \mathcal{X}$,

$$
\mathcal{F}\left(\begin{bmatrix} f \\ c \\ p \\ q \end{bmatrix}\right) = \begin{bmatrix} 0 \\ \beta f - \omega(p)c \\ -\rho(c,n)p - \sigma(c,n)p + \tau(c,n)q \\ -\nu(c,n)q + \sigma(c,n)p - \tau(c,n)q \end{bmatrix}.
$$

With an appropriate Lipschitz continuity condition on $\mathcal{F}$, the generalized solutions of (M.3.5) are given by the strongly continuous semigroup of nonlinear operators $\mathcal{U}(t), t \geq 0$ in $\mathcal{X}$ satisfying

$$
\mathcal{U}(t)\Phi = \mathcal{S}(t)\Phi + \int_0^t \mathcal{S}(t-u)\mathcal{F}(\mathcal{U}(t-u)\Phi)du, \Phi \in \mathcal{X}, t \geq 0,
$$

where $\Phi = [f_0, c_0, p_0, q_0]^T$ is the initial condition,

$$
\mathcal{S}(t)\begin{bmatrix} f_0 \\ c_0 \\ p_0 \\ q_0 \end{bmatrix} = \begin{bmatrix} e^{-\delta t}f_0 \\ T(\eta t)c_0 \\ \hat{S}(t)p_0 \\ \hat{T}(t)q_0 \end{bmatrix},
$$

$\hat{S}(t), t \geq 0$ is the strongly continuous linear semigroup in $\hat{Y}$ as in Theorem 8, and $\hat{T}(t), t \geq 0$ is the linear semigroup in $\hat{Y}$ defined by $(\hat{T}(t)\hat{\phi})(a,s) = T(\gamma t)\hat{\phi}(a,s), \hat{\phi} \in \hat{Y}, a \in (0, a_1), s \in (s_0, s_1), t \geq 0$.

The analysis of the general model (M.3.4) presents many difficulties. The haptotaxis term in (1.97), which is nonlinearly dependent on the proliferating cell density through the matrix degradative enzyme MDE, is particularly difficult to handle. Model (M.3.4) allows nonlinearities of the division moduli $\theta_i(x, a, s)$, $i = 1, 2, \ldots, n$ in (1.98) and (1.101), which are dependent on the oxygen concentration $c(x, t)$ and the total cell population density $n(x, t)$. The methods developed here can be extended to treat this nonlinear dependence, which is important in understanding tumor growth. In [119] a special case of (M.3.4), with haptotaxis, but without age or size dependence, is analyzed and the global existence of unique solutions is proved. In [16] a numerical treatment of (M.3.4) is given in the case with haptotaxis and age dependence in 2-dimensional spatial regions, but without size dependence, and without mutated phenotype classes. In [118] an analysis of (M.3.4) with age structure, but without size structure and with only one phenotype class is given. Treatment of the general age-size-space structured model (M.3.4) offers major challenges, both analytically and computationally, and is the objective of further study.

In this chapter population models involving continuum age, size, and spatial structuring have been developed. The analysis of these models requires a broad range of mathematical techniques to treat the effects of individual constituents on total population behavior. Models such as these, which incorporate detailed information about individual behavior, have great potential to inform scientific issues in biology. Modern biological research produces immense quantities of data, which is amenable to mathematical modeling and computer simulation. The models developed in the other chapters of this volume are also based on highly detailed, multi-scaled, and intricately structured formulations: discrete spatial patch-size structure and host parasite-load structure in models of macroparasitic disease transmission (Chap. 2); spatial structuring in models of microparasitic disease transmission (Chap. 3); physiological and age structure in models of the spatial spread of infectious diseases (Chap. 4); multi-variable models relating macro and microvariables in ecology (Chap. 5); and models of multiple-species competition in flow reactors (Chap. 6).

Acknowledgement

Supported by PHS-NIH grant 1P50CA113007-01.

References

1. J. Adam, N. Bellomo: *A Survey of Models for Tumor-Immune System Dynamics* (Birkhäuser, Boston 1997)
2. J. Adam, S. Maggelakis: Diffusion related growth characteristics of a spherical prevascular carcinoma, Bull. Math. Biol. **52** (1990) pp 549–582
3. B. Ainseba: Age-dependent population dynamics diffusive systems, Disc. Cont. Dyn. Sys.-Ser. B **4** (2004) pp 1233–1247
4. B. Ainseba, S. Anita: Local exact controllability of the age-dependent population dynamics with diffusion, Abst. Appl. Anal. **6** (2001) pp 357–368
5. B. Ainseba, B. M. Langlais: On a population control problem dynamics with age dependence and spatial structure, J. Math. Anal. Appl. **248** (2000) pp 455–474
6. J. Al-Omari, S. A. Gourley: Monotone travelling fronts in an age-structured reaction-diffusion model of a single species, J. Math. Biol. **45** (2002) pp 294–312
7. D. Ambrosi, L. Preziosi: On the closure of mass balance models for tumor growth, Math. Mod. Meth. Appl. Sci. **12** (2002) pp 737–754
8. A. Anderson: A hybrid mathematical model of solid tumour invasion: The importance of cell adhesion, Math. Med. Biol. **22** (2005) pp 163–186
9. A. Anderson, M. Chaplain: Mathematical modelling, simulation and prediction of tumour-induced angiogenesis, Invasion Metastasis **16** (1996) pp 222–234
10. A. Anderson, M. Chaplain: Continuous and discrete mathematical models of tumor-induced angiogenesis, Bull. Math. Biol. **60** (1998) pp 857–899
11. W. Arendt, A. Grabosch, G. Greiner, U. Groh, H. Lotz, U. Moustakas, R. Nagel, F. Neubrander, U. Schlotterbeck: *One-Parameter Semigroups of Positive Operators* (Springer Lecture Notes in Mathematics **1184**, New York 1986)
12. O. Arino: A survey of structured cell-population dynamics, Acta Biotheoret. **43** (1995) pp 3–25
13. B. Ayati: A variable time step method for an age-dependent population model with nonlinear diffusion, SIAM J. Num. Anal. **37** (2000) pp 1571–1589

14. B. Ayati: A structured-population model of *Proteus mirabilis* swarm-colony development, J. Math. Biol. **52** (2006) pp 93–114
15. B. Ayati, T. Dupont: Galerkin methods in age and space for a population model with nonlinear diffusion, SIAM J. Num. Anal. **40** (2000) pp 1064–1076
16. B. Ayati, G. Webb, R. Anderson: Computational methods and results for structured multiscale models of tumor invasion, SIAM J. Multsc. Mod. Simul. **5** (2006) pp 1–20
17. J. Banasiak, L. Arlotti: *Perturbations of Positive Semigroups with Applications* (Springer, Berlin Heidelberg New York, 2006)
18. N. Bellomo, L. Preziosi: Modelling and mathematical problems related to tumor evolution and its interaction with the immune system, Math. Comp. Mod. **32** (2000) pp 413–452
19. G. Bell, E. Anderson: Cell growth and division I. A mathematical model with applications to cell volume distributions in mammalian suspension cultures, Biophys. J. **7** (1967) pp 329–351
20. R. Bellman, K. Cooke: *Differential Difference Equations* (Academic, New York 1963)
21. A. Bertuzzi, A. Gandolfi: Cell kinetics in a tumor cord, J. Theor. Biol. **204** (2000) pp 587–599
22. A. Bertuzzi, A. d'Onofrio, A. Fasano, A. Gandolfi: Modeling cell populations with spatial structure: steady state and treatment-induced evolution of tumor cords, Discr. Cont. Dyn. Sys. B **4** (2004) pp 161–186
23. A. Bressan: *Hyperbolic Systems of Conservation Laws*, (Oxford University Press, Oxford 2000)
24. S. Busenberg, M. Iannelli: Separable models in age-dependent population dynamics, J. Math. Biol. **22** (1985) pp 145–173
25. S. Busenberg, M. Iannelli: A class of nonlinear diffusion problems in age-dependent population dynamics, Nonl. Anal. Theory. Meth. Appl. **7** (1983) pp 501–529
26. S. Busenberg, K. Cooke: *Vertically Transmitted Diseases*, (Springer Biomathematics **23**, New York 1992)
27. H. Byrne, M. Chaplain: Free boundary value problems associated with the growth and development of multicellular spheroids, Eur. J. Appl. Math. **8** (1997) pp 639–658
28. H. M. Byrne, S. A. Gourley: The role of growth in avascular tumor growth, Math. Comp. Mod. **26** (1997) 35–55
29. H. Byrne, L. Preziosi: Modelling solid tumour growth using the theory of mixtures, Math. Med. Biol. **20** (4) (2003) pp 341–366
30. C. Castillo-Chavez, Z. Feng: Global stability of an age-structure model for TB and its applications to optimal vaccination, Math. Biosci. **151** (1984) pp 135–154
31. W. Chan, B. Guo: On the semigroups of age-size dependent population dynamics with spatial diffusion, Manusc. Math. **66** (1989) pp 161–180
32. W. L. Chan, B. Z. Guo: On the semigroups for age dependent population dynamics with spatial diffusion source, J. Math. Anal. Appl. **184** (1994) pp 190–199
33. Ph. Clement, H. Heijmans, S. Angenent, C. van Duijn, B. de Pagter: *One-Parameter Semigroups* (North Holland, Amsterdam 1987)
34. A. Coale: *The Growth and Structure of Human Populations* (Princeton University Press, Princeton 1972)

35. S. Cui, A. Friedman: Analysis of a mathematical model of the growth of necrotic tumors, J. Math. Anal. Appl. **255** (2001) pp 636–677

36. J. Cushing: *An Introduction to Structured Population Dynamics* (SIAM, Philadelphia 1998)

37. C. Cusulin, M. Iannelli, G. Marinoschi: Age-structured diffusion in a multi-layer environment, Nonl. Anal. **6** (2005) pp 207–223

38. E. DeAngelis, L. Preziosi: Advection-dffusion models for solid tumour evolution *in vivo* and related free boundary problem, Math. Mod. Meth. Appl. Sci. **10** (2000) pp 379–407

39. M. Delgado, M. Molina-Becerra, A. Surez: A nonlinear age-dependent model with spatial diffusion, J. Math. Anal. Appl. **313** (2006) pp 366–380

40. Q. Deng, T. Hallam: An age structured population model in a spatially heterogeneous environment: Existence and uniqueness theory, Nonl. Anal. **65** (2206) pp 379–394

41. G. Diblasio: Non-linear age-dependent population diffusion, J. Math. Biol. **8** (1979) pp 265–284

42. O. Diekmann, Ph. Getto: Boundedness, global existence and continuous dependence for nonlinear dynamical systems describing physiologically structured populations, J. Dif. Eqs. **215** (2005) pp 268–319

43. O. Diekmann, M. Gyllenberg, H. Huang, M. Kirkilionis, J.A.J. Metz, H. Thieme: On the formulation and analysis of general deterministic structured population models II. Nonlinear theory, J. Math. Biol. **2** (2001) pp 157–189

44. O. Diekmann, H. Heijmans, H. Thieme: On the stability of the cell size distribution, J. Math. Biol. **19** (1984) pp 227–248

45. O. Diekmann, H. Heijmans, H. Thieme: On the stability of the cell size distribution II. Time-periodic development rates, Comp. Math. Appl. Part A **12** (1986) pp 491–512

46. J. Dyson, R. Villella-Bressan, G. Webb: Asymptotic behavior of solutions to abstract logistic equations, Math. Biosci. **206** (2007) pp 216–232

47. J. Dyson, R. Villella-Bressan, G. Webb: The evolution of a tumor cord cell population, Comm. Pure Appl. Anal. **3** (2004) pp 331–352

48. J. Dyson, R. Villella-Bressan, G. Webb: The steady state of a maturity structured tumor cord cell population, Discr. Cont. Dyn. Sys. B, **4** (2004) pp 115–134

49. K-J. Engel, R. Nagel: *One-Parameter Semigroups for Linear Evolution Equations*, K-J. Engel, R. Nagel, Eds., (Springer Graduate Texts in Mathematics **194**, New York, 2000)

50. G. Fragnelli, L. Tonetto: A population equation with diffusion, J. Math. Anal. Appl. **289** (2004) pp 90–99

51. W. Feller: On the integral equation of renewal theory, Ann. Math. Stat. **12** (1941) pp 243–267

52. Z. Feng, W. Huang, C. Castillo-Chavez: Global behavior of a multi-group SIS epidemic model with age structure, J. Diff. Eqs. **218**(2) (2005) pp 292–324

53. Z. Feng, C-C. Li, F. Milner: Schistosomiasis models with density dependence and age of infection in snail dynamics, Math. Biosci. **177–178** (2002) pp 271–286

54. W. Fitzgibbon, M. Parrott, G. Webb: Diffusion epidemic models with incubation and crisscross dynamics, Math. Biosci. **128** (1995) pp 131–155

55. W. Fitzgibbon, M. Parrott, G. Webb: A diffusive age-structured SEIRS epidemic model, Meth. Appl. Anal. **3** (1996) pp 358–369

56. A. Friedman: A hierarchy of cancer models and their mathematical challenges, Discr. Cont. Dyn. Sys. B **4** (2004) pp 147–160
57. A. Friedman, F. Reitich: On the existence of spatially patterned dormant malignancies in a model for the growth of non-necrotic vascular tumors, Math. Mod. Meth. Appl. Sci. **77** (2001) pp 1–25
58. M. Garroni, M. Langlais: Age-dependent population diffusion with external constraint, J. Math. Biol. **14** (1982) pp 77–94
59. J. Goldstein: *Semigroups of Linear Operators and Applications* (Oxford University Press, New York 1985)
60. H. Greenspan: Models for the growth of solid tumor by diffusion, Stud. Appl. Math. **51** (1972), pp 317–340
61. G. Greiner: A typical Perron-Frobenius theorem with applications to an age-dependent population equation, in *Infinite Dimensional Systems*, Eds. F. Kappel, W. Schappacher (Springer Lecture Notes in Mathematics **1076**, 1989) pp 786–100
62. G. Greiner, R. Nagel: Growth of cell populations via one-parameter semigroups of positive operators in Mathematics Applied to Science, Eds. J.A. Goldstein, S. Rosencrans, and G. Sod, Academic Press, New York (1988) pp 79–104
63. G. Grippenberg, S-O. Londen, O. Staffans: *Volterra Integral and Functional Equations* (Cambridge University Press, Cambridge 1990)
64. M. Gurtin, R. MacCamy: Nonlinear age-dependent population dynamics, Arch. Rat. Mech. Anal. **54** (1974) pp 281–300
65. M. Gurtin, R. MacCamy: Diffusion models for age-structured populations, Math. Biosci. **54** (1981) pp 49–59
66. M. Gyllenberg: Nonlinear age-dependent poulation dynamics in continuously propagated bacterial cultures, Math. Biosci. **62** (1982) pp 45–74
67. H. Heijmans: The dynamical behaviour of the age-size-distribution of a cell population, in *The Dynamics of Physiologically Structured Populations* (Springer Lecture Notes in Biomathematics **68**, 1986) pp 185–202
68. E. Hille, R. Phillips: *Functional Analysis and Semigroups*, (American Mathematical Society Colloquium Publications, Providence 1957)
69. F. Hoppensteadt: *Mathematical Theories of Populations: Demographics, Genetics, and Epidemics* (SIAM, Philadelphia 1975)
70. C. Huang: An age-dependent population model with nonlinear diffusion in R^n, Quart. Appl. Math. **52** (1994) pp 377–398
71. W. Huyer: A size-structured population model with dispersion, J. Math. Anal. Appl. **181** (1994) pp 716–754
72. W. Huyer: Semigroup formulation and approximation of a linear age-dependent population problem with spatial diffusion, Semigroup Forum **49** (1994) pp 99–114
73. M. Iannelli: *Mathematical Theory of Age-Structured Population Dynamics* (Giardini Editori e Stampatori, Pisa 1994)
74. M. Iannelli, M. Marcheva, F. A. Milner: *Gender-Structured Population Modeling*, (Mathematical Methods, Numerics, and Simulations, SIAM, Philadelphia 2005)
75. H. Inaba: *Mathematical Models for Demography and Epidemics*, (University of Tokyo Press, Tokyo 2002)
76. W. Kermack, A. McKendrick: Contributions to the mathematical theory of epidemics III. Further studies on the problem of endemicity, Proc. Roy. Soc. **141** (1943) pp 94–122

77. N. Keyfitz: *Introduction to the Mathematics of Population*, (Addison Wesley, Reading 1968)
78. K. Kubo, M. Langlais: Periodic solutions for a population dynamics problem with age-dependence and spatial structure, J. Math. Biol. **29** (1991) pp 363–378
79. K. Kunisch, W. Schappacher, G. Webb: Nonlinear age-dependent population dynamics with diffusion, Inter. J. Comput. Math. Appl. **11** (1985) pp 155–173
80. M. Langlais: A nonlinear problem in age-dependent population diffusion, SIAM J. Math. Anal. **16** (1985) pp 510–529
81. M. Langlais: Large time behavior in a nonlinear age-dependent population dynamics problem with spatial diffusion, J. Math. Biol. **26** (1988) pp 319–346
82. M. Langlais, F. A. Milner: Existence and uniqueness of solutions for a diffusion model of host-parasite dynamics, J. Math. Anal. Appl **279** (2003) pp 463–474
83. H. Levine, S. Pamuk, B. Sleeman, M. Nilsen-Hamilton: Mathematical modeling of capillary formation and development in tumor angiogenesis: Penetration into the stroma, Bull. Math. Biol. **63** (2001) pp 801–863
84. H. Levine, B. Sleeman: Modelling tumor-induced angiogenesis, in *Cancer Modelling and Simulation*, Ed. L. Preziosi (Chapman and Hall/CRC, Boca Raton, 2003)
85. A. Lotka: The stability of the normal age-distribution, Proc. Natl. Acad. Sci. USA **8** (1922) pp 339–345
86. A. Lotka: On an integral equation in population analysis, Ann. Math. Stat. **10** (1939) pp 1–35
87. R. MacCamy: A population problem with nonlinear diffusion, J. Diff. Eqs. **39** (21) (1981) pp 52–72
88. P. Magal: Compact attractors for time-periodic age-structured population models, Elect. J. Dif. Eqs. **65** (2001) pp 1–35
89. P. Magal, H. R. Thieme: Eventual compactness for semiflows generated by nonlinear age-structured models, Comm. Pure Appl. Anal. **3** (2004) pp 695–727
90. P. Marcati: Asymptotic behavior in age dependent population diffusion model, SIAM J. Math. Anal. **12** (1981) pp 904–914
91. R. Martin: *Nonlinear Operators and Differential Equations in Banach Spaces*, (Wiley-Interscience, New York 1976)
92. A. McKendrick: Applications of mathematics to medical problems, Proc. Edin. Math. Soc. **44** (1926) pp 98–130
93. J. Metz, O. Diekmann: *The Dynamics of Physiologically Structured Populations* (Springer Lecture Notes in Biomathematics **68**, New York 1986)
94. R. Nagel: *One-Parameter Semigroups of Positive Operators*, Ed. R. Nagel (Springer, Berlin Heidelberg New York 1986)
95. J. Nagy: The ecology and evolutionary biology of cancer: A review of mathematical models of necrosis and tumor cell diversity, Math. Biosci. Eng. **2** (2005) pp 381–418
96. A. Pazy: *Semigroups of Linear Operators and Applications to Partial Differential Equations* (Springer Series in Applied Mathematical Sciences, New York 1983)
97. J. Pollard: *Mathematical Models for the Growth of Human Populations*, (Cambridge University Press, Cambridge 1973)
98. J. Prüss: Equilibrium solutions of age-specific population dynamics of several species, J. Math. Biol. **11** (1981) pp 65–84
99. J. Prüss: On the qualitative behaviour of populations with age-specific interractions, Comp Math. Appl. **9** (1983) pp 327–339

100. J. Prüss: Stability analysis for equilibria in age-specific population dynamics, Nonl. Anal. **7** (1983) pp 1291–1313
101. A. Rhandi: Positivity and stability for a population equation with diffusion on L^1, Positivity **2** (1998) pp 101–113
102. A. Rhandi, R. Schnaubelt: Asymptotic behaviour of a non-autonomous population equation with diffusion in L^1, Discr. Cont. Dyn. Sys. **5** (1999) pp 663–683
103. I. Segal: Nonlinear semigroups, Ann. Math. **78** (1963) pp 339–364
104. F. Sharpe, A. Lotka: A problem in age-distribution, Philos. Mag. **6** (1911) pp 435–438
105. J. Sherratt, M. Chaplain: A new mathematical model for avascular tumour growth, J. Math. Biol. **43** (2001) pp 291–312
106. J. Sinko, W. Streifer: A new model for age-size structure of a population, Ecology **48** (1967) pp 910–918
107. B. Sleeman, A. Anderson, M. Chaplain: A mathematical analysis of a model for capillary network formation in the absence of endothelial cell proliferation, Appl. Math. Lett. **12** (1999) pp 121–127
108. J. W.-H So, J. Wu, X. Zou: A reaction–diffusion model for a single species with age structure. I Travelling wavefronts on unbounded domains, Proc. Roy. Soc. **457** (2001) pp 1841–1853
109. H. R. Thieme: Semiflows generated by Lipschitz perturbations of non-densely defined operators, Dif. Int. Eqs. **3** (1990) pp 1035–1066
110. H. R. Thieme: Analysis of age-structured population models with an additional structure, in *Mathematical Population Dynamics, Proceedings of the 2nd International Conference*, Rutgers University 1989, Eds. O. Arino, D.E. Axelrod, M. Kimmel (Lecture Notes in Pure and Applied Mathematics **131**, 1991) pp 115–126
111. H. R. Thieme: Quasi-compact semigroups via bounded perturbation, in *Advances in Mathematical Population Dynamics: Molecules, Cells and Man*, Eds. O. Arino, D. Axelrod, M. Kimmel (World Scientific, Singapore 1997) pp 691–711
112. H. R. Thieme: Positive perturbation of operator semigroups: Growth bounds, essential compactness, and asynchronous exponential growth, Disc. Cont. Dyn. Sys. **4** (1998) pp 735–764
113. H. Thieme: Balanced exponential growth of operator semigroups, J. Math. Anal. Appl. **223** (1998) pp 30–49
114. H. R. Thieme: *Mathematics in Populations Biology*, (Princeton University Press, Princeton 2003)
115. H. R. Thieme, J. I. Vrabie: Relatively compact orbits and compact attractors for a class of nonlinear evolution equations, J. Dyn. Diff. Eqs. **15** (2003) pp 731–750
116. S. Tucker, S. Zimmerman: A nonlinear model of population-dynamics containing an arbitrary number of continuous structure variables, SIAM J. Appl. Math. **48** (1988) pp 549–591
117. H. Von Foerster: Some remarks on changing populations, in *The Kinetcs of Cellular Proliferation*, Ed. F. Stohlman (Grune and Stratton, New York 1959)
118. C. Walker: Global well-posedness of a haptotaxis model with spatial and age structure, Differ. Integral Equations Athens **20** (9) (2007) pp 1053–1074
119. C. Walker, G. Webb: Global existence of classical solutions to a haptotaxis model, SIAM J. Math. Anal. **38** (5) (2007) pp 1694–1713

120. G. Webb: Lan age-dependent epidemic model with spatial diffusion, Arch. Rat. Mech. Anal. **75** (1980) pp 91–102
121. G. Webb: Diffusive age-dependent population models and a application to genetics, Math. Biosci. **61** (1982) pp 1–16
122. G. Webb: *Theory of Nonlinear Age-Dependent Population Dynamics* (Marcel Dekker **89**, New York 1985)
123. G. Webb: Logistic models of structured population growth, J. Comput. Appl. **12** (1986) pp 319–335
124. G. Webb: An operator-theoretic formulation of asynchronous exponential growth, Trans. Amer. Math. Soc. **303** (1987) pp 751–763
125. G. Webb: Periodic and chaotic behavior in structured models of cell population dynamics, in *Recent Developments in Evolution Equations*, Pitman Res. Notes Math. Ser. **324** (1995) pp 40–49
126. G. Webb: The steady state of a tumor cord cell popultion, J. Evol Eqs. **2** (2002) pp 425–438
127. G. Webb: Structured population dynamics, in *Mathmatical Modelling of Population Dynamics* (Banach Center Publications **63**, Institute of Mathematics, Polish Academy of Sciences, Warsaw 2004)

2

Infinite ODE Systems Modeling Size-Structured Metapopulations, Macroparasitic Diseases, and Prion Proliferation

M. Martcheva[1] and H.R. Thieme[2]

[1] Supported by NSF grants DMS-0137687 and DMS-0406119, Department of Mathematics, University of Florida, 358 Little Hall, P.O. Box 118105, Gainesville, FL 32611–8105, USA
maia@math.ufl.edu

[2] Supported by NSF grants DMS-9706787 and DMS-0314529, Department of Mathematics and Statistics, Arizona State University, P.O. Box 871804, Tempe, AZ 85287-1804, USA
thieme@math.asu.edu

Summary. Infinite systems of ordinary differential equations can describe:

- Spatially implicit metapopulation models with discrete patch-size structure
- Host-macroparasite models that distinguish hosts by their parasite loads
- Prion proliferation models that distinguish protease-resistant protein aggregates by the number of prion units they contain

It is the aim of this chapter to develop a theory for infinite ODE systems in sufficient generality (based on operator semigroups) and, besides well-posedness, to establish conditions for the solution semiflow to be dissipative and have a compact attractor for bounded sets. For metapopulations, we present conditions for uniform persistence on the one hand and prove on the other hand that a metapopulation dies out, if there is no emigration from birth patches or if empty patches are not colonized.

2.1 Introduction

Infinite systems of ordinary differential equations,

$$w' = f(t, w, x),$$

$$x'_j = \sum_{j=0}^{\infty} \alpha_{jk} x_k + g_j(t, w, x), \qquad j = 0, 1, 2, \ldots. \tag{2.1}$$

where $x(t)$ is the sequence of functions $(x_j(t))_{j=0}^{\infty}$, can describe:

- Spatially implicit metapopulation models with discrete patch-size structure [2, 5, 7, 38, 42]
- Host-macroparasite models which distinguish hosts by their parasite loads [6, 13, 24, 25, 33, 34, 48, 49]
- Prion proliferation models which distinguish protease-resistant protein aggregates by the number of prion units they contain [41, 45]

Spatially Implicit Metapopulation Models

A metapopulation is a group of populations of the same species which occupy separate areas (patches) and are connected by dispersal. Each separate population in the metapopulation is referred to as a local population. Metapopulations occur naturally or by human activity as a result of habitat loss and fragmentation.

In system (2.1), x_j denotes the number of patches with j occupants and w the average number of migrating individuals, or wanderers. The coefficients α_{jk} describe the transition from patches with k occupants to patches with j occupants due to deaths, births and emigration of occupants. The function f gives the rate of change of the number of dispersers due to patch emigration, immigration and disperser death. The functions g_j describe the rate of change of the numbers of patches with j occupants due to the immigration of dispersers. The coefficients α_{jk} have the properties typical for infinite transition matrices in stochastic processes with continuous time and discrete state (continuous-time birth and death chains, e.g., see [1] and the references therein). Since they form an unbounded set, existence and uniqueness of solutions to (2.1) is non-trivial. It is the aim of this chapter to develop a this-related theory in sufficient generality and also establish conditions for the solution semiflow to be dissipative [26], have a compact attractor for bounded sets [26, 52], and be uniformly persistent [5, 27, 56, 58]. We also prove that a metapopulation dies out, if nobody emigrates from its birth patch or if empty patches are not colonized.

It is worth mentioning that, though the linear special case $x_j' = \sum_{k=0}^{\infty} \alpha_{jk} x_k$ can be interpreted as a stochastic model for a population that is not distributed over patches [39], the model (2.1) is a deterministic model. It inherits the property though that subpopulations on individual patches can become extinct at finite time which is an important feature of real metapopulations. As a trade-off, the metapopulation model (2.1) is spatially implicit and not able to take spatial heterogeneities into account. A spatially explicit metapopulation model would be a finite system of ordinary differential equations $y_j' = \sum_{j=1}^{N} d_{jk} y_k + f_j(t, y)$, $j = 1, \ldots, N$, where N is the number of patches and y_j the size of the local population on patch j. The coefficients d_{jk} would describe the movement from patch k to patch j and the non-linearities f_j the local demographics on patch j due to births and deaths. An example of a spatially explicit metapopulation model (underlying an epidemic model) can be found in Chap. 4. Spatially explicit models can take account of how the patches are situated relatively to each other and of differences between the patches, but do not have the property that a local population can become extinct in finite time. The most basic spatially implicit metapopulation model is the Levins model [36,37] which only considers empty and occupied patches. Incorporating a structure which distinguishes between patches according to local population size makes it possible, e.g., to compare emigration strategies which are based on how crowded a patch is [38].

Alternatively, spatially implicit metapopulation models can be structured by a continuous rather than a discrete variable. This leads to non-local partial differential equations or integral equations [23]. The partial differential equations one obtains are similar to those considered in Chap. 1, but have non-linear terms in the derivative with respect to the size-structure variable. For general information on mathematical metapopulation theory we refer to [20,28,38].

Host-Macroparasite Models

The connection between metapopulation and host-macroparasite models is not incidental as a macroparasite population is a metapopulation with the hosts being the patches and the parasites in single hosts forming the local populations. In the epidemiology of infectious diseases, the Levins metapopulation model corresponds to a prevalence model that only considers susceptible and infective individuals. Such models (possibly after adding classes which take account of incubation and immunity) are quite adequate for microparasitic (viral, bacterial, fungal) diseases where the infectious agents multiply rapidly and it basically only matters whether a host is infectious or not. Macroparasitic (worm, e.g.) diseases, however, are characterized by highly variable parasite loads in individual hosts with very different effects on host health. Models like (2.1), called density models in [13], can take these into account with x_j denoting the number of hosts with j parasites and w denoting the average number of free-living parasites. The coefficients α_{jk} describe the transition from hosts with k parasites to hosts with j parasites due to deaths, births and release of parasites. The function f gives the rate of change of the number of free-living parasites due to death and the entry into or exit from hosts. The functions g_j describe the rate of change of the numbers of hosts with j parasites due to the acquisition of parasites from the pool of free-living parasites.

Since parasite loads often depend on the age of the host, host age has been included into density models [6,13,24,25,34,49]. This leads to an infinite system of partial differential equations. The analysis of these models uses moment generating functions. The use of a generating function would convert the system (2.1) into a single partial differential equation. An infinite system of partial differential equations incorporating age dependence would also be converted into a single partial differential equation, however with one more variable and partial derivative. This approach yields impressive and illuminating results, but requires the transition matrix to correspond to a simple birth and death process (possibly with catastrophes). Levins type metapopulation models with patch age have been considered in [19].

Models for Prion Proliferation

Prion proteins have been linked to fatal diseases called transmissible spongiform encephalopathies (TSE) including Creutzfeldt–Jakob disease (CJD), kuru, scrapie, and bovine spongiform encephalopathy (BSE, "mad cow disease"). The prion diseases in an individual host are associated with the accumulation of single prion proteins (monomers) into prion protein aggregates (polymers). An aggregate is a stringlike formation possibly containing several thousand units with each unit being a former monomer. Monomers are considered healthy because they can easily be degraded by proteinase while polymers are much more proteinase-resistant and are neurotoxic. The system (2.1), with some modification, covers the models of prion proliferation suggested in [41, 45]. Since a detailed derivation of a special metapopulation model can already be found in [38] (cf. Sect. 2.12), we explain the prion model in some more detail here.

The amount of aggregates which contain j prion units (former monomers) is represented by x_j while the amount of (healthy) prion monomers is w. We assume that aggregates grow by adding one monomer at a time, the respective rate is σ_j for an aggregate to grow from j to $j+1$ units. This process is sometimes called polymerization. An aggregate of size k can break into two pieces of sizes j and $k-j$: the respective per unit rate is b_{jk} if $j \leq k-j$ and $b_{k-j,k}$ if $j \geq k-j$. Aggregates of size j are chemically degraded at a rate κ_j while single monomers are degraded at a rate δ. Monomers are produced at a constant rate Λ. The model in [41] has the form

$$w' = \Lambda - w \sum_{k=1}^{\infty} \sigma_k x_k - \delta w,$$

$$x_j' = w(\sigma_{j-1} x_{j-1} - \sigma_j x_j) - \kappa_j x_j + \sum_{k=j+1}^{\infty} (b_{jk} + b_{k-j,k}) x_k - x_j \sum_{k=1}^{j-1} b_{kj}, \qquad (2.2)$$

$$j = 1, 2, \ldots, \qquad \sigma_0 = 0.$$

Notice that the polymerization rate is of mass action type as it involves the product of the amount of monomers w and the amount of polymers containing $j-1$ or j units respectively. There is no equation for x_0 because aggregates containing 0 units do not exist, differently from empty patches in metapopulations or hosts without worms in macroparasite diseases. To fit (2.2) into (2.1), without an x_0-equation, we set

$$\alpha_{jk} = \begin{cases} b_{jk} + b_{k-j,k}, & 1 \le j \le k-1, \\ -\kappa_k - \displaystyle\sum_{i=1}^{k-1} b_{ik}, & 1 \le j = k, \\ 0, & j > k \ge 1. \end{cases} \tag{2.3}$$

The model in [45] (see also [41, App.A]) allows for the fact that small aggregates below a certain minimum size, m, are unstable. So, if an aggregate splits and one of the pieces has a size less than m, it immediately disintegrates into monomers,

$$w' = \Lambda - w \sum_{k=1}^{\infty} \sigma_k x_k - \delta w + \sum_{j=1}^{m-1} j \sum_{k=m}^{\infty} (b_{jk} + b_{j-k,k}) x_k,$$

$$x'_j = w(\sigma_{j-1} x_{j-1} - \sigma_j x_j) - \kappa_j x_j + \sum_{k=j+1}^{\infty} (b_{jk} + b_{k-j,k}) x_k - x_j \sum_{k=1}^{j-1} b_{kj}, \tag{2.4}$$

$$j = m, m+1, \ldots, \qquad \sigma_{m-1} = 0.$$

The system (2.1) can be adapted to this model by striking the equations for $x_0, \ldots, x_{m-1}$ and defining the coefficients $(\alpha_{jk})_{j,k=m}^{\infty}$ as in (2.3) with the modification that $j \ge m$ and $k \ge m$. Analogous models where the amount of units in an aggregate are modeled by a continuous rather than a discrete variable have been considered in [15, 21, 35, 47, 53, 61]. Saturation effects in polymerization have been incorporated in [22].

The system (2.4) includes the special case that b_{jk} is constant for $1 < m \le j < k$ which may be a reasonable approximation of reality, while the assumption that b_{jk} is constant for $1 \le j < k$ is clearly unrealistic. This special case allows a moment closure which reduces the infinite system of ODEs to a system of three ODEs which can been completely analyzed (cf. [47]). Since we consider the case of variable b_{jk} here, it is not clear whether to favor system (2.2) or system (2.4). Notice that there is another conceptual difference between the systems. System (2.2) distinguishes between monomers which have been part of an aggregate before (in other words aggregates consisting of one unit), represented by the variable x_1, and the "virgin" monomers represented by w. Only virgin monomers are attached to the aggregates. Such a distinction between monomers and single unit aggregates is not made in system (2.4) where the monomers resulting from aggregate disintegration return to the monomer class represented by w. A drawback of system (2.4) may be that it could be very difficult to assign a specific value to m. Thinking along the lines that polymer splitting can result in complete disintegration, it seems to be more realistic (and mathematically more difficult) to assume that for each $j \in \mathbb{N}$ there is a probability $q_j \in [0, 1]$ of a piece of j units to disintegrate into monomers after polymer splitting,

$$w' = \Lambda - w \sum_{k=1}^{\infty} \sigma_k x_k - \delta w + \sum_{j=1}^{\infty} j q_j \sum_{k=j+1}^{\infty} (b_{jk} + b_{j-k,k}) x_k,$$

$$x'_j = w(\sigma_{j-1} x_{j-1} - \sigma_j x_j) - \kappa_j x_j$$

$$+ (1 - q_j) \sum_{k=j+1}^{\infty} (b_{jk} + b_{k-j,k}) x_k - x_j \sum_{k=1}^{j-1} b_k,$$

$$j = 1, 2, \ldots, \qquad \sigma_0 = 0.$$

$\qquad$ (2.5)

Obviously this system encompasses the two previous ones.

Outline of the Mathematical Approach

For the mathematical treatment of (2.1), we choose a somewhat more abstract approach than the ones in [2] and [5] from which we have received much inspiration in order to include a variety of models (in Sect. 2.12 and [38] we assume that only juveniles migrate) and to include state transitions which are not of nearest-neighbor type like in the prion proliferation models. The biological interpretation (restricted here to metapopulation and host-macroparasite systems) gives us guidance how to choose the appropriate state space. Assuming that meaningful solutions are non-negative, the number of patches (hosts) is given by $\sum_{j=0}^{\infty} x_j$ and the number of occupants (in-host parasites) by $\sum_{j=1}^{\infty} j x_j$. Recall the sequence space

$$\ell^1 = \left\{ (x_j)_{j=0}^{\infty}; \ x_j \in \mathbb{R}, \sum_{j=0}^{\infty} |x_j| < \infty \right\} \qquad (2.6)$$

with norm

$$\|x\| = \sum_{j=0}^{\infty} |x_j|, \qquad x = (x_j)_{j=0}^{\infty}. \qquad (2.7)$$

We introduce the subspace

$$\ell^{11} = \left\{ (x_j)_{j=0}^{\infty}; \ x_j \in \mathbb{R}, \sum_{j=0}^{\infty} j|x_j| < \infty \right\} \qquad (2.8)$$

which becomes a Banach space of its own under the norm

$$\|x\|_1 = \sum_{j=0}^{\infty} (1 + j)|x_j|, \qquad x = (x_j)_{j=0}^{\infty}. \qquad (2.9)$$

Other, equivalent, choices are possible, of course. We treat (2.1) as a semilinear operator differential equation

$$w' = f(t, w, x), \qquad x' = A_1 x + g(t, w, x)$$

on the non-negative cone of the Banach space $\mathbb{R} \times \ell^{11}$ where A_1 is the infinitesimal generator of a positive C_0-semigroup on ℓ^{11} and the functions f and $g(\sqcup) = \big(g_j(\sqcup)\big)_{j=0}^{\infty}$ are locally Lipschitz continuous.

The main body of the chapter is structured as follows.

2.2 The Homogeneous Linear System: Kolmogorov's Differential Equation

The linear special case of (2.1),

$$x'_j = \sum_{k=0}^{\infty} \alpha_{jk} x_k, \qquad j \in \mathbb{Z}_+, \tag{2.10}$$

is known as Kolmogorov's differential equation [31] and has been widely studied [17, XVII.9] [18, XIV.7] [29, Sects. 23.10–23.12] [16, 30, 50, 51]. See [4] and [59] for more references. We write $\mathbb{Z}_+$ for the set of non-negative integers and $\mathbb{N}$ for the natural numbers starting at 1, $\mathbb{Z}_+ = \mathbb{N} \cup \{0\}$. We review results proved in [39].

Assumption 1 We make the following assumptions concerning the coefficients α_{jk}, $j, k \in \mathbb{Z}_+$.

(a) $\alpha_{jj} \leq 0 \leq \alpha_{jk}$, $k \neq j$.

(b) $\alpha^{\diamond} := \sup_{k=0}^{\infty} \sum_{j=0}^{\infty} \alpha_{jk} < \infty$.

(c) There exist constants $c_0, c_1 > 0, \epsilon > 0$ such that

$$\sum_{j=1}^{\infty} j\alpha_{jk} \leq c_0 + c_1 k - \epsilon|\alpha_{kk}| \qquad \forall k \in \mathbb{Z}_+.$$

Notice that the sequence $|\alpha_{jj}|$ may be unbounded and is so in many applications. Let ℓ^1 denote the Banach space of real sequences $x = (x_k)_{k=0}^{\infty}$ with $\|x\| := \sum_{k=0}^{\infty} |x_k| < \infty$. ℓ_+^1 denotes the cone of non-negative sequences in ℓ^1.

Recall that a C_0-semigroup on a Banach space X is a family of bounded linear operators on X, $\{S(t); t \geq 0\}$, such that $S(t + s) = S(t)S(s)$ for all $t, s \geq 0$ and $S(t)x \xrightarrow{t \to 0} x = S(0)x$ for all $x \in X$. It follows that $S(t)x$ is a continuous function of $t \geq 0$ for all $x \in X$.

The *infinitesimal generator* of the C_0-semigroup S, A, is defined by

$$A = \lim_{h \to 0+} \frac{1}{h}(S(h)x - x), \qquad x \in D(A),$$

where $D(A)$ is the subspace of elements x where this limit exists. $D(A)$ is dense in X and A is a closed operator. If $x \in D(A)$, then $S(t)x$ is differentiable in $t \geq 0$ and

$$\frac{d}{dt}S(t)x = AS(t)x = S(t)Ax.$$

Notice that the first equation can be interpreted as an abstract linear differential equation. For this and more see the textbooks [4, 9, 14, 29, 32, 40, 46, 52].

Theorem 2. *Let* $x^{[n]} = (x_j^{[n]})_{j=0}^{\infty}$ *be the unique componentwise solution of the (essentially finite) linear system of ordinary differential equations*

$$\begin{aligned} \frac{d}{dt}x_j^{[n]} &= \sum_{k=0}^{n} \alpha_{jk}x_k^{[n]}, & j &= 0, \ldots, n, \\ \frac{d}{dt}x_j^{[n]} &= \alpha_{jj}x_j^{[n]}, & j &> n, \end{aligned} \qquad (2.11)$$

with initial data $x^{[n]}(0) = \check{x}$. *Then* $S^{[n]}(t)\check{x} = x^{[n]}(t)$ *defines a sequence of C_0-semigroups* $S^{[n]}$ *on* ℓ^1. *There exists a C_0-semigroup S on* ℓ^1 *such that* $S^{[n]}(t)\check{x} \to S(t)\check{x}$ *in* ℓ^1 *for every* $\check{x} \in \ell^1$, $t \geq 0$. *If* $\check{x} \in \ell_+^1$, $S^{[n]}(t)\check{x} \in \ell_+^1$, $S(t)\check{x} \in \ell_+^1$, *and the convergence of $S^{[n]}(t)\check{x}$ to $S(t)\check{x}$ as $n \to \infty$ is monotone increasing. The domain of the infinitesimal generator $A^{[n]}$ of $S^{[n]}$ is*

$$D(A^{[n]}) = \left\{ x \in \ell^1; \sum_{j=0}^{\infty} |\alpha_{jj}|\,|x_j| < \infty \right\} =: D_0, \qquad (2.12)$$

and $A^{[n]}x = \left(\sum_{k=0}^{\infty} \alpha_{jk}^{[n]}x_k\right)_{j=0}^{\infty}$, $x = (x_k)_{k=0}^{\infty}$, *with*

$$\alpha_{jk}^{[n]} = \left\{ \begin{array}{l} \alpha_{jk}; \ \ j,k \leq n \\ \alpha_{jj}; \ j = k > n \\ 0; \ \ otherwise \end{array} \right\}, \qquad j,k \in \mathbb{Z}_+. \tag{2.13}$$

The following estimates hold

$$\|S^{[n]}(t)\| \leq \|S(t)\| \leq e^{\alpha^\circ t}, \qquad t \geq 0.$$

On the subspace D_0 introduced in (2.12) we define a linear operator $\breve{A}$,

$$\breve{A}x = \Big(\sum_{k=0}^{\infty} \alpha_{jk} x_k\Big)_{j=0}^{\infty}, \qquad x \in D_0. \tag{2.14}$$

Lemma 1. *Let the Assumption 1 be satisfied. Then D_0 is dense in ℓ^1, $\breve{A} : D_0 \to \ell^1$ is well-defined and linear and $\|(\lambda - \breve{A})x\| \geq (\lambda - \alpha^\circ)\|x\|$ for all $x \in D_0$, $\lambda \in \mathbb{R}$. The closure of $\breve{A}$ is the infinitesimal generator of the semigroup S in Theorem 2 and $\sum_{j=0}^{\infty} (\breve{A}x)_j = \sum_{k=0}^{\infty} \Big(\sum_{j=0}^{\infty} \alpha_{jk}\Big) x_k$ for all $x \in D_0$.*

Remark 1. That S is generated by the closure of $\breve{A}$ is proved in [59]. Without part (c) in Assumption 1, the semigroup S still exists and its infinitesimal generator extends $\breve{A}$ [59] but it may no longer coincide with the closure of $\breve{A}$ [4, Theorem 7.11]. Further, without (c), solutions to (2.10) may no longer be uniquely determined by their initial data [50, Sect. 6].

In our context, the space of main interest is

$$\ell^{11} = \Big\{ x \in \ell^1; \sum_{j=1}^{\infty} j|x_j| < \infty \Big\}$$

with norm $\|x\|_1 = \|x\| + \sum_{j=1}^{\infty} j|x_j|$ which allows us to address the total number of patch occupants in the context of metapopulations or the total number of in-host parasites in the context of macroparasitic diseases.

Theorem 3. *Let the Assumption 1 be satisfied. Then the following hold:*

(a) *The semigroup S in Theorem 2 leaves ℓ^{11} invariant and the restrictions of $S(t)$ to ℓ^{11}, $S_1(t)$, form a C_0-semigroup on ℓ^{11} which is generated by the part of $\breve{A}$ in ℓ^{11}, denoted by A_1, i.e. A_1 is the restriction of $\breve{A}$ to*

$$D(A_1) = \{x \in \ell^{11} \cap D_0; \breve{A}x \in \ell^{11}\}.$$

Further $\|S_1(t)\|_1 \leq e^{\omega t}$ for all $t \geq 0$, with $\omega = \max\{c_1, \alpha^\circ + c_0\}$.

(b) *The semigroups $S^{[n]}$ in Theorem 2 leave ℓ^{11} invariant. Their restrictions to ℓ^{11}, $S_1^{[n]}$, form C_0-semigroups on ℓ^{11} and also satisfy the estimate $\|S_1^{[n]}(t)\|_1 \leq e^{\omega t}$ for all $t \geq 0$. Their infinitesimal generators, $A_1^{[n]}$, have the same domain*

$$D\big(A_1^{[n]}\big) = \Big\{x \in \ell^{11}; \sum_{j=1}^{\infty} j|a_{jj}||x_j| < \infty\Big\} =: D_1.$$

Finally, for all $\check{x} \in \ell^{11}$, $S_1^{[n]}(t)\check{x} \to S_1(t)\check{x}$ in ℓ^{11}, with the convergence being uniform in bounded intervals in $\mathbb{R}_+$.

Lemma 2. *Let the Assumption 1 be satisfied. Then $D_1 \subseteq D(A_1)$ and $\sum_{k=0}^{\infty}\big(\sum_{j=1}^{\infty} j|\alpha_{jk}|\big)|x_k| < \infty$ for all $x \in D_1$.*

Several other approximations of the semigroups S and S_1 have been suggested [4, 16, 30, 50, 60]; the one used here has the advantage that it is easy to show that the approximating semigroups are differentiable. It is closely related to the approach in [51], but the construction there does not really yield approximating semigroups on the same Banach space.

Lemma 3. *Let the Assumption 1 be satisfied. Then the semigroups $S^{[n]}(t)$ on ℓ^1 and $S_1^{[n]}(t)$ on ℓ^{11} are differentiable for $t > 0$. Further there exist constants $c_n > 0$ such that*

$$\left.\begin{aligned}\|\tfrac{d}{dt}S^{[n]}(t)\| &\le c_n + (et)^{-1}, \\ \|\tfrac{d}{dt}S_1^{[n]}(t)\|_1 &\le c_n + (et)^{-1}\end{aligned}\right\} \qquad \forall t \in (0,1). \qquad (2.15)$$

Proof. We only give the proof for $S^{[n]}$; the proof for $S_1^{[n]}$ is completely analogous. Equivalently we show that $S^{[n]}(t)$ maps ℓ^1 into $D(A^{[n]})$. By construction, (2.11), $[S^{[n]}(t)x]_j = e^{\alpha_{jj}t}x_j$ for $j > n$. Hence, with appropriate constants $c_n > 0$, for $t > 0$,

$$\sum_{j=n+1}^{\infty} |\alpha_{jj}|\big[S^{[n]}(t)x\big]_j \le \sum_{j=n+1}^{\infty} |\alpha_{jj}|e^{-|\alpha_{jj}|t}x_j \le \frac{1}{et}\|x\|.$$

By (2.12) and (2.13), $S(t)$ maps ℓ^1 into $D(A^{[n]})$ for $t > 0$ and

$$\|A^{[n]}S^{[n]}(t)x\| \le \sum_{j,k=1}^{n} |\alpha_{jk}||S^{[n]}(t)x| + \sum_{j=n+1}^{\infty} |\alpha_{jj}|\big[S^{[n]}(t)x\big]_j$$

$$\le c_n\|x\| + (et)^{-1}\|x\| \qquad \forall t \in (0,1),$$

with appropriate constants $c_n > 0$. $\square$

We conjecture that the semigroups in Lemma 3 are analytic, but the estimate in the proof does not completely match [46, Chap. 2, Theorem 5.2(d)]. In general, neither the semigroup S nor the semigroup S_1 are differentiable for all $t > 0$. As an example we consider the simple death process,

$$\alpha_{k-1,k} = k = -\alpha_{kk}, \quad k \in \mathbb{N}, \qquad \alpha_{jk} = 0 \quad \text{otherwise}. \qquad (2.16)$$

Let $e^{[n]} = (\delta_{kn})_{k=0}^{\infty}$ (the Kronecker symbols) be the sequence where all terms are 0 except the nth term which is 1. It is well-known [1, 6.4.2] that $[S(t)e^{[n]}]_j = 0$ for $j > n$ and

$$[S(t)e^{[n]}]_j = \binom{n}{j}e^{-jt}(1 - e^{-t})^{n-j}, \qquad j = 0, \ldots, n. \qquad (2.17)$$

Since $e^{[n]}$ is an element of both D_0 and $D(A_1)$ (notice $\|e^{[n]}\| = 1$ and $\|e^{[n]}\|_1 = n + 1$), we can differentiate $S(t)e^{[n]}$ and $S_1(t)e^{[n]}$ and

$$\left[\frac{d}{dt}S(t)e^{[n]}\right]_j = \begin{cases} 0; & j > n, \\ [S(t)e^{[n]}]_j \dfrac{ne^{-t} - j}{1 - e^{-t}}; & j = 0, \ldots n. \end{cases} \qquad (2.18)$$

We choose $\bar{t} = \ln 2$ such that $e^{-t} = 1/2$ for $t = \bar{t}$. As we show in the appendix

$$2\frac{\left\|\frac{d}{dt}S_1(\bar{t})e^{[2n]}\right\|_1}{\|e^{2n}\|_1} \geq \left\|\frac{d}{dt}S(\bar{t})e^{[2n]}\right\| = 2n\binom{2n}{n}2^{-2n} \qquad (2.19)$$

$$\geq \sqrt{n-1}e^{-1/2}, \qquad \bar{t} = \ln 2, n \geq 2. \qquad (2.20)$$

This implies that $S(t)$ is not strongly differentiable at any $t \leq \ln 2$. Otherwise $S'(\bar{t}) = AS(\bar{t})$ would be a bounded linear operator [46, 2.4] contradicting this estimate because $\|e^{[2n]}\| = 1$. Similarly, $S_1(t)$ is not differentiable at any $t \leq \ln 2$.

2.3 Solution to the Semilinear System

We can formally rewrite (2.1) as a semilinear Cauchy problem

$$\begin{aligned} w' &= f(t, w, x), \\ x' &= A_1 x + g(t, w, x), \end{aligned} \qquad (2.21)$$

where

$$g(t, w, x) = (g_j(t, w, x))_{j=0}^{\infty}$$

and A_1 is the infinitesimal generator of the semigroup S_1 considered in Theorem 3. Since in general the semigroup S_1 is not differentiable (see the discussion at the end of the previous section), we cannot expect to find a solution of (2.21) in the strict sense if $x(0) = \check{x} \in \ell_+^{11}$ rather than $x(0) \in D(A_1)$. The pair of continuous functions $w : [0, \tau) \to \mathbb{R}_+$ and $x : [0, \tau) \to \ell_+^{11}$ is called an *integral solution* of (2.21) with initial condition $w(0) = \check{w}$, $x(0) = \check{x}$ if

$$w' = f(t, w, x), \quad t \in [0, \tau), \qquad w(0) = \check{w},$$

$$x(t) = \check{x} + A_1 \int_0^t x(s)ds + \int_0^t g(s, w(s), x(s))ds, \qquad t \in [0, \tau), \qquad (2.22)$$

with the understanding that $\int_0^t x(s)ds \in D(A_1)$ for all $t \in [0,\tau)$. Equivalently to the second equation in (2.22), x is a *mild solution* of $x' = A_1 x + g(t,w,x)$, if it satisfies the integral equation

$$x(t) = S_1(t)\check{x} + \int_0^t S_1(t-s)g(s,w(s),x(s))ds, \qquad t \in [0,\tau), \qquad (2.23)$$

where S_1 is the C_0-semigroup generated by A_1 on ℓ^{11} [4, Proposition 3.31].

2.3.1 Local Existence

A standard approach to local existence of solutions consists in assuming that the non-linearities satisfy a Lipschitz condition. We also need assumptions which make the solutions preserve positivity.

Assumption 4 $f : \mathbb{R}_+^2 \times \ell^{11} \to \mathbb{R}$ and $g : \mathbb{R}_+^2 \times \ell^{11} \to \ell^{11}$ are continuous and have the following properties:

(a) $f(t,0,x) \geq 0$ for all $x \in \ell_+^{11}$, $t \geq 0$.
(b) For every $j \in \mathbb{Z}_+$, $g_j(t,w,x) \geq 0$ whenever $w \geq 0$, $x \in \ell_+^{11}$, $x_j = 0$, $t \geq 0$.
(c) For every $r > 0$ there exists a Lipschitz constant Λ_r such that

$$\left. \begin{array}{r} |f(t,w,x) - f(t,\tilde{w},\tilde{x})| \\ \|g(t,w,x) - g(t,\tilde{w},\tilde{x})\|_1 \end{array} \right\} \leq \Lambda_r(|w - \tilde{w}| + \|x - \tilde{x}\|_1),$$

whenever $t \in [0,r]$, $w, \tilde{w} \in [0,r]$, $x, \tilde{x} \in \ell_+^{11}$, $\|x\|_1, \|\tilde{x}\|_1 \leq r$.

Theorem 5. *Under the Assumptions 1, and 4, for every $\check{w} \in \mathbb{R}_+$ and $\check{x} \in \ell_+^{11}$, there exists some $\tau \in [0,\infty]$ and a unique continuous solution $w : [0,\tau) \to [0,\infty)$, $x : [0,\tau) \to \ell_+^{11}$ of (2.22).*

Remark 2. $\tau \in [0,\infty)$ can be chosen in such a way that the solution (w,x) cannot be extended to a solution on a larger open interval.

Since we want our solutions to preserve positivity, we do not refer for the proof to general results which use Banach's fixed point theorem [46, Chap. 6 Theorem 1.2] [52, Theorem 46.1], but to results which use generalizations of the explicit Euler approximation to solve ordinary differential equations [11, Chap. 1 Theorem 1.1].

Proof. We apply [40, VIII.2, Theorem 2.1] (or [55, Sect. 2]). We set $D = \mathbb{R}_+ \times \ell_+^{11}$. Notice that (2.23) can be rewritten in terms of $y(t) = (w(t), x(t))$ and $\check{y} = (\check{w}, \check{x})$ as

$$y(t) = \bar{S}(t)\check{y} + \int_0^t \bar{S}(t-s)\bar{f}(s,y(s))ds$$

with the C_0-semigroup $\bar{S}(t)\breve{y} = (\breve{w}, S_1(t)\breve{x})$ on $\mathbb{R} \times \ell^{11}$ and the non-linearity $\bar{f}(s, \breve{y}) = (f(s, \breve{y}), g(s, \breve{y}))$. The local existence of solutions with values in D follows once we have checked the subtangential condition

$$\frac{1}{h}d\big(y + h\bar{f}(t, y), D\big) \to 0, \qquad h \to 0+, y \in D,$$

where $d(z, D)$ is the distance from the point z to the set D. This subtangential condition can be broken up into two tangential conditions,

$$\left.\begin{array}{l} \frac{1}{h}d\big(w + hf(t, w, x), \mathbb{R}_+\big) \to 0 \\[2mm] \frac{1}{h}d\big(x + hg(t, w, x), \ell_+^{11}\big) \to 0 \end{array}\right\} \qquad h \to 0+, w \in \mathbb{R}_+, x \in \ell_+^{11}.$$

In a Banach lattice Z with positive cone Z_+,

$$d(z, Z_+) \leq \|z - z^+\| = \|z^-\|,$$

where z^+ and z^- are the positive and negative part of the vector z. Since, for $z \in Z_+$, $\|z^-\| = 0$,

$$\limsup_{h \to 0+} \frac{1}{h}d(z + h\tilde{z}, Z_+) \leq \limsup_{h \to 0+} \frac{1}{h}\|[z + h\tilde{z}]^-\|$$
$$= \lim_{h \to 0+} \frac{1}{h}\Big(\|[z + h\tilde{z}]^-\| - \|z^-\|\Big)$$
$$=: D_+\|z^-\|\tilde{z},$$

which is the right derivative of the convex functional $z \mapsto \|z^-\|$ at z in the direction of $\tilde{z}$ [40, II.5]. If $Z = \mathbb{R}$,

$$D_+\|z^-\|\,\tilde{z} = D_+z^-\tilde{z} = \lim_{h \to 0+} \frac{1}{h}\big([z + h\tilde{z}]^- - z^-\big)$$
$$= \left\{\begin{array}{l} \lim_{h \to 0+} \dfrac{1}{h}(0 - 0); \ z > 0 \\[2mm] \lim_{h \to 0+} \dfrac{1}{h}[h\tilde{z}]^-; \ z = 0 \\[2mm] \lim_{h \to 0+} \dfrac{1}{h}(-z - h\tilde{z} + z); \ z < 0 \end{array}\right\} = \left\{\begin{array}{l} 0; \ z > 0 \\ \tilde{z}^-; \ z = 0 \\ -\tilde{z}; \ z < 0 \end{array}\right\}.$$

For $z \in \mathbb{R}_+$ $\tilde{z} \in \mathbb{R}$,

$$\limsup_{h \to 0+} \frac{1}{h}d(z + h\tilde{z}, \mathbb{R}_+) \leq \left\{\begin{array}{ll} 0; & z > 0, \\ \tilde{z}^-; & z = 0. \end{array}\right.$$

For $z, \tilde{z} \in \ell^1$,

$$D_+\|z^-\|\tilde{z} = \sum_{j=0}^{\infty} \lim_{h \to 0+} \frac{1}{h}\big([z_j + h\tilde{z}_j]^- - z_j^-\big) = \sum_{j=0}^{\infty} D_+z_j^-\,\tilde{z}_j.$$

For $z, \tilde{z} \in \ell^{11}$,

$$D_+ \|z^-\|_1 \tilde{z} = \sum_{j=0}^{\infty} (1+j) D_+ z_j^- \, \tilde{z}_j.$$

We summarize. For $w \in \mathbb{R}_+$ and $x \in \ell_+^{11}$,

$$\lim_{h\to 0+} \frac{1}{h} d(w + hf(t,w,x), \mathbb{R}^+) \le \left\{ \begin{array}{c} 0; \ w > 0 \\ [f(t,w,x)]^- ; \ w = 0 \end{array} \right\}$$

and

$$\lim_{h\to 0+} \frac{1}{h} d(x + hg(t,w,x), \ell_+^{11}) \le \sum_{j=0}^{\infty} (1+j) \left\{ \begin{array}{c} 0; \ x_j > 0 \\ [g_j(t,w,x)]^- ; \ x_j = 0 \end{array} \right\}.$$

By Assumption 4 (a) and (b), these expressions are 0 and the subtangential condition is satisfied.   □

2.3.2 Global Existence

In order to establish global existence we make the following additional assumptions. Recall D_1 in Theorem 3,

$$D_1 = \left\{ x \in \ell^{11}; \sum_{j=1}^{\infty} j|\alpha_{jj}| \, |x_j| < \infty \right\}.$$

Assumption 6 There exist constants $c_2, c_3 \ge 0$ such that for all $t \ge 0$, $w \ge 0$ and $x \in \ell_+^{11} \cap D_1$ the following hold:

- $\sum_{j=0}^{\infty} g_j(t,w,x) \le c_3 \|x\|.$

- $f(t,w,x) + \sum_{j=1}^{\infty} j g_j(t,w,x) \le c_2(w + \|x\|_1).$

Theorem 7. *Let Assumptions 1, 4, and 6 be satisfied. Then, for every $\breve{w} \in [0,\infty)$ and $\breve{x} \in \ell_+^{11}$, there exists a unique continuous solution $w : [0,\infty) \to \mathbb{R}_+$, $x : [0,\infty) \to \ell_+^{11}$ of (2.22). The solution satisfies the estimates*

$$\|x(t)\| \le \|\breve{x}\| e^{\alpha^\circ t}, \qquad |w(t)| + \|x(t)\|_1 \le (\breve{w} + \|\breve{x}\|_1) e^{\omega_2 t}$$

where α° is from Assumption 1 and $\omega_2 \in \mathbb{R}$ is an appropriate constant.

Remark 3. On every bounded subinterval of $[0,\infty)$, the solution x is the uniform limit of solutions $x^{[n]}$ on $[0,\infty)$ with values in ℓ_+^{11} which solve the system

$$\frac{d}{dt}w^{[n]}(t) = f\big(t, w^{[n]}(t), x^{[n]}(t)\big), \qquad w^{[n]}(0) = \breve{w},$$

$$\frac{d}{dt}x_j^{[n]}(t) - g_j\big(t, w^{[n]}(t), x^{[n]}(t)\big) = \begin{cases} \displaystyle\sum_{k=0}^{n} \alpha_{jk} x_k^{[n]}(t), & j = 0, \ldots, n, \\ \alpha_{jj} x_k^{[n]}(t), & j > n, \end{cases} \qquad (2.24)$$

$$x^{[n]}(0) = \breve{x},$$

and satisfy

$$\sum_{j=1}^{\infty} j|a_{jj}| \int_0^t x_j^{[n]}(s)ds < \infty. \qquad (2.25)$$

The following estimates will be used frequently for $m = 0, 1$ and $0^0 := 1$,

$$\sum_{j=0}^{\infty} j^m x_j^{[n]}(t) \leq \sum_{j=0}^{\infty} j^m \breve{x}_j + \sum_{k=0}^{\infty} \bigg(\sum_{j=0}^{\infty} j^m \alpha_{jk}\bigg) \int_0^t x_k^{[n]}(s)ds$$
$$+ \int_0^t \sum_{j=0}^{\infty} j^m g_j\big(s, w^{[n]}(s), x^{[n]}(s)\big)ds. \qquad (2.26)$$

Proof. In order to derive the estimates which imply global existence of solutions we consider the approximating problems where the infinite matrix (α_{jk}) is replaced by the infinite matrices $(\alpha_{jk}^{[n]})$ in (2.13) because this will allow us to interchange the order of summation freely. The matrices also satisfy the assumptions of Theorem 5. So, for every $n \in \mathbb{N}$, there exists some $\tau_n \in [0, \infty]$ and a solution on $[0, \tau_n)$ of

$$\frac{d}{dt}w^{[n]} = f\big(t, w^{[n]}, x^{[n]}\big), \qquad w^{[n]}(0) = \breve{w},$$

$$x^{[n]}(t) = \breve{x} + A_1^{[n]} \int_0^t x^{[n]}(s)ds + \int_0^t g\big(s, w^{[n]}(s), x^{[n]}(s)\big)ds,$$

with the understanding that $\int_0^t x^{[n]}(s)ds \in D(A_1^{[n]}) = D_1$. (2.25) follows from the definition of D_1.

Again $\tau_n \in [0, \infty]$ can be chosen such that the solution $(w^{[n]}, x^{[n]})$ cannot be extended to a solution on a larger interval.

If we spell the equation for $x^{[n]}$ out componentwise for x, we see that $w^{[n]}$ and $x_j^{[n]}$ can be differentiated and satisfy (2.24). Since the $x^{[n]}$ are non-negative, for $m = 0, 1$,

$$\sum_{j=0}^{\infty} j^m x_j^{[n]}(t)$$

$$= \sum_{j=0}^{\infty} j^m \breve{x}_j + \sum_{j=0}^{n} j^m \bigg(\sum_{k=0}^{n} \alpha_{jk} \int_0^t x_k^{[n]}(s)ds\bigg) + \sum_{j=n+1}^{\infty} j^m \alpha_{jj} \int_0^t x_j^{[n]}(s)ds$$

$$+ \sum_{j=0}^{\infty} j^m \int_0^t g_j(s, w^{[n]}, x^{[n]})ds.$$

We can change the order of summations, use that $\alpha_{jk} \geq 0$ for $j \neq k$, and obtain the estimate (2.26) which implies

$$w^{[n]}(t) + \left\| x^{[n]}(t) \right\|_1$$
$$\leq \breve{w} + \|\breve{x}\|_1 + \int_0^t f(s, w^{[n]}(s), x^{[n]}(s))ds$$
$$+ \sum_{k=0}^{\infty} \left(\sum_{j=0}^{\infty} (1+j)\alpha_{jk} \right) \int_0^t x_k^{[n]}(s)ds$$
$$+ \int_0^t \sum_{j=0}^{\infty} (1+j)g_j(s, w^{[n]}, x^{[n]})ds.$$

By Assumptions 1, 4, and 6,

$$w^{[n]}(t) + \left\| x^{[n]}(t) \right\|_1$$
$$\leq \breve{w} + \|\breve{x}\|_1 + (\alpha^\circ + c_0 + c_3) \int_0^t \|x^{[n]}(s)\|ds$$
$$+ c_2 \int_0^t w^{[n]}(s)ds + (c_1 + c_2) \int_0^t \|x^{[n]}(s)\|_1 ds \Big)$$
$$\leq \omega_2 \left(\int_0^t w^{[n]}(s)ds + \int_0^t \|x^{[n]}(s)\|_1 ds \right)$$

with an appropriate $\omega_2 > 0$. By Gronwall's inequality,

$$w^{[n]}(t) + \left\| x^{[n]}(t) \right\|_1 \leq e^{\omega_2 t} \left(\breve{w} + \|\breve{x}\|_1 \right).$$

Suppose $\tau_n < \infty$. The growth bounds in Assumption 6 imply that $g(t, w^{[n]}, x^{[n]})$ and $f(t, w^{[n]}, x^{[n]})$ are bounded on $[0, \tau_n)$. It follows from the variation of parameters formula,

$$w^{[n]}(t) = \breve{w} + \int_0^t f(s, w^{[n]}(s), x^{[n]}(s))ds,$$
$$x^{[n]}(t) = S_1^{[n]}(t)\breve{x} + \int_0^t S_1^{[n]}(t-s)g(s, w^{[n]}(s), x^{[n]}(s))ds,$$

that $w^{[n]}$ and $x^{[n]}$ can be continuously extended to $[0, \tau_n]$. By the local existence theorem they can be extended to an interval larger than $[0, \tau_n]$, contradicting the maximality of the solution.

We return to the solution (w, x) in Theorem 5 which, by (2.23), is given by

$$w(t) = \breve{w} + \int_0^t f(s, w(s), x(s))ds,$$
$$x(t) = S_1(t)\breve{x} + \int_0^t S_1(t-s)g(s, w(s), x(s))ds.$$

We subtract this system of equations from the previous,

$$\left|w(t) - w^{[n]}(t)\right| \leq \int_0^t \left|f(s, w(s), x(s)) - f(s, w^{[n]}(s), x^{[n]}(s))\right| ds$$

and

$$\left\|x(t) - x^{[n]}(t)\right\|$$
$$\leq \left\|\left[S_1(t) - S_1^{[n]}(t)\right]\check{x}\right\|_1$$
$$+ \int_0^t \left\|\left[S_1(t - s) - S_1^{[n]}(t - s)\right]g(s, w(s), x(s))\right\|_1 ds$$
$$+ \int_0^t \left\|S_1^{[n]}(t - s)\right\|_1 \left\|g(s, w(s), x(s)) - g(s, w^{[n]}(s), x^{[n]}(s))\right\|_1 ds.$$

We use $\left\|S_1^{[n]}(t)\right\|_1 \leq e^{\omega t}$ (Theorem 3) and the Lipschitz conditions for f and g in Assumption 4. For every $r \in (0, \tau)$, we find a Lipschitz constant Λ_r such that

$$\left|w(t) - w^{[n]}(t)\right| + \left\|x(t) - x^{[n]}(t)\right\|_1$$
$$\leq \left\|\left[S_1(t) - S_1^{[n]}(t)\right]\check{x}\right\|_1$$
$$+ \int_0^t \left\|\left[S_1(t - s) - S_1^{[n]}(t - s)\right]g(s, w(s), x(s))\right\|_1 ds$$
$$+ \Lambda_r \int_0^t e^{\omega(t-s)}\left(\left|w(s) - w^{[n]}(s)\right| + \left\|x(s) - x^{[n]}(s)\right\|_1\right) ds.$$

Since

$$\left\|\left[S^{[n]}(t) - S(t)\right]\check{x}\right\|_1 \to 0, \qquad n \to \infty, \ t \geq 0, \ \check{x} \in \ell_+^{11}$$

by Theorem 3, Lebesgue's theorem of dominated convergence implies that second summand on the right hand side of the last inequality converges to 0 for all $t \geq 0$. A Gronwall argument implies that

$$\left|w(t) - w^{[n]}(t)\right| + \left\|x(t) - x^{[n]}(t)\right\|_1 \to 0, \qquad n \to \infty,$$

uniformly for t in every compact subinterval of $[0, \tau)$. This implies that

$$w(t) + \|x(t)\|_1 \leq e^{\omega_2 t}(\check{w} + \|\check{x}\|_1), \qquad t \in [0, \tau).$$

A similar argument as before implies that $\tau = \infty$. □

2.3.3 A Semiflow

A map $\Phi : \mathbb{R}_+ \times \ell_+^{11} \to \ell_+^{11}$ is called a *semiflow* on ℓ_+^{11} if $\Phi(t+s, \check{x}) = \Phi(t, \Phi(s, \check{x}))$ for all $t, s \geq 0$ and $\Phi(0, \check{x}) = \check{x}$ whenever $\check{x} \in \ell_+^{11}$. If Φ is continuous, it is called a *continuous semiflow*. The following theorem is essentially proved in the same way as the continuous dependence of solutions of ODEs on initial data with the Gronwall inequality playing a crucial role [52, Theorem 46.4] [55, Sect. 3].

Theorem 8. *Let the assumptions of Theorem 7 be satisfied and f and g_j not depend on time t. Then the map $\Phi : \mathbb{R}_+ \times \mathbb{R}_+ \times \ell_+^{11} \to \mathbb{R}_+ \times \ell_+^{11}$ defined by $\Phi(t, (\breve{w}, \breve{x})) = (w(t), x(t))$ with x being a solution of (2.22) is a continuous semiflow.*

2.4 General Metapopulation Models and Boundedness of Solutions

In the following we concentrate on metapopulations to derive boundedness results. A special feature of a certain class of metapopulation models is that the number of patches (islands) does not increase.

2.4.1 Decrease or Constancy of Patch Number

We formulate the Assumption that guarantees this feature.

Assumption 9 (a) $\quad \sum_{j=0}^{\infty} \alpha_{jk} \leq 0$ for all $k \in \mathbb{Z}_+$

(b) $\quad \sum_{j=0}^{\infty} g_j(t, w, x) \leq 0$ for all $t \geq 0$, $w \geq 0$, $x \in \ell_+^{11}$.

Proposition 1. *Let the Assumptions of Theorem 7 and Assumption 9 be satisfied. Then $\|x(t)\| \leq \|\breve{x}\|$ for all $t \geq 0$ for every non-negative solution of (2.22).*

Proof. Recall that x solves

$$x(t) = \breve{x} + A_1 \int_0^t x(s)ds + \int_0^t g(s, w(s), x(s))ds.$$

Since $x(t) \in \ell_+^{11}$,

$$\|x(t)\| = \sum_{j=0}^{\infty} x_j(t)$$

$$= \|\breve{x}\| + \sum_{j=0}^{\infty} \left(A_1 \int_0^t x(s)ds \right)_j + \int_0^t \sum_{j=0}^{\infty} g_j(s, w(s), x(s))ds.$$

By Lemma 1 and Assumption 9, $\|x(t)\| \leq \|\breve{x}\|$ because $\alpha^{\diamond} \leq 0$. $\quad\square$

The same proof yields the following result.

Corollary 1. *Let the Assumptions of Theorem 7 be satisfied and $\sum_{j=0}^{\infty} \alpha_{jk} = 0$ for all $k \in \mathbb{Z}_+$ and $\sum_{j=0}^{\infty} g_j(t, w, x) = 0$ for all $t, w \geq 0$, $x \in \ell_+^{11}$. Then $\|x(t)\| = \|x(0)\|$ for all $t > 0$ and all solutions of (2.22).*

2.4.2 Uniform Eventual Boundedness of Solutions

Assumption 10 *There exist constants $c_4, c_5, \epsilon_4 > 0$ such that, for all $w \geq 0$, $x \in D_1 \cap \ell_+^{11}$,*

$$f(t, w, x) + \sum_{k=0}^{\infty} \left(\sum_{j=1}^{\infty} j\alpha_{jk} \right) x_k + \sum_{j=1}^{\infty} jg_j(t, w, x)$$

$$\leq c_4 \|x\| + c_5 - \epsilon_4 \left(w + \sum_{j=1}^{\infty} jx_j \right).$$

By Lemma 2, the series in the second term exist. If the previous assumptions are added, the solutions of the model equations are uniformly eventually bounded and the solution semiflow is called *dissipative*.

Theorem 11. *Let Assumptions 1, 4, 9, and 10 be satisfied. Then, with the constants c_4, c_5, ϵ_4 from Assumption 10,*

$$w(t) + \sum_{j=1}^{\infty} jx_j(t) \leq \left(\check{w} + \sum_{j=1}^{\infty} j\check{x}_j \right) e^{-\epsilon_4 t} + \frac{c_4 \|\check{x}\| + c_5}{\epsilon_4}$$

for all solutions (w, x) of (2.1) with initial data $\check{w} \geq 0$, $\check{x} \in \ell_+^{11}$.

Proof. The Assumptions 9 and 10 imply Assumption 6, and we have global solutions for the initial values in question by Theorem 7. By Proposition 1, $\|x(t)\| \leq \|\check{x}\|$ for all $t \geq 0$. We consider the functions $x^{[n]}$ on $[0, \infty)$ in (2.24) which approximate x by Remark 3. By estimate (2.26),

$$\left(w^{[n]}(t) + \sum_{j=1}^{\infty} jx_j^{[n]}(t) \right)$$

$$\leq \check{w} + \sum_{j=1}^{\infty} j\check{x}_j + \int_0^t f\big(t, w^{[n]}(s), x^{[m]}(s)\big) ds$$

$$+ \sum_{k=0}^{\infty} \left(\sum_{j=0}^{\infty} j\alpha_{jk} \right) \int_0^t x_k^{[n]}(s) ds + \int_0^t \sum_{j=0}^{\infty} jg_j\big(s, w^{[n]}(s), x^{[n]}(s)\big) ds.$$

By Lemma 2, the double series exists absolutely. Since the functions $x_k^{[n]}$ are non-negative, we can interchange the series and the integral. By Assumption 10 (notice that $\int_0^t x^{[n]}(s) ds \in D_1$),

$$w^{[n]}(t) + \sum_{j=1}^{\infty} jx_j^{[n]}(t) \leq \check{w} + \sum_{j=1}^{\infty} j\check{x}_j + c_4 \int_0^t \|x^{[n]}(s)\| ds + c_5 t$$

$$- \epsilon_4 \left(\int_0^t w^{[n]}(s) ds + \int_0^t \left(\sum_{j=1}^{\infty} jx_j^{[n]}(s) \right) ds \right).$$

By Gronwall's inequality,

$$w^{[n]}(t) + \sum_{j=1}^{\infty} j x^{[n]}(t) \le e^{-\epsilon_4 t}\left(\breve{w} + \sum_{j=1}^{\infty} j \breve{x}_j\right) + \frac{c_5}{\epsilon_4}$$

$$+ c_4 \int_0^t \|x^{[n]}(t-s)\| e^{-\epsilon_4 s} ds.$$

We take the limit $n \to \infty$, use $\|x(t)\| \le \|\breve{x}\|$ and obtain the statement of this theorem. $\square$

2.5 Extinction Without Migration or Colonization of Empty Patches

If there is no emigration from the patches, we can assume that the average number of migrating individuals (wanderers), $w(t)$, is exponentially decreasing, more generally, w is bounded on $[0, \infty)$, $\int_0^\infty w(t)dt < \infty$. In this section we derive conditions such that this implies that the solutions of (2.22) satisfy

$$\sum_{j=1}^{\infty} j x_j(t) \to 0 \qquad \text{as} \qquad t \to \infty,$$

i.e., the occupant part of the population goes extinct together with its migrating part. We also show that the occupant population goes extinct if empty patches are not colonized.

Assumption 12 (a) $\displaystyle\sum_{j=0}^{\infty} \alpha_{jk} \le 0, \qquad k = 0, 1, 2, \dots.$

(b) For all $k \in \mathbb{N}$ there is some $j \in \mathbb{Z}_+$, $j < k$, such that $\alpha_{jk} > 0$.

(c) $g_j(0, x) = 0$ for all $x \in \ell_+^{11}$, $j = 0, 1, \dots.$

(d) $\displaystyle\sum_{j=0}^{\infty} g_j(w, x) \le 0$ for all $w \ge 0, x \in \ell_+^{11}$.

(e) There exists a constant $c > 0$ such that

$$\sum_{j=1}^{\infty} j g_j(w, x) \le cw\|x\| \text{ for all } w \ge 0,\ x \in \ell_+^{11}.$$

(f) $\displaystyle\limsup_{k \to \infty} \sum_{j=0}^{\infty} \frac{j \alpha_{jk}}{k} < 0.$

If the Assumptions 1, 4 (for g), 10, and 12 are satisfied, then also the Assumption 9 is satisfied and unique solutions exist to (2.22) which are defined and bounded on $[0, \infty)$.

Proposition 2. *Let the Assumptions 1, 4, 10, and 12 be satisfied. Let $c > 0$ be the number in Assumption 12. Then there exist $m \in \mathbb{N}$ and $\epsilon_1 > 0$ such that for every solution x of (2.22) with $\check{x} \in \ell_+^{11}$,*

$$\sum_{j=1}^{\infty} j x_j(t) \le e^{-\epsilon_1 t} \sum_{j=1}^{\infty} j \check{x}_j + \int_0^t e^{-\epsilon_1(t-s)} \sum_{k=0}^{m-1} \xi_k x_k(s) ds$$

$$+ c \|\check{x}\| \int_0^t e^{-\epsilon_1(t-s)} w(s) ds,$$

where $\xi_k = \sum_{j=1}^{\infty} \alpha_{jk} + \epsilon_1 k$.

Proof. Let $x^{[n]}$ be the solutions of (2.24) which approximate x. By (2.26),

$$\|x^{[n]}(t)\| \le \|\check{x}\|,$$

$$\sum_{j=1}^{\infty} j x_j^{[n]}(t) \le \sum_{j=1}^{\infty} j \check{x}_j + \sum_{k=0}^{\infty} \tilde{\xi}_k \int_0^t x_k^{[n]}(s) ds$$

$$+ \int_0^t \left(\sum_{j=1}^{\infty} j g_j\big(w(s), x^{[n]}(s)\big) \right) ds.$$

By part (f) of Assumption 12, $\tilde{\xi}_k \le -\epsilon_1 k$ for $k \ge m$ with appropriate $\epsilon_1 > 0$, $m \in \mathbb{N}$, and, by part (e),

$$\sum_{j=1}^{\infty} j x_j^{[n]}(t) \le \sum_{j=1}^{\infty} j \check{x}_j + \sum_{k=0}^{m-1} \tilde{\xi}_k \int_0^t x_k^{[n]}(s) ds - \epsilon_1 \sum_{k=m}^{\infty} k \int_0^t x_k^{[n]}(s) ds$$

$$+ \int_0^t \left(\sum_{j=0}^{\infty} c w(s) x_j^{[n]}(s) \right) ds$$

$$= \sum_{j=1}^{\infty} j \check{x}_j + \sum_{k=0}^{m-1} \xi_k \int_0^t x_k^{[n]}(s) ds - \epsilon_1 \sum_{k=0}^{\infty} k \int_0^t x_k^{[n]}(s) ds$$

$$+ \int_0^t c w(s) \|x^{[n]}(s)\| ds,$$

with $\xi_k = \tilde{\xi}_k + \epsilon_1 k$. We take the limit as $n \to \infty$, obtain $\|x(t)\| \le \|\check{x}\|$ and

$$\sum_{j=1}^{\infty} j x_j(t) \le \sum_{j=1}^{\infty} j \check{x}_j + \sum_{k=0}^{m-1} \xi_k \int_0^t x_k(s) ds - \epsilon_1 \int_0^t \sum_{k=1}^{\infty} k x_k(s) ds$$

$$+ \int_0^t c w(s) \|\check{x}\| ds.$$

Gronwall's inequality implies the assertion. $\square$

Next we show that the size of the occupant population tends to zero as $t \to \infty$ if there is no emigration from patches and the migrating part of the metapopulation decreases exponentially as a result.

Theorem 13. *Let Assumptions 1, 4, 10, and 12 be valid. Further let $\alpha_{00} = 0$. Let w, x be a solution of (2.22) on $[0, \infty)$ such that w is bounded on $\mathbb{R}_+$ and $\int_0^\infty w(t)dt < \infty$. Then*

$$\sum_{k=1}^\infty k \int_0^\infty x_k(s)ds < \infty \qquad and \qquad \sum_{k=1}^\infty kx_k(t) \to 0, \qquad t \to \infty.$$

We mention that the assumption $\alpha_{00} = 0$ together with the other assumptions on the coefficients α_{jk} implies that $\alpha_{j0} = 0$ for all $j \in \mathbb{Z}_+$.

Proof. Recall that x is an integral solution,

$$x(t) - \check{x} = A_1 \int_0^t x(s)ds + \int_0^t g(w(s), x(s))ds.$$

For the single terms this means that

$$x_j(t) - \check{x}_j = \sum_{k=1}^\infty \alpha_{jk} \int_0^t x_k(s)ds + \int_0^t g_j(w(s), x(s))ds.$$

Recall that x_k is non-negative, $\alpha_{jk} \geq 0$ for $j \neq k$, and $x_j(t) \leq \|\check{x}\|$. By Theorem 11, the functions w and x (with values in ℓ_+^{11}) are bounded. Since $g_j(0, x) = 0$ and g_j are locally Lipschitz continuous, there exist constants $\Lambda_j > 0$ such that, for all $j, k \in \mathbb{Z}_+$, $j \neq k$, $t \geq 0$,

$$\alpha_{j,k} \int_0^t x_k(s)ds \leq |\alpha_{jj}| \int_0^t x_j(s)ds + \|\check{x}\| + \Lambda_j \int_0^t w(s)ds.$$

Let $k \in \mathbb{N}$ be arbitrary. By successive application of Assumption 12 (b) we find numbers $k_0 < \cdots < k_m$ such that $k_0 = 0$, $k_m = k$ and $\alpha_{k_i,k_{i+1}} > 0$ for $i = 0, \ldots, m-1$. Since $\alpha_{00} = 0$,

$$\alpha_{0k_1} \int_0^t x_{k_1}(s)ds \leq \|\check{x}\| + \Lambda_0 \int_0^t w(s)ds.$$

Since $\int_0^\infty w(s)ds < \infty$, also $\int_0^\infty x_{k_1}(s)ds < \infty$. Since $\alpha_{k_i,k_{i+1}} > 0$, we obtain step by step that

$$\int_0^\infty x_{k_i}(s)ds < \infty \qquad \forall i = 1, 2, \ldots, m,$$

in particular $\int_0^\infty x_k(s)ds < \infty$ where $k \in \mathbb{N}$ has been arbitrary. The claims now follow from the inequality in Proposition 2, the first by integrating it, the second by applying Lebesgue's theorem of dominated convergence. Notice that $\xi_0 = 0$ because $\alpha_{j0} = 0$ for all $j \in \mathbb{Z}_+$. $\square$

We turn to the case that empty patches are not colonized. This is mathematically captured in the assumption that the function g_0 is non-negative.

Theorem 14. *Let Assumptions 1, 4, 10, and 12 be valid. Further let $\alpha_{00} = 0$ and $g_0(w, x) \geq 0$ for all $w \geq 0$, $x \in \ell_+^{11}$. Let x be a solution of (2.22) on $[0, \infty)$ with values in ℓ_+^{11}. Then*

$$\sum_{k=1}^{\infty} k \int_0^{\infty} x_k(s)ds < \infty \qquad and \qquad \sum_{k=1}^{\infty} kx_k(t) \to 0, \qquad t \to \infty.$$

Proof. We revisit the proof of Theorem 13. From the integral equation for x_0, we obtain the inequality,

$$\alpha_{01} \int_0^t x_1(s)ds \leq \|\breve{x}\| - \int_0^t g_0(w(s), x(s))ds \leq \|\breve{x}\| \qquad \forall t \geq 0.$$

Except for this modification, the proof proceeds in exactly the same way. $\square$

2.6 A More Specific Metapopulation Model

For the rest of the chapter we restrict our considerations which concern qualitative aspects of metapopulation models (compact attractors, (in)stability of equilibria, persistence) to a somewhat more specific model framework in order to cut down on obscuring technicalities,

$$\begin{aligned}
w' &= \sum_{k=1}^{\infty} \eta_k x_k - w \sum_{k=0}^{\infty} \sigma_k x_k - \delta w, \\
x_j' &= \sum_{k=0}^{\infty} \alpha_{jk} x_k + w \sum_{k=0}^{\infty} \gamma_{jk} x_k, \qquad j = 0, 1, \dots.
\end{aligned} \tag{2.27}$$

The coefficients γ_{jk} describe the transition from patches with k occupants to patches with j occupants due to immigrating dispersers. The terms σ_k describe the average loss rate of dispersers due to settlement on a patch with k occupants. Below we will impose a balance equation or inequality linking γ_{jk} and σ_k. The coefficients η_k describe the rate at which individuals emigrate from a patch with k occupants. $\delta > 0$ is the per capita mortality rate of dispersers. We assume the following.

Assumption 15 (a) $\alpha_{jj}, \gamma_{jj} \leq 0 \leq \alpha_{jk}, \gamma_{jk}$ for $j \neq k$, $j, k \in \mathbb{Z}_+$. Further

$$\sum_{j=0}^{\infty} \alpha_{jk} \leq 0 \quad \text{and} \quad \sum_{j=0}^{\infty} \gamma_{jk} \leq 0 \quad \text{for all } k \in \mathbb{Z}_+.$$

(b) There exist constants $c_0, c_1 > 0, \epsilon > 0$ such that

$$\sum_{j=1}^{\infty} j\alpha_{jk} \leq c_0 + c_1 k - \epsilon|\alpha_{kk}| \qquad \forall k \in \mathbb{Z}_+.$$

(c) There exists a constant $c_7 > 0$ such that $0 \le \eta_k, \sigma_k \le c_7 k$ for all $k \in \mathbb{N}$.

(d) There exists a constant $c_8 > 0$ such that $\displaystyle\sum_{j=1}^{\infty} j|\gamma_{jk}| \le c_8(1+k)$ for all $k \in \mathbb{Z}_+$.

(e) $\displaystyle\sum_{j=1}^{\infty} j\gamma_{jk} \le \sigma_k$ for all $k \in \mathbb{Z}_+$.

Part (e) of the last assumption expresses a balance law which guarantees that the rate at which a patch with k occupants gains new occupants through immigration of dispersers does not exceed the rate at which dispersers leave the disperser pool to settle on a patch with k occupants. A strict inequality means that some dispersers die during the immigration. Mathematically part (e), together with part (c), implies that the second part of Assumption 6 is satisfied. The first part of that assumption, with $c_3 = 0$, is satisfied by Assumption 15 (a). The other parts of Assumption 15 either repeat the Assumption 1 or make sure that the functions f and g in Assumption 4 are well-defined and satisfy the Lipschitz conditions. Theorem 7 implies the following result.

Theorem 16. *Let the Assumption 15 be satisfied. Then, for every $\breve{w} \in [0, \infty)$ and $\breve{x} \in \ell_+^{11}$, there exists a unique integral solution $w : [0, \infty) \to \mathbb{R}_+$, $x : [0, \infty) \to \ell_+^{11}$ of (2.27),*

$$w' = \sum_{k=1}^{\infty} \eta_k x_k - w \sum_{k=0}^{\infty} \sigma_k x_k - \delta w,$$

$$x_j(t) - \breve{x}_j = \sum_{k=0}^{\infty} \alpha_{jk} \int_0^t x_k(s)ds + \sum_{k=0}^{\infty} \gamma_{jk} \int_0^t w(s)x_k(s)ds, \qquad (2.28)$$

$$j = 0, 1, \dots$$

The solution satisfies the estimates

$$\|x(t)\| \le \|\breve{x}\|, \qquad |w(t)| + \|x(t)\|_1 \le (\breve{w} + \|\breve{x}\|_1)e^{\omega_2 t}$$

with some $\omega_2 > 0$.

We add an assumption to obtain uniform eventual boundedness of solutions.

Assumption 17 There exists constants $c_4 > 0$ and $\epsilon_4 > 0$ such that

$$\eta_k + \sum_{j=1}^{\infty} j\alpha_{jk} \le c_4 - \epsilon_4 k \qquad \forall k \in \mathbb{Z}_+.$$

In order to check Assumption 10, we observe that, by Lemma 2, for $x \in D_1$,

$$\sum_{k=0}^{\infty} \eta_k x_k + \sum_{k=0}^{\infty} \left(\sum_{j=1}^{\infty} j \alpha_{jk} \right) x_k = \sum_{k=0}^{\infty} \left(\eta_k + \sum_{j=1}^{\infty} j \alpha_{jk} \right) x_k$$

$$\leq c_4 \|x\| - \epsilon_4 \sum_{k=1}^{\infty} k x_k.$$

If we combine this inequality with the one in Assumptions 15(e), 10 follows with $c_5 = 0$. One readily checks that the other assumptions of Theorem 11 are satisfied.

Theorem 18. *Let the Assumptions 15 and 17 be satisfied. Then, with the constants c_4 and $\epsilon_4 > 0$ from Assumption 17,*

$$w(t) + \sum_{j=1}^{\infty} j x_j(t) \leq \left(\breve{w} + \sum_{j=1}^{\infty} j \breve{x}_j \right) e^{-\epsilon_4 t} + \frac{c_4 \|\breve{x}\|}{\epsilon_4}$$

for all solutions (w, x) of (2.1) with initial data $\breve{w} \geq 0$, $\breve{x} \in \ell_+^{11}$. Further $\|x(t)\| \leq \|\breve{x}\|$ for all $t \geq 0$.

2.6.1 Extinction Without Migration or Colonization

The metapopulation in system (2.27) dies out, if there is no emigration from the patches or if empty patches are not colonized.

Corollary 2. *Let Assumptions 15 and 17 be valid. Assume that $\alpha_{00} = 0$ and that for all $k \in \mathbb{N}$ there is some $j \in \mathbb{Z}_+$, $j < k$, such that $\alpha_{jk} > 0$. Further let (σ_j) be a bounded sequence and $\limsup_{k \to \infty} \sum_{j=1}^{\infty} \frac{j \alpha_{jk}}{k} < 0$. Finally and most importantly let $\gamma_{00} = 0$ or $\eta_j = 0$ for all $j \in \mathbb{N}$. Then, for model (2.27),*

$$\sum_{k=1}^{\infty} k \int_0^{\infty} x_k(s) ds < \infty \qquad and \qquad \lim_{t \to \infty} \sum_{k=1}^{\infty} k x_k(t) = 0.$$

Proof. If $\gamma_{00} = 0$, the statement follows from Theorem 14. If $\eta_j = 0$ for all $j \in \mathbb{N}$, then $w' \leq -\delta w$ by (2.27) and $\int_0^{\infty} w(t) dt < \infty$ and w is bounded on $\mathbb{R}_+$. The statement now follows from Theorem 13. $\square$

2.6.2 An a Priori Estimate for Equilibria

An equilibrium of (2.27) is a time-independent solution of (2.27). Equivalently it is a time-independent solution of (2.28). In either case, an equilibrium (w, x), $w \geq 0$, $x = (x_k) \in \ell_+^{11}$ satisfies $x \in D(A_1)$ and

$$\delta w = \sum_{k=1}^{\infty} \eta_k x_k - w \sum_{k=0}^{\infty} \sigma_k x_k,$$

$$0 = A_1 x + w \Gamma x,$$

(2.29)

where $[A_1 x]_j = \sum_{k=0}^{\infty} \alpha_{jk} x_k$, $x \in D(A_1)$, and $[\Gamma x]_j = \sum_{k=0}^{\infty} \gamma_{jk} x_k$, $x \in \ell^{11}$. Γ maps ℓ_+^{11} into ℓ_+^{11} by Assumption 15.

Theorem 19. *Let the Assumptions 15 and 17 be satisfied. Then, for every solution $x \in D(A_1) \cap \ell_+^{11}$ of $0 = A_1 x + w \Gamma x$, where $w \geq 0$ is given, we have the estimate*

$$\sum_{k=1}^{\infty} \eta_k x_k - w \sum_{k=0}^{\infty} \sigma_k x_k \leq c_4,$$

with c_4 from Assumption 17. If (w,x) is an equilibrium of (2.27), we also have $\delta w \leq c_4$.

Proof. Let $x \in \ell_+^{11} \cap D(A_1)$ satisfy $A_1 x + w \Gamma x = 0$. Then

$$\int_0^t S_1(s) w \Gamma x \, ds = - \int_0^t S_1(s) A_1 x \, ds = x - S_1(t) x.$$

By Theorem 3, for every $t \geq 0$, $x = \lim_{n \to \infty} x^{[n]}(t)$ where

$$x^{[n]}(t) = S_1^{[n]}(t) x + w \int_0^t S_1^{[n]}(s) \Gamma x \, ds.$$

Since the semigroups $S_1^{[n]}$ are differentiable (Lemma 3), we can differentiate $x^{[n]}(t)$ in ℓ^{11} for $t > 0$ and

$$\frac{d}{dt} x^{[n]}(t) = A_1^{[n]} x^{[n]}(t) + w \Gamma x.$$

By (2.11), Assumptions 15 and 17,

$$\frac{d}{dt} \sum_{j=1}^{\infty} j x_j^{[n]}(t)$$

$$= \sum_{k=0}^{n} \left(\sum_{j=1}^{n} j \alpha_{jk} \right) x_k^{[n]}(t) + \sum_{j=n+1}^{\infty} j \alpha_{jj} x_j^{[n]}(t) + w \sum_{k=0}^{\infty} \left(\sum_{j=1}^{\infty} j \gamma_{jk} \right) x_k$$

$$\leq \sum_{k=0}^{\infty} \left(\sum_{j=1}^{\infty} j \alpha_{jk} \right) x_k^{[n]}(t) + w \sum_{k=0}^{\infty} \sigma_k x_k$$

$$\leq \sum_{k=0}^{\infty} (c_4 - \epsilon_4 k - \eta_k) x_k^{[n]}(t) + w \sum_{k=0}^{\infty} \sigma_k x_k.$$

We integrate this inequality,

$$\sum_{j=1}^{n} j x_j^{[n]}(t) \le \sum_{j=1}^{n} j x_j e^{-\epsilon_4 t} + \frac{c_4}{\epsilon_4} - \int_0^t \sum_{k=0}^{n} \eta_k x_k^{[n]}(t-s) e^{-\epsilon_4 s} ds$$

$$+ w \int_0^t \sum_{k=0}^{\infty} \sigma_k x_k e^{-\epsilon_4 s} ds.$$

We first take the limit $n \to \infty$ and then the limit $t \to \infty$,

$$\sum_{j=1}^{\infty} j x_j \le \frac{c_4}{\epsilon_4} - \frac{1}{\epsilon_4} \sum_{k=0}^{\infty} \eta_k x_k + \frac{w}{\epsilon_4} \sum_{k=0}^{\infty} \sigma_k x_k.$$

In particular,

$$\sum_{k=1}^{\infty} \eta_k x_k - w \sum_{k=0}^{\infty} \sigma_k x_k \le c_4.$$

$\square$

2.7 Compact Attractors

We continue to study the metapopulation model (2.27) under the Assumptions 15 and 17. We now fix the number of initial patches to be $N \in \mathbb{N}$ and choose the state space

$$X_N = \{(w, x) \in \mathbb{R}_+ \times \ell_+^{11}; \|x\| \le N\}.$$

We let f and g be independent of time. By Theorems 8 and 16,

$$\Phi(t, (\breve{w}, \breve{x})) = (w(t), x(t)), \qquad t \ge 0,$$

is a continuous semiflow on X_N. In the following it is convenient to introduce the notation $\Phi_t(x) = \Phi(t, x)$ for $t \ge 0$, $x \in X_N$. This way we obtain a family of maps $\{\Phi_t; t \ge 0\}$ on X_N with the property $\Phi_t \circ \Phi_s = \Phi_{t+s}$ in non-linear analogy to operator semigroups.

Let $B \subseteq X_N$. A non-empty compact invariant subset C of X_N is called a *compact attractor* of B if for every open set U, $C \subseteq U \subseteq X_N$, there exists some $r \ge 0$ such that $\Phi_t(B) \subseteq U$ for all $t \ge r$.

Equivalently, $d(\Phi_t(x), C) \to 0$ as $t \to \infty$, uniformly in $x \in B$. Here $d(y, B) = \inf\{d(y, z); z \in B\}$ is the distance from the point y to the set B.

A non-empty compact invariant subset C of X_N is called the *compact attractor of bounded subsets* of X_N if C is a compact attractor of every bounded subset B of X_N. Obviously, by its invariance, a compact attractor of bounded subsets is uniquely determined.

General results concerning compact attractors of bounded sets can be found in [26] and [52]. They involve concepts like dissipativity and asymptotic smoothness of the semiflow. For this particular semiflow a more direct approach seems to work better. We need some additional assumptions.

Assumption 20 (a) $\sup\limits_{k\in\mathbb{N}} \dfrac{|\alpha_{jk}|}{k} < \infty$ for all $j \in \mathbb{Z}_+$.

(b) $\sup\limits_{k\in\mathbb{N}} \dfrac{|\gamma_{jk}|}{k} < \infty$ for all $j \in \mathbb{Z}_+$.

(c) $\sup\limits_{k\in\mathbb{N}} \sigma_k < \infty$.

Our main tool is the *separation measure of non-compactness* [3, II.3], α_s, which has the following sequential characterization in a metric space (X, d). If $Y \subseteq X$,

$$\alpha_s(Y) = \inf\big\{c > 0; \text{ each sequence } (x_n) \text{ in } Y \text{ has a}$$
$$\text{subsequence } (x_{n_j}) \text{ with } \limsup_{j,k\to\infty} d(x_{n_j}, x_{n_k}) \le c\big\}. \qquad (2.30)$$

It is related to the Kuratowski and the Hausdorff measures of non-compactness, α_K and α_H, by

$$\alpha_H(Y) \le \alpha_s(Y) \le \alpha_K(Y) \le 2\alpha_H(Y), \qquad Y \subseteq X. \qquad (2.31)$$

We will use the following two of its properties:

Lemma 4. *(a)* $\alpha_s(B) = \alpha_s(\bar{B})$ *for any bounded subset B of X and its closure $\bar{B}$.*
(b) Let (X, d) be a complete metric space. If B_t is a family of non-empty, closed, bounded sets defined for $t > r$ that satisfy $B_t \subseteq B_s$ whenever $s \le t$ and $\alpha_s(B_t) \to 0$ as $t \to \infty$, then $\cap_{t>r} B_t$ is a non-empty compact set.

(a) follows from (2.30), while (b) is a consequence of the inequality (2.31) and the fact that α_H and α_K satisfy (b) [3, II.2].

Lemma 5. *Let Φ be a semiflow and B a bounded set and $r \ge 0$ such $\Phi_t(B) \subseteq B$ for all $t \ge r$, $\alpha_s(\Phi_t(B)) \to 0$ for $r \le t \to \infty$. Then B has a compact attractor, namely*

$$\omega(B) = \bigcap_{t\ge 0} \overline{\bigcup_{s\ge t} \Phi_s(B)}.$$

This result holds for any measure of non-compactness.

Proof. Let $B_t = \overline{\bigcup_{s\ge t} \Phi_s(B)}$. Then $B_t \subseteq B$ for $t \ge r$. By definition B_t is a decreasing family of subsets of B_0. For $t \ge r$,

$$B_t = \overline{\Phi_{t-r}\Big(\bigcup_{s\ge t} \Phi_{s+r-t}(B)\Big)} \subseteq \overline{\Phi_{t-r}(B)}.$$

By Lemma 4 (a),

$$\alpha_s(B_t) \le \alpha_s(\Phi_{t-r}(B)) \to 0, \qquad r \le t \to \infty.$$

By Lemma 4 (b), $\omega(B) = \bigcap_{t \geq 0} B_t$ is non-empty and compact. Suppose that $\omega(B)$ does not attract B. Then there exist sequences $x_n \in B$ and $t_n \to \infty$ as $n \to \infty$ and $\epsilon > 0$ such that $d(\Phi(t_n, x_n), \omega(B)) > \epsilon$. Define

$$C_m = \overline{\{\Phi(t_n, x_n); n \geq m\}}.$$

Then $C_{m+1} \subseteq C_m$ for all $m \in \mathbb{N}$. Further

$$C_m = \overline{\Phi_{t_m - r}(\{\Phi(t_n + r - t_m, x_n); n \geq m\})} \subseteq \overline{\Phi_{t_m - r}(B)}.$$

By assumption, $\alpha_s(C_m) \to 0$ as $m \to \infty$. So $\bigcap_{m \in \mathbb{N}} C_m$ is non-empty and compact. Choose z in this intersection. Then $z \in \omega(B)$ and $d(\Phi(t_n, x_n), z) < \epsilon$ for some $n \in \mathbb{N}$, a contradiction. Since $\omega(B)$ is compact and attracts B, it is invariant [26, Lemma 3.3.1]. □

Theorem 21. *Let the Assumptions 15, 17, and 20 be satisfied. Then the semi-flow Φ on X_N induced by the solutions of (2.27) has a compact attractor of all bounded subsets of X_N.*

Proof. Let B_0 be the following bounded set.

$$B_0 = \left\{ (w, x) \in X_N; w + \sum_{j=1}^{\infty} j x_j \leq \frac{c_4 N}{\epsilon_4} + 1 \right\},$$

where c_4 and ϵ_4 are the constants from Theorem 18. By Theorem 18, for every bounded set B there exists some $r > 0$ such that $\Phi_t(B) \subseteq B_0$ for all $t \geq r$. So it is sufficient to prove that the set B_0 has a compact attractor. There exists some $r_0 > 0$ such that $\Phi_t(B_0) \subseteq B_0$ for all $t \geq r_0$. By Lemma 5 it is sufficient to show that $\alpha_s(\Phi_t(B_0)) \to 0$.

Let $y, \tilde{y} \in \mathbb{R}$. Then, for sufficiently small $|h|$,

$$|y + h\tilde{y}| - |y| = \begin{cases} h\tilde{y}, & y > 0, \\ |h||\tilde{y}|, & y = 0, \\ -h\tilde{y}, & y < 0. \end{cases}$$

We divide by h and take the limit $h \to 0$ either from the right or the left,

$$D_{\pm}|y|\tilde{y} := \lim_{h \to 0\pm} \frac{|y + h\tilde{y}| - |y|}{h} = \begin{cases} \tilde{y}, & y > 0, \\ \pm|\tilde{y}|, & y = 0, \\ -\tilde{y}, & y < 0. \end{cases}$$

In particular,

$$D_{-}|y|\tilde{y} \leq \tilde{y}\, \mathrm{sign}_0(y) \qquad \text{where} \qquad \mathrm{sign}_0(y) = \begin{cases} 1, & y > 0, \\ 0, & y = 0, \\ -1, & y < 0. \end{cases}$$

Let $\check{x}, \tilde{x} \in B_0$ and $x^{[n]}$ and $\tilde{x}^{[n]}$ be the approximating solutions of $\Phi(\sqcup, \check{x})$ and $\Phi(\sqcup, \check{x})$ as in Remark 3. By [40, VI.4],

$$\frac{d_-}{dt}\left|x_j^{[n]}(t) - \tilde{x}_j^{[n]}(t)\right| = D_-\left|x_j^{[n]}(t) - \tilde{x}_j^{[n]}(t)\right|\left(\frac{d}{dt}x_j^{[n]}(t) - \frac{d}{dt}\tilde{x}_j^{[n]}(t)\right)$$

$$\leq \left(\frac{d}{dt}x_j^{[n]}(t) - \frac{d}{dt}\tilde{x}_j^{[n]}(t)\right)\mathrm{sign}_0\left(x_j^{[n]}(t) - \tilde{x}_j^{[n]}(t)\right).$$

Here $\frac{d_-}{dt}$ denotes the left derivative. Notice that $y\,\mathrm{sign}_0(y) = |y|$. By (2.24) and (2.27), for $j = 1, \ldots, n$,

$$\frac{d_-}{dt}\left|x_j^{[n]}(t) - \tilde{x}_j^{[n]}(t)\right| \leq \sum_{k=0}^{n}\alpha_{jk}\left|x_k^{[n]}(t) - \tilde{x}_k^{[n]}(t)\right|$$

$$+ w^{[n]}(t)\sum_{k=0}^{\infty}\gamma_{jk}\left|x_k^{[n]}(t) - \tilde{x}_k^{[n]}(t)\right|$$

$$+ \left|w^{[n]}(t) - \tilde{w}^{[n]}(t)\right|\sum_{k=0}^{\infty}|\gamma_{jk}|\left|\tilde{x}_k^{[n]}(t)\right|.$$

We multiply this inequality by j, add over $j = 1, \ldots, n$, change the order of summation and use $\alpha_{jk} \geq 0$ for $j \neq k$,

$$\frac{d_-}{dt}\sum_{j=1}^{n}j\left|x_j^{[n]}(t) - \tilde{x}_j^{[n]}(t)\right|$$

$$\leq \sum_{k=0}^{\infty}\left(\sum_{j=1}^{n}j\alpha_{jk}\right)\left|x_k^{[n]}(t) - \tilde{x}_k^{[n]}(t)\right|$$

$$+ w^{[n]}(t)\sum_{k=0}^{\infty}\left(\sum_{j=1}^{n}j\gamma_{jk}\right)\left|x_k^{[n]}(t) - \tilde{x}_k^{[n]}(t)\right|$$

$$+ \left|w^{[n]}(t) - \tilde{w}^{[n]}(t)\right|\sum_{k=0}^{\infty}\left(\sum_{j=1}^{\infty}j|\gamma_{jk}|\right)\left|\tilde{x}_k^{[n]}(t)\right|.$$

Notice that $\sum_{j=1}^{\infty}j|\gamma_{jk}| \leq c_8(k+1)$ for all $k \in \mathbb{Z}_+$ by Assumption 15(d). By Assumption 17, we can choose $\epsilon > 0$ and $m \in \mathbb{N}$ such that $\sum_{j=1}^{\infty}j\alpha_{jk} \leq -\epsilon k$ for all $k > m$. Set $\xi_k = \sum_{j=1}^{\infty}j\alpha_{jk} + \epsilon k$. For $n > m$,

$$\frac{d_-}{dt} \sum_{j=1}^{n} j \big| x_j^{[n]}(t) - \tilde{x}_j^{[n]}(t) \big|$$

$$\leq \sum_{k=0}^{m} \xi_k \big| x_k^{[n]}(t) - \tilde{x}_k^{[n]}(t) \big| - \epsilon \sum_{j=1}^{n} j \big| x_j^{[n]}(t) - \tilde{x}_j^{[n]}(t) \big|$$

$$+ w^{[n]}(t) \sum_{k=0}^{\infty} \Big(\sum_{j=1}^{n} j \gamma_{jk} \Big) \big| x_k^{[n]}(t) - \tilde{x}_k^{[n]}(t) \big|$$

$$+ \big| w^{[n]}(t) - \tilde{w}^{[n]}(t) \big| \sum_{k=0}^{\infty} c_8 (1+k) \big| \tilde{x}_k^{[n]}(t) \big|.$$

We integrate this differential inequality,

$$\sum_{j=1}^{n} j \big| x_j^{[n]}(t) - \tilde{x}_j^{[n]}(t) \big|$$

$$\leq e^{-\epsilon t} \sum_{j=1}^{n} j \big| x_j^{[n]}(0) - \tilde{x}_j^{[n]}(0) \big| + \int_0^t e^{-\epsilon(t-s)} \sum_{k=0}^{m} \xi_k \big| x_k^{[n]}(s) - \tilde{x}_k^{[n]}(s) \big| ds$$

$$+ \int_0^t e^{-\epsilon(t-s)} w^{[n]}(s) \sum_{k=0}^{\infty} \Big(\sum_{j=1}^{n} j \gamma_{jk} \Big) \big| x_k^{[n]}(s) - \tilde{x}_k^{[n]}(s) \big| ds$$

$$+ \int_0^t e^{-\epsilon(t-s)} \big| w^{[n]}(s) - \tilde{w}^{[n]}(s) \big| c_8 \big\| \tilde{x}^{[n]}(s) \big\|_1 ds.$$

The infinite matrices $(\alpha_{jk}^{[n]})$ satisfy the same assumptions as the infinite matrix (α_{jk}) with the same constants. So $w^{[n]}, x^{[n]}$ satisfy the estimates in Theorem 18 with the same constants as w, x. By Lebesgue's theorem of dominated convergence (first applied to the sum and then to the integral), we can take the limit as $n \to \infty$,

$$\sum_{j=1}^{\infty} j \big| x_j(t) - \tilde{x}_j(t) \big|$$

$$\leq e^{-\epsilon t} \sum_{j=1}^{\infty} j \big| x_j(0) - \tilde{x}_j(0) \big| + \int_0^t e^{-\epsilon(t-s)} \sum_{k=0}^{m} \xi_k \big| x_k(s) - \tilde{x}_k(s) \big| ds$$

$$+ \int_0^t e^{-\epsilon(t-s)} w(s) \sum_{k=0}^{\infty} \Big(\sum_{j=1}^{\infty} j \gamma_{jk} \Big) \big| x_k(s) - \tilde{x}_k(s) \big| ds$$

$$+ \int_0^t e^{-\epsilon(t-s)} \big| w(s) - \tilde{w}(s) \big| c_8 \big\| \tilde{x}(s) \big\|_1 ds.$$

By Assumptions 15(e) and 20(c), there exists some $c_9 > 0$ such that $\sum_{j=1}^{\infty} j \gamma_{jk} \leq c_9$ for all $k \in \mathbb{Z}_+$. We split up the last but one sum in the last inequality at $k = i$ where $i \in \mathbb{N}$ is arbitrary. Then

$$\sum_{j=1}^{\infty} j|x_j(t) - \tilde{x}_j(t)|$$

$$\leq e^{-\epsilon t}\|x(0) - \tilde{x}(0)\|_1 + \int_0^t e^{-\epsilon(t-s)} \sum_{k=0}^{m} \xi_k |x_k(s) - \tilde{x}_k(s)| ds$$

$$+ \int_0^t e^{-\epsilon(t-s)} w(s) \sum_{k=0}^{i} c_9 |x_k(s) - \tilde{x}_k(s)| ds$$

$$+ \frac{c_9}{i} \int_0^t e^{-\epsilon(t-s)} w(s)\|x(s) - \tilde{x}(s)\|_1 ds$$

$$+ \int_0^t e^{-\epsilon(t-s)}|w(s) - \tilde{w}(s)|c_8\|\tilde{x}(s)\|_1 ds.$$

Let $((\breve{w}^{\{n\}}, \breve{x}^{\{n\}}))$ be a sequence in B_0 and $(w^{\{n\}}(t), x^{\{n\}}(t)) = \Phi(t, (\breve{w}^{\{n\}}, \breve{x}^{\{n\}}))$. It follows from (2.28), Assumption 20(a, b), and Theorem 18 that $w^{\{n\}}$ and, for each $j \in \mathbb{Z}_+$, $x_j^{\{n\}}$ are equi-bounded and equi-continuous with respect to n on every finite interval in $\mathbb{R}_+$. By the Arzela–Ascoli theorem and a diagonalization procedure, after choosing subsequences, $w^{\{n\}}$, $x_j^{\{n\}}$ are Cauchy sequences for each j uniformly on every finite interval in $\mathbb{R}_+$. We set $x = x^{\{l\}}$ and $\tilde{x} = x^{\{n\}}$ in the inequality above. Then

$$\limsup_{l,n\to\infty} \sum_{j=1}^{\infty} j|x_j^{\{l\}}(t) - x_j^{\{n\}}(t)|$$

$$\leq e^{-\epsilon t} \limsup_{l,n\to\infty}\|x^{\{l\}}(0) - x^{\{n\}}(0)\|_1$$

$$+ \frac{c_9}{i} \limsup_{l,n\to\infty} \int_0^t e^{-\epsilon(t-s)} w^{\{l\}}(s)\|x^{\{l\}}(s) - x^{\{n\}}(s)\|_1 ds.$$

Since this estimate holds for every $i \in \mathbb{N}$ and each $x^{\{n\}}$ satisfies the estimates in Theorem 18, with the same constants, we can take the limit $i \to \infty$ and

$$\limsup_{l,n\to\infty} \sum_{j=1}^{\infty} j|x_j^{\{l\}}(t) - x_j^{\{n\}}(t)| \leq e^{-\epsilon t} \limsup_{l,n\to\infty}\|x^{\{l\}}(0) - x^{\{n\}}(0)\|_1.$$

Since $\|x\|_1 \leq |x_0| + 2\sum_{j=1}^{\infty} j|x_j|$,

$$\limsup_{l,n\to\infty} \|\Phi_t(\breve{x}^{\{l\}}) - \Phi_t(\breve{x}^{\{n\}})\|_1 \leq 2e^{-\epsilon t}\|\breve{x}^{\{l\}} - \breve{x}^{\{n\}}\|_1 \leq 4e^{-\epsilon t}\|B_0\|_1$$

where $\|B_0\|_1 = \sup_{\breve{x} \in B_0} \|\breve{x}\|_1$. By (2.30), $\alpha_s(\Phi_t(B_0)) \leq 4e^{-\epsilon t}\|B_0\|_1 \to 0$ as $t \to \infty$. This finishes the proof. $\square$

2.8 Towards the Stability of Equilibria

For the metapopulation model (2.27) we make assumptions which guarantee that the number of patches does not change in time.

Assumption 22 Assume

(a) For all $k \in \mathbb{N}$ there is some $j \in \mathbb{Z}_+$, $j < k$, with $\alpha_{jk} > 0$, $\alpha_{00} = 0$, and

$$\sum_{j=0}^{\infty} \alpha_{jk} = 0 = \sum_{j=0}^{\infty} \gamma_{jk} \qquad \forall k \in \mathbb{Z}_+.$$

(b) $\displaystyle\limsup_{k \to \infty} \sum_{j=0}^{\infty} \frac{j\alpha_{jk}}{k} < 0.$

Occasionally we will also assume the following.

Assumption 23 (a) The sequence (σ_n) is bounded.

(b) There exist positive constants c_0, c_1, ϵ such that

$$\sum_{j=1}^{\infty} j\gamma_{jk} \le c_0 + c_1 k - \epsilon|\gamma_{kk}| \qquad \text{for all } k \in \mathbb{N}.$$

By Corollary 1, $\|x(t)\| = \|\check{x}\|$. We fix the initial patch number to be N and obtain $\sum_{j=0}^{\infty} x_j(t) = N$. We will use this equality to eliminate x_0. Notice that Assumption 15(a) and 22(a) imply that $\alpha_{j0} = 0$ for all $j \in \mathbb{Z}_+$. We equivalently rewrite (2.27) as

$$w' = \sum_{k=1}^{\infty} \eta_k x_k - w \sum_{k=1}^{\infty} (\sigma_k - \sigma_0) x_k - \left(N\sigma_0 + \delta\right)w,$$

$$x'_j = \sum_{k=1}^{\infty} \alpha_{jk} x_k + w\left(\gamma_{j0} N + \sum_{k=1}^{\infty} (\gamma_{jk} - \gamma_{j0}) x_k\right), \qquad j = 1, 2, \ldots$$

(2.32)

This system can be cast in more condensed notation,

$$w' = \langle x, x^* \rangle - \xi w + w\langle x, y^* \rangle, \qquad x' = \tilde{A}x + wz + w\Gamma_0 x, \qquad (2.33)$$

with $x(t) = (x_j(t))_{j=1}^{\infty}$.

Remark 4. $x(t)$ takes values in $\tilde{\ell}^{11}$, the space of sequences $x = (x_j)_{j=1}^{\infty}$ with norm $\|x\|_1^{\sim} = \sum_{j=1}^{\infty} j|x_j|$. Further

$$\xi = N\sigma_0 + \delta, \qquad u = (\gamma_{j0})_{j=1}^{\infty}, \qquad z = Nu,$$

x^* and y^* in the dual space of $\tilde{\ell}^{11}$,

$$\langle x, x^* \rangle = \sum_{k=1}^{\infty} \eta_k x_k, \qquad \langle x, y^* \rangle = \sum_{k=1}^{\infty} (\sigma_0 - \sigma_k) x_k.$$

Finally

$$\tilde{A}x = \left(\sum_{k=1}^{\infty} \alpha_{jk} x_k\right)_{j=1}^{\infty}, \qquad \Gamma_0 x = \tilde{\Gamma}x - \langle x, z^* \rangle u$$

with

$$\langle x, z^* \rangle = \sum_{k=1}^{\infty} x_k, \qquad \tilde{\Gamma}x = \left(\sum_{k=1}^{\infty} \gamma_{jk} x_k\right)_{j=1}^{\infty}.$$

2.8.1 Stability of Equilibria

Let $(\tilde{w}, \tilde{x})$ be an equilibrium, i.e. a constant solution of (2.33). $(\tilde{w}, \tilde{x}) = (0,0)$ is an equilibrium, e.g., called the *extinction equilibrium*. Any other equilibrium in $\mathbb{R}_+ \times \tilde{\ell}_+^{11}$ is called a *persistence equilibrium*. To study the stability of the equilibrium $(\tilde{w}, \tilde{x})$, we expand the system about the equilibrium. We set $w = \tilde{w} + v$ and $x = \tilde{x} + y$ and obtain the following equation for v and y, where we have replaced $\tilde{x}$ by x and $\tilde{w}$ by w,

$$\begin{aligned} v' &= \langle y, x^* \rangle - \xi v + v\langle x, y^* \rangle + w\langle y, y^* \rangle + v\langle y, y^* \rangle, \\ y' &= \tilde{A}y + vz + w\Gamma_0 y + v\Gamma_0 x + v\Gamma_0 y. \end{aligned} \tag{2.34}$$

This is an abstract Cauchy problem (evolution equation)

$$(v, y)' = \mathcal{A}(v, y) + g(v, y), \tag{2.35}$$

where $\mathcal{A}$ is the linear operator defined in ℓ^{11} by

$$\mathcal{A}(v, y) = \big(\langle y, x^* \rangle - \xi v + v\langle x, y^* \rangle + w\langle y, y^* \rangle, \tilde{A}y + vz + w\Gamma_0 y + v\Gamma_0 x\big), \tag{2.36}$$
$$v \in \mathbb{R}, y \in \tilde{\ell}^{11},$$

and g the non-linear map on ℓ^{11} defined by

$$g(v, y) = \big(v\langle y, y^* \rangle, v\Gamma_0 y\big). \tag{2.37}$$

Proposition 3. *Let the Assumptions 15 and 22 be satisfied. Let $\tilde{A}$ and $\tilde{\Gamma}$ be as in Remark 4. Let $w \geq 0$. If $w > 0$ also assume Assumption 23. Then $\tilde{A}_w = \tilde{A} + w\tilde{\Gamma}$, with appropriate domain, is the generator of a positive C_0-semigroup $\tilde{S}$ on $\tilde{\ell}^{11}$ with strictly negative growth bound.*

Proof. Define $\beta_{jk} = \alpha_{jk} + w\gamma_{jk}$ for $j, k \in \mathbb{Z}_+$. The operator $\tilde{A}_w$ is associated with the infinite matrix $(\beta_{jk})_{j,k=1}^{\infty}$. For $k \in \mathbb{N}$,

$$\sum_{j=1}^{\infty} \beta_{jk} = -\alpha_{0k} - w\gamma_{0k}$$

which is non-positive for $k \in \mathbb{N}$ and strictly negative for $k = 1$. Also the other assumptions of [39, Proposition 6.3] are satisfied. It follows that $\tilde{A}_w$ with domain $\{x \in \tilde{\ell}^{11}; \sum_{j=1}^{\infty} |\beta_{jj}||x_j| < \infty, \tilde{A}_w x \in \tilde{\ell}^{11}\}$ is the generator of a C_0-semigroup $\tilde{S}(t)$ on $\tilde{\ell}^{11}$ and there exist $\epsilon > 0$, $M \geq 1$ such that $\|\tilde{S}(t)\|_1^{\tilde{\,}} \leq Me^{-\epsilon t}$. □

Proposition 4. *Let $w \geq 0$ and $x \in \tilde{\ell}^{11}$. Let $\Gamma_0 y = \tilde{\Gamma} y + \langle y, z^* \rangle u$ with the ingredients as in Remark 4. Let the Assumptions 15 and 22 be satisfied and, if $w > 0$, also Assumption 23. Then $\mathcal{A}$ is the generator of a C_0-semigroup T with strictly negative essential growth bound (essential type).*

Proof. $\mathcal{A} = B + C$ where

$$B(v, y) = (-\xi v, \tilde{A}_w y),$$
$$C(v, y) = (\langle y, x^* \rangle + v \langle x, y^* \rangle, vz - w \langle y, z^* \rangle u + v \Gamma_0 x)$$

By Proposition 3, $\tilde{A}_w$ is the generator of a C_0-semigroup $\tilde{S}$ on ℓ^{11} with strictly negative growth bound. B is the generator of the semigroup $S(t)(v, y) = (e^{-\xi t}v, \tilde{S}(t)y)$. S also has a strictly negative growth bound. The linear operator C on ℓ^{11} has finite-dimensional range and therefore is compact. The perturbation $\mathcal{A} = B + C$ generates a C_0-semigroup T such that $T(t) - S(t)$ is compact for every $t \geq 0$. So the essential growth bound of T does not exceed the growth bound of S and is strictly negative [14, Chap. 4 Proposition 2.12]. $\qquad\square$

Theorem 24. *Let the Assumptions 15 and 22 be satisfied and $\tilde{w}, \tilde{x}$ be an equilibrium of (2.32). If $\tilde{w} \neq 0$, also make Assumption 23.*
Then the following hold:
(a) If all eigenvalues of $\mathcal{A} = B + C$ have strictly negative real part, then the equilibrium $(\tilde{w}, \tilde{x})$ is locally asymptotically stable in the following sense. There exist $M \geq 1$ and $r > 0$ such that

$$\left\| (w(t), x(t)) - (\tilde{w}, \tilde{x}) \right\|_1 \leq M e^{-rt} \left\| (w(0), x(0)) - (\tilde{w}, \tilde{x}) \right\|_1 \qquad \forall t \geq 0,$$

for all solutions of (2.32).
(b) If $\mathcal{A} = B + C$ has at least one eigenvalue with strictly positive real part, then the equilibrium $(\tilde{w}, \tilde{x})$ is unstable in the following sense: there exist some $\epsilon > 0$ and a sequence $0 < t_n \to \infty$ as $n \to \infty$ and a sequence of solutions w^n, x^n of (2.32) such that $w^n(0) \to \tilde{w}, x^n(0) \to \tilde{x}$ as $n \to \infty$ and $\left\| (w^n(t_n), x^n(t_n)) - (\tilde{w}, \tilde{x}) \right\|_1 \geq \epsilon$ for all $n \in \mathbb{N}$.

Proof. We notice that the non-linearity g in (2.35) and (2.37) satisfies $\frac{\|g(v,y)\|_1}{\|(v,y)\|_1} \to 0$ as $v \to 0, y \to 0$. Let $\Phi(t, \check{w}, \check{x})$ be the semiflow induced by the solutions of (2.32) with initial data $\check{w}$ and $\check{x}$. It follows from standard arguments (essentially from Gronwall's inequality, cf. [55, Sect. 3], e.g.) that, for each $t \geq 0$, $\Phi(t, \cdot)$ is differentiable at $(\tilde{w}, \tilde{x})$ with derivative $T(t)$ from Proposition 4. The results now follow from [12] along the lines of [55, Sect. 4]. $\qquad\square$

2.9 Instability of Every Other Equilibrium: General Result

The following derivation of an instability condition for equilibria is more efficiently done on a somewhat more abstract level and may apply to other situations where an unstructured (part of the) population [in our case the dispersers] is paired with a structured (part of the) population [in our case the occupants]. We consider the system

$$w' = f(w, x), \qquad x' = \Lambda x + g(w, x). \tag{2.38}$$

Here Λ is a closed linear operator in an ordered Banach space X with cone X_+ and $f : \mathbb{R}_+ \times X_+ \to \mathbb{R}$, $g : \mathbb{R}_+ \times X_+ \to X$ are continuously differentiable.

We assume that $\mathbb{R}_+$ is contained in the resolvent set of Λ and also in the resolvent set of $\Lambda + g_x(w, x)$ for each $w \geq 0$ and $x \in X_+$.

g_w and g_x denote the partial derivatives of $g(w, x)$ with respect to w and x. Since Λ^{-1} exists and is bounded, -1 is in the resolvent set of $\Lambda^{-1} g_x(w, x)$ and

$$(\mathbb{I} + \Lambda^{-1} g_x(w, x))^{-1} \Lambda^{-1} = (\Lambda + g_x(w, x))^{-1}. \tag{2.39}$$

2.9.1 The Equilibria

A pair (w, x) is an equilibrium solution of (2.38) if and only if $0 = f(w, x)$ and x satisfies the fixed point equation

$$x = -\Lambda^{-1} g(w, x). \tag{2.40}$$

Assume that for every $w > 0$ there exists a solution $x = \phi(w)$ of (2.40). If follows from our assumptions and the implicit function theorem [8, Chap. 2, Theorem 2.3] that ϕ is differentiable (analytic if g is analytic) and

$$\phi'(w) = -\Lambda^{-1} \Big(g_w(w, \phi(w)) + g_x(w, \phi(w)) \phi'(w) \Big). \tag{2.41}$$

By our assumptions and (2.39),

$$\begin{aligned} \phi'(w) &= - \big(\mathbb{I} + \Lambda^{-1} g_x(w, \phi(w)) \big)^{-1} \Lambda^{-1} g_w(w, \phi(w)) \\ &= - \big(\Lambda + g_x(w, \phi(w)) \big)^{-1} g_w(w, \phi(w)). \end{aligned} \tag{2.42}$$

We substitute the solution $x = \phi(w)$ of (2.40) into $0 = f(w, x)$,

$$0 = f(w, \phi(w)) =: F(w). \tag{2.43}$$

Theorem 25. *A pair (w, x) with $w \in \mathbb{R}_+$ and $x \in X_+$ is an equilibrium if and only if $F(w) = 0$ and $x = \phi(w)$. In particular there is a one-to-one correspondence between equilibria and zeros of F. F is analytic if f and g are analytic.*

For later use we differentiate the function F,

$$F'(w) = f_w(w, \phi(w)) + f_x(w, \phi(w)) \phi'(w).$$

We substitute (2.42),

$$F'(w) = f_w(w, x) - f_x(w, x) \big(\Lambda + g_x(w, x) \big)^{-1} g_w(w, x). \tag{2.44}$$

2.9.2 The Eigenvalue Problem of the Linearized System

We linearize (2.38) around an equilibrium (w, x),

$$v' = f_w(w, x)v + f_x(w, x)y, \qquad y' = \Lambda y + g_w(w, x)v + g_x(w, x)y. \qquad (2.45)$$

The associated eigenvalue problem has the form

$$\lambda v = f_w(w, x)v + f_x(w, x)y,$$
$$\lambda y = \Lambda y + g_w(w, x)v + g_x(w, x)y. \qquad (2.46)$$

Consider $\lambda \geq 0$. We solve the second equation for y,

$$y = (\lambda - \Lambda - g_x(w, x))^{-1} g_w(w, x)v = v(\lambda - \Lambda - g_w(w, x))^{-1} g_w(w, x).$$

We notice that $(v, y) \neq (0, 0)$ if and only if $v \neq 0$. We substitute the expression for y into the first equation of (2.46) and divide by v,

$$\lambda = f_w(w, x) + f_x(w, x)(\lambda - \Lambda - g_w(w, x))^{-1} g_w(w, x).$$

This leads to the following characteristic equation,

$$0 = Q(\lambda) := \lambda - f_w(w, x) - f_x(w, x)\big(\lambda - \Lambda - g_w(w, x)\big)^{-1} g_w(w, x).$$

We evaluate $Q(\lambda)$ for $\lambda = 0$ and compare it to (2.44),

$$Q(0) = -f_w(w, x) + f_x(w, x)\big(\Lambda + g_w(w, x)\big)^{-1} g_w(w, x) = -F'(w).$$

Notice that $Q(\lambda) \to \infty$ as $\lambda \to \infty$. If $Q(0) < 0$, the characteristic equation has a root $\lambda > 0$ by the intermediate value theorem.

Theorem 26. *Let (w, x) be an equilibrium of (2.38) and $F'(w) > 0$. Then the associated linear operator has a strictly positive eigenvalue.*

By Theorem 25, we can order the equilibria (w, x) according to their w-component provided that the zeros of F are isolated which is the case, e.g., if f and g and so F are analytic.

Corollary 3. *Assume that the zeros of F are isolated and there is no $w > 0$ with both $F(w) = 0$ and $F'(w) = 0$. Then, for every other equilibrium, the associated linear operator has a strictly positive eigenvalue.*

Proof. If the zeros of F are isolated, then, for every $b > 0$, then we have finitely many equilibria (w_j, x_j) with $0 \leq w_j \leq b$ and can order them like $w_1 < w_2 < \cdots$. Since $F'(w_j) \neq 0$, F changes sign at each w_j. So $F'(w_j) > 0$ for every other j and the associated linear operator has a strictly positive eigenvalue by Theorem 26. $\square$

2.10 Existence of Equilibria and Instability of Every Other Equilibrium

After eliminating the equation for the empty patches, our system can be rewritten in the form (2.32) and then in a more condensed form for $w(t) \in \mathbb{R}_+$ and $x(t) = (x_j(t))_{j=1}^{\infty} \in \tilde{\ell}_+^{11}$,

$$
\begin{aligned}
w' &= \langle x, x^* \rangle - w \langle x, y^* \rangle - w(\sigma_0 N + \delta), \\
x' &= \tilde{A}x + w\tilde{\Gamma}x + w(N - \langle x, z^* \rangle)u,
\end{aligned}
\tag{2.47}
$$

which is the same as (2.33). Here u and x^*, y^*, z^* are as in Remark 4 as are the bounded linear quasi-positive operator $\tilde{\Gamma}$ on ℓ^{11} and $\tilde{A}$, the generator of a positive C_0-semigroup on $\tilde{\ell}^{11}$. The system (2.33) fits into the framework of (2.38) by setting $\Lambda = \tilde{A}$ and

$$
\begin{aligned}
f(w, x) &= \langle x, x^* \rangle - w \langle x, y^* \rangle - w(\sigma_0 N + \delta), \\
g(w, x) &= w\tilde{\Gamma}x + w(N - \langle x, z^* \rangle)u.
\end{aligned}
\tag{2.48}
$$

For each $w \geq 0$, $\tilde{A}_w = \tilde{A} + w\tilde{\Gamma}$ is also the infinitesimal generator of a positive C_0-semigroup $\tilde{S}$ on $\tilde{\ell}^{11}$. Notice that we obtain the operator $\mathcal{A}$ in (2.36) when we linearize (2.47) about an equilibrium.

We make the Assumptions 15, 22 and 23. By Proposition 3, $\tilde{S}$ has strictly negative growth bound and so, for each $w \geq 0$, $\tilde{A}_w$ has positive resolvents $(\lambda - \tilde{A}_w)^{-1}$ for all $\lambda \geq 0$. We take the partial derivative of g in (2.48) with respect to x,

$$
g_x(w, x)y = w\tilde{\Gamma}y - w\langle y, z^* \rangle z, \qquad z = Nu.
\tag{2.49}
$$

Lemma 6. *If $\lambda - \tilde{A}_w$ has a bounded positive inverse for $\lambda \geq 0$, $\lambda - \tilde{A} - g_x(w, x)$ has a bounded inverse and*

$$
(\lambda - \tilde{A} - g_x(w, x))^{-1}\tilde{x} = (\lambda - \tilde{A}_w)^{-1}\tilde{x} - w\zeta(\lambda - \tilde{A}_w)^{-1}z
$$

where

$$
\zeta = \frac{\langle (\lambda - \tilde{A}_w)^{-1}\tilde{x}, z^* \rangle}{1 + w\langle (\lambda - \tilde{A}_w)^{-1}z, z^* \rangle}.
$$

Proof. In order to find $\hat{x} = (\lambda - \tilde{A} - g_x(w, x))^{-1}\tilde{x}$, we solve the equation

$$
\lambda\hat{x} - \tilde{A}\hat{x} - w\tilde{\Gamma} + w\langle \hat{x}, z^* \rangle z = \tilde{x}.
$$

See (2.48). This can be rewritten as

$$
(\lambda - \tilde{A}_w)\hat{x} = \tilde{x} - w\langle \hat{x}, z^* \rangle z.
$$

Since the resolvent exists for $\tilde{A}_w$,

$$\hat{x} = (\lambda - \tilde{A}_w)^{-1}\tilde{x} - w\langle\hat{x}, z^*\rangle(\lambda - \tilde{A}_w)^{-1}z.$$

We apply the functional z^*,

$$\langle x, z^*\rangle = \langle(\lambda - \tilde{A}_w)^{-1}\tilde{x}_w, z^*\rangle - w\langle x, z^*\rangle\langle(\lambda - \tilde{A}_w)^{-1}z, z^*\rangle.$$

We solve for $\zeta := \langle x, z^*\rangle$ and substitute ζ into the equation for x. This yields the assertion. $\square$

Since f and g are analytic, (w, x) is an equilibrium of (2.47) if and only if $x = \phi(w)$ and $F(w) = 0$ where ϕ and F are analytic functions on $\mathbb{R}_+$ (see Theorem 25).

2.10.1 Equilibria

To find a concrete expression for the solutions $x = \phi(w)$ of the equation $\Lambda x + g(w, x) = 0$, which is identical to

$$0 = \tilde{A}x + w\tilde{\Gamma}x + w\big(N - \langle x, z^*\rangle\big)u = \tilde{A}_w x + w\big(N - \langle x, z^*\rangle\big)u,$$

we apply the inverse of $\tilde{A}_w$ to the second equation in (2.33),

$$x = -w\big(N - \langle x, z^*\rangle\big)\tilde{A}_w^{-1}u. \tag{2.50}$$

In order to calculate $\langle x, z^*\rangle$, we apply the functional z^* to this equation,

$$\langle x, z^*\rangle = -w\big(N - \langle x, z^*\rangle\big)\langle\tilde{A}_w^{-1}u, z^*\rangle.$$

We solve for $\langle x, z^*\rangle$,

$$\langle x, z^*\rangle = -wN\frac{\langle\tilde{A}_w^{-1}u, z^*\rangle}{1 - w\langle\tilde{A}_w^{-1}u, z^*\rangle}.$$

Notice that the denominator is positive because $-\tilde{A}_w^{-1}$ is a positive operator. Further

$$\langle x, z^*\rangle \in [0, N). \tag{2.51}$$

We rewrite

$$\langle x, z^*\rangle = N\Big(1 - \frac{1}{1 - w\langle\tilde{A}_w^{-1}u, z^*\rangle}\Big). \tag{2.52}$$

We substitute this expression into the one for $x = \phi(w)$, recall $Nu = z$ from Remark 4, and find

$$\phi(w) = w\psi(w),$$

$$\psi(w) = -\frac{1}{1 - w\langle\tilde{A}_w^{-1}z, z^*\rangle}\tilde{A}_w^{-1}u \in X_+. \tag{2.53}$$

By (2.43),

$$F(w) = f(w, \phi(w))$$
$$= \langle \phi(w), x^* \rangle - w \langle \phi(w), y^* \rangle - \sigma_0 w N - \delta w.$$

(2.54)

At this point, we need an estimate for $\phi(w)$. We recall that there is a one-to-one correspondence between equilibria of (2.32) and equilibria of the original system (2.27) with $\|x\| = 1$. This means that $\phi(w) = (x_j)_{j=1}^\infty$ where $x \in \ell_+^{11}$, $A_1 x + w\Gamma x = 0$ and $x_0 = N - \sum_{j=1}^\infty x_j$. By Theorem 19,

$$\sum_{k=1}^\infty \eta_k x_k - w \sum_{k=0}^\infty \sigma_k x_k \le c_4.$$

After eliminating $x_0 = N - \sum_{k=1}^\infty x_j$ this reads

$$\sum_{k=1}^\infty \eta_k \phi_k(w) - w \sum_{k=1}^\infty [\sigma_k - \sigma_0] \phi_k(w) - w\sigma_0 N \le c_4.$$

By Remark 4 and (2.54), $F(w) \le c_4 - \delta w$ and $F(w) < 0$ for large $w > 0$.

We substitute $\phi(w) = w\psi(w)$ into F. For $w > 0$, equation $F(w) = 0$ then takes the form

$$\tilde{F}(w) = \delta$$

with

$$\tilde{F}(w) = \frac{F(w)}{w} + \delta$$

being analytic in $w > 0$ and $\tilde{F}(w) < \delta$ for large $w > 0$ and

$$\tilde{F}(0) = \langle \psi(0), x^* \rangle - \sigma_0 N, \qquad \psi(0) = -\tilde{A}^{-1} z.$$

We combine Theorems 24 and 26. The associated linear operator in Theorem 26 coincides with the operator $\mathcal{A}$ in Theorem 24. Notice that, for $w > 0$, $\tilde{F}(w) = \delta$ and $\tilde{F}'(w) = 0$ is equivalent to $F(w) = 0$ and $F'(w) = 0$.

Theorem 27. *Let the Assumptions 15, 22, and 23 be satisfied, $\xi = \sigma_0 N + \delta$.*

(a) If $\xi < -\langle \tilde{A}^{-1} z, x^* \rangle$, the extinction equilibrium is unstable and there exists a persistence equilibrium. For all but finitely many $\xi < -\langle \tilde{A}^{-1} z, x^* \rangle$, there exists an odd number of persistence equilibria (w_j, x_j), $w_1 < w_2 < \cdots$. Every even-indexed persistence equilibrium is unstable.

(b) If $\xi > -\langle \tilde{A}^{-1} z, x^* \rangle$, the extinction equilibrium is stable. For all but finitely many $\xi > -\langle \tilde{A}^{-1} z, x^* \rangle$, there exists no persistence equilibrium or an even number of persistence equilibria (w_j, x_j), $w_1 < w_2 < \cdots$. Every odd-indexed persistence equilibrium is unstable.

Proof. Assume that $\tilde{F}(w) = \delta$ has a solution. Since $\tilde{F}(w) < \delta$ for large $w > 0$, F is not constant. F is analytic and so is F'. Since F' is not zero everywhere, there is no accumulation of arguments w with $\tilde{F}'(w) = 0$. Since $\tilde{F}(w) < \delta$ for large $w > 0$ there are only finitely many $w > 0$ such that $\tilde{F}(w) = -\delta$ and

$\tilde{F}'(w) = 0$. So for all but finitely many δ, we have $\tilde{F}'(w) \neq 0$ for all $w > 0$ with $\tilde{F}(w) = \delta$.

(a) Here we consider the case $\delta < \tilde{F}(0)$.

As $\tilde{F}(w) < \delta$ for large w, there exists an $w > 0$ such that $\tilde{F}(w) = \delta$ by the intermediate value theorem. For all but finitely many δ, $\tilde{F}'(w) \neq 0$ for all w with $\tilde{F}(w) = \delta$. Choose such a δ. Since $\tilde{F}(w) < \delta$ for sufficiently large $w > 0$, $\tilde{F}(w)$ crosses the line $\tilde{F} = \delta$ an odd number of times, the first time with a negative derivative, the second time with a positive derivative etc. By Theorems 24 and 26, every w with $\tilde{F}'(w) > 0$, i.e., every even-indexed equilibrium, is unstable. (b) is proved similarly. The stability proof for the extinction equilibrium is postponed to Theorem 33. $\square$

Application of these results to special metapopulation models can be found in [38].

2.11 Stability of the Extinction Equilibrium Versus Metapopulation Persistence

The total population size of the metapopulation is given by the sum of the number of dispersers and the total number of patch occupants,

$$P(t) = w(t) + \sum_{j=1}^{\infty} j x_j(t).$$

The extinction equilibrium is characterized by $P = 0$. The stability of the extinction equilibrium can be formulated in terms of the total population size.

The extinction equilibrium is locally stable if, for every $\epsilon > 0$, there exists some $\delta > 0$ such that $P(t) \leq \epsilon$ whenever $P(0) < \delta$. The extinction equilibrium is locally asymptotically stable, if in addition there exists some $\delta_0 > 0$ such that $P(t) \to 0$ as $t \to \infty$ whenever $P(0) < \delta_0$.

The following two concepts imply the instability of the extinction equilibrium.

The metapopulation is called *weakly uniformly persistent* if there exists some $\epsilon > 0$ (independent of the initial conditions) such that

$$\limsup_{t \to \infty} P(t) > \epsilon \qquad \text{whenever} \quad P(0) > 0.$$

The metapopulation is called *(strongly) uniformly persistent* if there exists some $\epsilon > 0$ (independent of the initial conditions) such that

$$\liminf_{t \to \infty} P(t) > \epsilon \qquad \text{whenever} \quad P(0) > 0.$$

Obviously uniform persistence implies weak uniform persistence. The converse holds under additional assumptions the most crucial of which is the existence of a compact attractor. Actually we will establish uniform persistence in a stronger sense. Material on persistence theory for semiflows on infinite dimensional spaces can be found in [27, 56, 58, 62].

2.11.1 Local Asymptotic Stability of the Extinction Equilibrium

We turn to the stability of the extinction equilibrium for the specific metapopulation model (2.27). After elimination of the empty patches, this is the equilibrium $\tilde{w} = 0$, $\tilde{x} = 0$ for (2.32) or rather its abstract formulation (2.33). Throughout this section, we make the Assumptions 15 and 22. We define a linear operator B_0 (on appropriate domain in ℓ^{11}) and a bounded linear operator C on ℓ^{11} by

$$B_0(w, x) = (-\xi w, \tilde{A}x), \qquad C(w, x) = (\langle x, x^* \rangle, wz). \qquad (2.55)$$

and a non-linear map g on ℓ^{11} by

$$g(w, x) = (w \langle x, y^* \rangle, w \Gamma_0 x), \qquad \Gamma_0 x = \tilde{\Gamma}x - \langle x, z^* \rangle \frac{1}{N} z. \qquad (2.56)$$

Then (2.33) can be written as $(w, x)' = (B_0 + C)(w, x) + g(w, x)$. The domain of B_0 is the same as the one of the operator A_1, $D(B_0) = D(A_1)$, $D_1 \subseteq D(B_0) \subseteq D_0$. For each $\epsilon \geq 0$, (2.32) can be written as the Cauchy problem

$$(w, x)' = \mathcal{A}_\epsilon(w, x) + g_\epsilon(w, x),$$

with

$$\mathcal{A}_\epsilon = B_0 - \epsilon \mathbb{I} + (1 - \epsilon)C, \qquad g_\epsilon = \epsilon \mathbb{I} + \epsilon C + g. \qquad (2.57)$$

Differently from g, the modified non-linearity g_ϵ, for $\epsilon > 0$, is positivity preserving in a neighborhood of the origin (the size of which depends on ϵ).

Lemma 7. *Let the Assumptions 15 and 22 hold. Then, for any $\epsilon > 0$, there exists some $\epsilon_0 > 0$ such that $g_\epsilon(w, x) \geq 0$ whenever $w \in [0, \epsilon_0]$, $x \in \tilde{\ell}^{11}_+$ $\|x\|^\sim_1 \leq \epsilon_0$.*

Proof. Let $w \in [0, \epsilon_0]$, $x \in \tilde{\ell}^{11}_+$, $\|x\|^\sim_1 \leq \epsilon_0$. We look at the first component of $g_\epsilon(w, x)$. By (2.56), (2.57), Remark 4, and Assumption 15(c),

$$\epsilon w + \epsilon \langle x, x^* \rangle + w \langle x, y^* \rangle$$

$$\geq \epsilon w - w \sum_{k=1}^{\infty} \sigma_k x_k \geq w(\epsilon - c_7 \|x\|^\sim_1) \geq w(\epsilon - c_7 \epsilon_0) \geq 0,$$

if ϵ_0 is chosen small enough. We look at the second component of g_ϵ. By (2.57) (2.56), and (2.55),

$$\epsilon x + \epsilon w z + w\Gamma_0 x = \epsilon x + w\tilde{\Gamma} x + w\Big(\epsilon - \langle x, z^*\rangle \frac{1}{N}\Big)z.$$

The term in $(\cdot)$ can be estimated by

$$\geq \epsilon - \|x\|_{\tilde{1}} \frac{1}{N} \geq \epsilon - \epsilon_0 \frac{1}{N} \geq 0,$$

if $\epsilon_0 > 0$ is chosen small enough. As for the other term,

$$(\epsilon x + w\tilde{\Gamma} x)_j \geq (\epsilon - w\gamma_{jj})x_j \geq (\epsilon - wc_8)x_j$$

where c_8 is the constant in Assumption 15(d). The last expression is non-negative if $w \leq \epsilon_0$ and $\epsilon_0 > 0$ is chosen small enough.   □

The operators $(1 - \epsilon)C$ are compact for every $\epsilon \geq 0$. By Proposition 3, $\tilde{A}$ is the generator of a C_0-semigroup $\tilde{S}$ on $\tilde{\ell}^{11}$ with strictly negative growth bound. The operators $B_0 - \epsilon\mathbb{I}$ generate C_0-semigroups S^ϵ on ℓ^{11} which have the form

$$S^\epsilon(t)(w, x) = \Big(e^{-(\epsilon+\xi)t}w, e^{-\epsilon t}\tilde{S}(t)\tilde{x}\Big).$$

Obviously the semigroups S^ϵ have strictly negative growth bounds. For each $\epsilon \geq 0$, the operator $\mathcal{A}_\epsilon = B_0 - \epsilon\mathbb{I} + (1-\epsilon)C$ generates a C_0-semigroup $\{T^\epsilon(t); t \geq 0\}$ on ℓ^{11}. Since $(1 - \epsilon)C$ is compact, $T^\epsilon(t) - S^\epsilon(t)$ is compact for every $t \geq 0$ and the essential growth bound of T^ϵ equals the essential growth bound of S^ϵ [14, Chap. 4, Proposition 2.12] and is strictly negative. For all $\epsilon \in [0,1]$, the operators $(1 - \epsilon)C$ are positive, i.e, they map ℓ^{11}_+ into itself. Since the semigroup S^ϵ is positive, the standard perturbation formula implies that the semigroup T^ϵ is positive.

Proposition 5. *Let the Assumptions 15 and 22 be satisfied. Assume that there is a spectral value of $\mathcal{A}_0$ with non-negative real part. Then there exists some $\lambda_0 \geq 0$ with the following properties:*

(i) *λ_0 is a pole of the resolvent of $\mathcal{A}_0$, is isolated in the spectrum of $\mathcal{A}_0$ and an eigenvalue of $\mathcal{A}_0$ with finite algebraic multiplicity.*
(ii) *$\lambda_0 \geq \Re\tilde{\lambda}$ for every $\tilde{\lambda}$ in the spectrum of $\mathcal{A}_0$.*
(iii) *λ_0 is associated with positive eigenvectors of $\mathcal{A}_0$ and $\mathcal{A}_0^*$.*

Proof. By assumption, the spectral bound of $\mathcal{A}_0$,

$$\lambda_0 = \sup\{\Re\lambda; \lambda \in \sigma(\mathcal{A}_0)\},$$

is non-negative. Since $T(t) - S(t)$ is compact for every $t > 0$, $(\lambda - \mathcal{A}_0)^{-1} - (\lambda - B_0)^{-1}$ is compact for sufficiently large $\lambda > 0$, i.e., $\mathcal{A}_0$ is resolvent compact relatively to B_0 [57, Def. 3.7]. Then the spectral bound λ_0 is non-negative and has the asserted properties [57, Proposition 3.10].   □

Theorem 28. *Let the Assumptions 22 and 15 be satisfied. Assume that there is no element $v \in \ell_+^{11} \cap D(\mathcal{A}_0)$ such that $v \neq 0$ and $\mathcal{A}_0 v \geq 0$. Then the extinction equilibrium is locally asymptotically stable.*

Proof. It follows from the assumptions and (2.56) that g is continuously differentiable in ℓ_+^{11} and $g'(0) = 0$. Suppose that the spectral bound of $\mathcal{A}_0$,

$$\lambda_0 = \sup\{\Re\lambda; \lambda \in \sigma(\mathcal{A}_0)\},$$

is non-negative. Then the same arguments as in the proof of Proposition 5 imply that λ_0 has the properties (i), (ii), (iii) asserted in Proposition 5, in particular $\mathcal{A}_0 v = \lambda_0 v \geq 0$ with some $v \in \ell_+^{11} \cap D(\mathcal{A}_0)$, in contradiction to our assumption. Hence $\lambda_0 < 0$ and all eigenvalues of $\mathcal{A}_0 + g'(0) = \mathcal{A}_0$ have strictly negative real parts. The assertion follows from Theorem 24. Recall $\tilde{w} = 0$. $\square$

2.11.2 Instability of the Extinction Equilibrium

Theorem 29. *Let the Assumptions 22 and 15 be satisfied. Assume that there is an element $v \in \ell_+^{11} \cap D(\mathcal{A}_0)$ such that $v \neq 0$ and $\mathcal{A}_0 v \geq 0$. Further assume that there is no element $v \in \ell_+^{11} \cap D(\mathcal{A}_0)$ such that $v \neq 0$ and $\mathcal{A}_0 v = 0$. Then the extinction equilibrium is unstable.*

Remark 5. Under the assumptions of Theorem 29, there exists an eigenvalue $\lambda_0 > 0$ of $\mathcal{A}_0$ which is associated with positive eigenvectors of $\mathcal{A}_0$ and $\mathcal{A}_0^*$.

Proof. We choose some $v \in \ell_+^{11}$, $v \neq 0$, such that $\mathcal{A}_0 v \geq 0$. For $\lambda > 0$, $(\lambda - \mathcal{A}_0)v \leq \lambda v$. For sufficiently large λ, $(\lambda - \mathcal{A}_0)^{-1}$ exists and is a bounded positive operator. We apply it to the previous inequality arbitrarily many times, $v \leq \lambda^n(\lambda - \mathcal{A}_0)^{-n}v$. This implies that the spectral radius of $\lambda(\lambda - \mathcal{A}_0)^{-1}$ is greater than or equal to 1. Hence the spectral bound of $\mathcal{A}_0$, λ_0, satisfies $\lambda_0 \in [0, \infty)$ [57, Cor. 3.6]. By Proposition 5, λ_0 is an eigenvalue of $\mathcal{A}_0$ associated with an eigenvector $v \in \ell_+^{11}$ of $\mathcal{A}_0$ and a positive eigenvector of $\mathcal{A}_0^*$. Since $\mathcal{A}_0 v \neq 0$ for all $v \in \ell_+^{11}$, $v \neq 0$, $\lambda_0 > 0$. So $\mathcal{A}_0$ has a positive eigenvalue and, by Theorem 28 (notice that $\mathcal{A} = \mathcal{A}_0$ because $w = \tilde{w} = 0$), the extinction equilibrium $(0, 0)$ is unstable. $\square$

2.11.3 Persistence of the Metapopulation

Since persistence is a stronger property than instability of the extinction equilibrium, it is not surprising that we uphold the assumptions of Theorem 29. Then the operator $\mathcal{A}_0 = B_0 + C$ has a positive eigenvalue which is associated with a positive eigenvector of $\mathcal{A}_0^*$. We need this eigenvector to be strictly positive in an appropriate sense. To this end we make irreducibility assumptions for the transition matrix (α_{jk}).

Definition 1. The infinite matrix $(\alpha_{jk})_{j,k\in\mathbb{N}}$ is called *irreducible* if, for every $j, k \in \mathbb{N}$, $j \neq k$, there exist $n \in \mathbb{N}$ and $i_1, \ldots, i_n \in \mathbb{N}$ such that $i_1 = k$, $i_n = j$ and $\alpha_{i_{l+1},i_l} > 0$ for $l = 1, \ldots, n-1$;

If $k_0 \in \mathbb{N}$, the finite matrix $(\alpha_{jk})_{j,k=1}^{k_0}$ is called *irreducible* if the analogous statement holds with the set $\mathbb{N}$ be replaced by $\{0, \ldots, k_0\}$.

A number $k_0 \in \mathbb{N}$ is called the *irreducibility bound* of the infinite matrix (α_{jk}), if the matrix $(\alpha_{jk})_{j,k=0}^{k_0}$ is irreducible, $\alpha_{jk} = 0$ whenever $j > k_0$ and $k = 0, \ldots, j-1$, and $\alpha_{kk} < 0$ for $k > k_0$.

Analogously the irreducibility of an infinite matrix $(\breve{\alpha}_{jk})_{j,k\in\mathbb{Z}_+}$ or its irreducibility bound are defined.

Notice that the irreducibility together with the assumptions $\sum_{j=0}^{\infty} \alpha_{jk} \leq 0$, $\alpha_{jk} \geq 0$ for $j \neq k$, implies that $\alpha_{kk} < 0$ for all $k \in \mathbb{N}$. It is easy to see that the irreducibility bound (if there is one) is uniquely determined.

Assumption 30 Let one of the following be satisfied:

(a) The infinite matrix $(\alpha_{jk})_{j,k\in\mathbb{N}}$ is irreducible and $\gamma_{j0} > 0$ for some $j \in \mathbb{N}$ and $\eta_k > 0$ for some $k \in \mathbb{N}$.

or

(b) The matrix $(\alpha_{jk})_{j,k\in\mathbb{N}}$ has the irreducibility bound k_0, $\gamma_{j0} > 0$ for some $j \in \{1, \ldots, k_0\}$ and $\eta_k > 0$ for some $k \in \{1, \ldots, k_0\}$.

Proposition 6. *Let Assumption 22, 15 and 30 be satisfied. Then the eigenvalue λ_0 of $\mathcal{A}_0$ in Proposition 5 is associated with a strictly positive eigenvector v^* of $\mathcal{A}_0^*$, $\langle \mathbf{x}, v^* \rangle > 0$ for all $\mathbf{x} \in \ell_+^{11}$, $\mathbf{x} \neq 0$.*

Proof. Let us first assume (a) in the Assumption 30. The operator $\mathcal{A}_0 = B_0 + C$, with B_0 and C in (2.55) and x^*, z in Remark 4, is associated with the infinite matrix

$$(\beta_{jk})_{j,k=0}^{\infty} = \begin{pmatrix} -\xi & \eta_1 & \eta_2 & \cdots \\ \gamma_{10}N & \alpha_{11} & \alpha_{12} & \cdots \\ \gamma_{20}N & \alpha_{21} & \alpha_{22} & \cdots \\ \vdots & \vdots & \vdots & \vdots \end{pmatrix}. \tag{2.58}$$

By Assumption 30(a) this infinite matrix is irreducible and the semigroup T generated by $\mathcal{A}_0$ is strictly positive on ℓ_+^{11}, i.e., $[T(t)\mathbf{x}]_j > 0$ for every $t > 0$, $j \in \mathbb{Z}_+$, $\mathbf{x} \in \ell_+^{11}$, $\mathbf{x} \neq 0$. This implies that the eigenvector v^* of $\mathcal{A}_0$ associated with λ_0 is strictly positive, i.e. $\langle \mathbf{x}, v^* \rangle > 0$ for all $\mathbf{x} \in \ell_+^{11}$, $\mathbf{x} \neq 0$. Let us now assume (b) in the Assumption 30. v^* can be identified with a sequence $(y_j)_{j=0}^{\infty}$ with $y_j = \langle e_j, v^* \rangle \geq 0$ for all $j \in \mathbb{Z}_+$. Here e_j is the sequence which has 1 in the jth term and only zeros otherwise. Suppose that $y_j = 0$ for $j = 0, \ldots, k_0$. Let $k > k_0$ be the smallest natural number for which $y_k > 0$. Since $k > 1$, by the form of (2.58),

$$\langle e_k, \mathcal{A}^* v^* \rangle = \sum_{j=k}^{\infty} y_j \alpha_{jk}.$$

Since $\alpha_{jk} = 0$ for $j > k > k_0$ by Definition 1,

$$\langle e_k, \mathcal{A}^* v^* \rangle = \alpha_{kk} y_k.$$

But also $\langle e_k, \mathcal{A}^* v^* \rangle = \lambda_0 y_k$ which implies $0 < \lambda_0 = \alpha_{kk} \leq 0$, a contradiction. Hence $y_j > 0$ for at least one $j \in \{0, \dots, k_0\}$. Since the matrix $(\beta_{jk})_{j,k=0}^{k_0}$ is irreducible, $[T(t)\mathbf{x}]_j > 0$ for all $t > 0$, $j = 0, \dots, k_0$, $\mathbf{x} \in \ell_+^{11}$, $\mathbf{x} \neq 0$. Hence, for each $\mathbf{x} \in \ell_+^{11}$, $\mathbf{x} \neq 0$,

$$0 < \langle T(t)\mathbf{x}, v^* \rangle = e^{\lambda_0 t} \langle \mathbf{x}, v^* \rangle.$$

$\square$

Theorem 31. *Let Assumptions 22, 15, and 30 be satisfied. Assume that there is an element $v \in \ell_+^{11}$, $v \neq 0$, such that $\mathcal{A}_0 v = (B_0 + C)v \geq 0$. Further assume that there is no element $v \in \ell_+^{11}$, $v \neq 0$, such that $(B_0 + C)v = 0$.*

Then the metapopulation is uniformly weakly persistent, i.e., there exists some $\epsilon_0 > 0$ such that

$$\limsup_{t \to \infty} \left(w(t) + \sum_{j=1}^{\infty} j x_j(t) \right) \geq \epsilon_0$$

for all solutions of (2.27) with $\breve{w} \geq 0$, $\breve{x} \in \ell_+^{11}$, $\breve{w} + \sum_{j=1}^{\infty} j \breve{x}_j > 0$.

Proof. By Remark 5, $\mathcal{A}_0$ has an eigenvalue $\lambda_0 > 0$. We first show that the operators $\mathcal{A}_\epsilon$ also have positive eigenvalues provided that $\epsilon > 0$ is small enough. Let λ be a resolvent value of B_0. Then

$$\lambda - \mathcal{A}_\epsilon = \left[\mathbb{I} + \epsilon(\lambda - B_0)^{-1} - (1 - \epsilon)C(\lambda - B_0)^{-1} \right](\lambda - B_0).$$

If $\lambda > 0$ is chosen large enough,

$$\left\| \epsilon(\lambda - B_0)^{-1} - (1 - \epsilon)C(\lambda - B_0)^{-1} \right\| < 1$$

for all $\epsilon \in [0, 1]$ and the operator in $[\ ]$ has a bounded inverse. Thus $\lambda - \mathcal{A}_\epsilon$ has a bounded inverse and

$$(\lambda - \mathcal{A}_\epsilon)^{-1} = (\lambda - B_0)^{-1} \left[\mathbb{I} + \epsilon(\lambda - B_0)^{-1} - (1 - \epsilon)C(\lambda - B_0)^{-1} \right]^{-1}$$

$$\xrightarrow{\epsilon \to 0} (\lambda - B_0)^{-1} \left[\mathbb{I} - C(\lambda - B_0)^{-1} \right]^{-1} = (\lambda - \mathcal{A}_0)^{-1}.$$

As $\lambda_0 > 0$ is an eigenvalue of $\mathcal{A}_0$ and an isolated point of the spectrum of $\mathcal{A}_0$ by Proposition 5, we can choose $\epsilon > 0$ so small that $\lambda_\epsilon > 0$ for the spectral bound λ_ϵ of $\mathcal{A}_\epsilon$ [32, Chap. 4, Theorem 2.25 and Sect. 3.5]. Then Propositions 5 and 6 hold for $\mathcal{A}_\epsilon$ and λ_ϵ rather than $\mathcal{A}_0$ and λ_0. Once $\epsilon > 0$ has been chosen, by Lemma 7 there exists some $\epsilon_0 > 0$ such that $g_\epsilon(\mathbf{x}) := \epsilon \mathbf{x} + \epsilon C \mathbf{x} + g(\mathbf{x}) \geq 0$ for all $\mathbf{x} \in \ell_+^{11}$, $\|\mathbf{x}\|_1 \leq \epsilon_0$. Assume that there exists a non-negative solution $w, (x_j)_{j=0}^{\infty}$ of (2.27) with $\breve{w} \geq 0$, $\breve{x} \in \ell_+^{11}$, $\breve{w} + \sum_{j=1}^{\infty} j \breve{x}_j > 0$ and

$$\limsup_{t\to\infty}\Big(w(t) + \sum_{j=1}^{\infty} jx_j(t)\Big) < \epsilon_0.$$

If we set $x(t) = (x_j(t))_{j=1}^{\infty}$, w and x satisfy (2.33). Then $\mathbf{x} = (w, x)$ in ℓ_+^{11} with $\mathbf{x}(0) \neq 0$ and $\limsup_{t\to\infty} \|\mathbf{x}(t)\|_1 < \epsilon_0$. By Propositions 5 and 6, $\lambda_\epsilon = s(\mathcal{A}_\epsilon)$ is an eigenvalue of $\mathcal{A}_\epsilon$ and there exists $v_\epsilon^* \in X_+^*$, $X = \ell^{11}$, such that $\langle x, v_\epsilon^* \rangle > 0$ for all $x \in \ell_+^{11}$, $x \neq 0$. By making a time shift forward and using the semiflow property, we can assume that $\langle \mathbf{x}(t), v_\epsilon^* \rangle > 0$ and $\|\mathbf{x}(t)\|_1 \leq \epsilon_0$ for all $t \geq 0$. Then, for all $t \geq 0$,

$$\mathbf{x}(t) = \mathbf{x}(0) + \mathcal{A}_\epsilon \int_0^t \mathbf{x}(s)ds + \int_0^t g_\epsilon(\mathbf{x}(s))ds \geq \mathbf{x}(0) + \mathcal{A}_\epsilon \int_0^t \mathbf{x}(s)ds.$$

Let $\hat{\mathbf{x}}(\lambda)$ denote the Laplace transform of $\mathbf{x}$,

$$\hat{\mathbf{x}}(\lambda) = \int_0^{\infty} e^{-\lambda t}\mathbf{x}(t)dt.$$

We take the Laplace transform of the equation above,

$$\hat{\mathbf{x}}(\lambda) \geq \frac{1}{\lambda}\mathbf{x}(0) + \frac{1}{\lambda}\mathcal{A}_\epsilon\hat{\mathbf{x}}(\lambda).$$

We multiply by λ and apply the functional v_ϵ^*,

$$\lambda\langle\hat{\mathbf{x}}(\lambda), v_\epsilon^*\rangle \geq \langle\mathbf{x}(0), v_\epsilon^*\rangle + \lambda_\epsilon\langle\hat{\mathbf{x}}(\lambda), v_\epsilon^*\rangle.$$

For $\lambda = \lambda_\epsilon$ we obtain the contradiction, $0 \geq \langle\mathbf{x}(0), v_\epsilon^*\rangle > 0$. $\square$

If the solution semiflow has a compact attract, a stronger persistence results can be obtained.

Theorem 32. *Let the Assumptions 15, 17, 20, 22, and 30 be satisfied.*

Assume that there is an element $v \in \ell_+^{11}$, $v \neq 0$, such that $(B_0 + C)v \geq 0$. Further assume that there is no element $v \in \ell_+^{11}$, $v \neq 0$, such that $(B_0 + C) v = 0$.

Then the metapopulation is uniformly strongly persistent in the following sense: Under Assumption 30(a), for every $j \in \mathbb{Z}_+$, there exists some $\epsilon_j > 0$ such that

$$\liminf_{t\to\infty} w(t) \geq \epsilon_0, \qquad \liminf_{t\to\infty} x_j(t) \geq \epsilon_j \qquad \forall j \in \mathbb{N}$$

for all integral solutions of (2.27) with $\check{w} \geq 0$, $\check{x} \in \ell_+^{11}$, $\check{w} + \sum_{j=1}^{\infty} j\check{x}_j > 0$. Under Assumption 30(b), such a result holds for w and $x_1 \ldots, x_{k_0}$.

Proof. We define $\rho : \mathbb{R}_+ \times \ell_+^{11} \to \mathbb{R}_+$ by $\rho(w, x) = w + \sum_{j=1}^{\infty} jx_j$, $x = (x_j)_{j=0}^{\infty}$. By Theorem 31, the semiflow induced by the solutions of (2.27) is uniformly weakly ρ-persistent in the language of [58, A.5] and has a compact attractor by Theorem 21. We apply [58, Theorem A.34]. In order to show the persistence result for x_j, fix $j \in \mathbb{N}$ for (a) and $j \in \{1, \ldots, k_0\}$ for (b) and define $\tilde{\rho}(x) = x_j$, $x = (x_j)_{j=0}^{\infty}$. In order to show the persistence result for w define $\tilde{\rho}$ by $\tilde{\rho}(w, x) = w$. Let Φ be the semiflow induced by the solutions of (2.28), $\Phi_t(\breve{w}, \breve{x}) = (w(t), x(t))$ with $w, x = (x_j)_{j=0}^{\infty}$ satisfying (2.28). A total orbit $(w(t), x(t))$ of Φ is defined for all $t \in \mathbb{R}$ and satisfies $(w(t), x(t)) = \Phi_{t-r}(w(r), x(r))$ for all $t, r \in \mathbb{R}$, $t \geq r$. This is equivalent to

$$
\begin{aligned}
w' &= \sum_{k=1}^{\infty} \eta_k x_k - w \sum_{k=0}^{\infty} \sigma_k x_k - \delta w \quad \text{on } \mathbb{R}, \\
x_j(t) - x_j(r) &= \sum_{k=0}^{\infty} \alpha_{jk} \int_r^t x_k(s)ds + \sum_{k=0}^{\infty} \gamma_{jk} \int_r^t w(s)x_k(s)ds, \\
&\qquad j \in \mathbb{Z}_+, r, t \in \mathbb{R}, t > r.
\end{aligned}
\tag{2.59}
$$

Cf. (2.28). The assumptions of [58, Theorem A.34] are satisfied by the following Lemma. $\square$

Lemma 8. *Let the assumptions of Theorem 32 be satisfied. Let $w(t), x(t) = (x_j(t))_{j=0}^{\infty}$ be a non-negative solution of (2.59) which exists on $\mathbb{R}$ such that $w(t) + \|x(t)\|_1 \leq c$ for all $t \in \mathbb{R}$ with some constant $c > 0$ and $\|x(t)\| = N$ for all $t \in \mathbb{R}$.*

Then $w(t) > 0$ and $x_j(t) > 0$ for all $t \in \mathbb{R}$ and all $j \in \mathbb{N}$, whenever $w(t) + \sum_{k=1}^{\infty} kx_k(t) > 0$ for all $t \in \mathbb{R}$.

Proof. By (2.59), integrating the equation for w, for $t > r$,

$$
\begin{aligned}
w(t) &= w(r)\frac{\phi(t)}{\phi(r)} + \int_r^t \sum_{k=1}^{\infty} \eta_k x_k(s)\frac{\phi(t)}{\phi(s)}ds, \\
\phi(t) &= \exp\left(\int_0^t \left[\sum_{k=0}^{\infty} \sigma_k x_k(s) - \delta\right]ds\right) > 0, \\
x_j(t) &= x_j(r)\frac{\phi_j(t)}{\phi_j(r)} + \int_r^t \sum_{k \neq j, k=1}^{\infty} [\alpha_{jk} + \gamma_{jk}w(s)]x_k(s)\frac{\phi_j(t)}{\phi_j(s)}ds, \\
\phi_j(t) &= \exp\left(\int_0^t \left[\alpha_{jj} + \gamma_{jj}w(s)\right]ds\right) > 0.
\end{aligned}
\tag{2.60}
$$

The irreducibility assumptions are now combined with the following kind of arguments.

Case 1: Suppose that $x_k(r) > 0$ for some r, $k \in \mathbb{N}$. By (2.60), $x_k(t) \geq x_k(r)\frac{\phi_j(t)}{\phi(r)} > 0$ for all $t \geq r$. Now let $j \in \mathbb{N}$, $\alpha_{jk} > 0$. By (2.60),

$$
x_j(t) \geq \int_r^t \alpha_{jk}x_k(s)\frac{\phi_j(t)}{\phi_j(s)}ds > 0 \qquad \forall t > r.
$$

If we combine this argument with the respective irreducibility properties of the matrix $(\alpha_{jk})_{j,k\in\mathbb{N}}$ we obtain that $x_j(t) > 0$ for $t > r$ and $j \in \mathbb{N}$ or $j = 1, \ldots, k_0$ respectively.

By Assumption 30, there exists some $k \in \mathbb{N}$ such that $\eta_k > 0$. Then

$$w(t) \geq \int_r^t \eta_k x_k(s) \frac{\phi(t)}{\phi(s)} ds > 0 \qquad \forall t > r.$$

Case 2: Now assume that $w(r) > 0$ for some $r \in \mathbb{R}$. By (2.60), $w(t) > 0$ for all $t > r$. Since $\sum_{k=0}^{\infty} x_k(r) = N$ there are two cases, $x_0(r) > 0$ or $x_k(r) > 0$ for some $k \in \mathbb{N}$. If the second is the case, the considerations for case 1 imply that $x_j(t) > 0$ for all $t > r$ and all $j \in \mathbb{N}$ or $j = 1, \ldots, k_0$ respectively. So let us assume that $x_0(r) > 0$. Then $x_0(t) > 0$ for all $t \geq r$. By Assumption 30, there exists some $j \in \mathbb{N}$ (or $j \in \{1, \ldots, k_0\}$) such that $\gamma_{j0} > 0$. By (2.60),

$$x_j(t) \geq \int_r^t \gamma_{j0} w(s) x_0(s) \frac{\phi_j(t)}{\phi_j(s)} ds > 0 \qquad \forall t > r.$$

By Case 1, $x_j(t) > r$ for all $t > r$, $j \in \mathbb{N}$. □

We conclude this section by emphasizing that there is a distinct threshold condition (though we can only express it in abstract terms) which separates local stability of the extinction equilibrium on the one hand from existence of a persistence equilibrium and (weak or strong) persistence of the metapopulation on the other hand.

Theorem 33. *Let the Assumptions 15 and 22 be satisfied. Let z, x^*, ξ and the operator $\tilde{A}$ be as in Remark 4. Then the following hold:*

(a) Let $\xi > -\langle \tilde{A}^{-1} z, x^* \rangle$. Then the extinction equilibrium is locally asymptotically stable.

(b) Let $\xi < -\langle \tilde{A}^{-1} z, x^* \rangle$. Then the extinction equilibrium is unstable and there exists a persistence equilibrium. If in addition, Assumption 30 holds, the metapopulation is uniformly weakly persistent in the sense of Theorem 31. If we also add Assumptions 17 and 20, then the metapopulation is uniformly strongly persistent in the sense of Theorem 32.

Proof. (a) We apply Theorem 28. Suppose that the assumptions of this theorem are not satisfied. Then there exists an element $v \in \ell_+^{11} \cap D(\mathcal{A}_0)$, $v \neq 0$, such that $\mathcal{A}_0 v \geq 0$. By definition of $\mathcal{A}_0$ in (2.57) and by (2.55), $v = (w, x)$ with $w \geq 0$, $x \in \ell_+^{11}$, with

$$0 \leq -\xi w + \langle x, x^* \rangle, \qquad 0 \leq \tilde{A}x + wz. \tag{2.61}$$

By Proposition 3, $-\tilde{A}^{-1}$ exist and is a positive bounded linear operator. We apply it to the second inequality in (2.61), $x \leq -w\tilde{A}^{-1}z$. If $w = 0$, $x \in -\tilde{\ell}_+^{11}$ and so $x = 0$ and $v = 0$. Since $v \neq 0$, $w > 0$. We substitute $x \leq -w\tilde{A}^{-1}z$ in the first inequality in (2.61), $0 \leq -\xi w - w \langle \tilde{A}^{-1} z, x^* \rangle$. We divide by $w > 0$ and obtain a contradiction to the assumption $\xi > -\langle \tilde{A}^{-1} z, x^* \rangle$. So the assumptions of Theorem 28 are satisfied and the local asymptotic stability of the extinction equilibrium follows.

(b) The existence of a persistence equilibrium has already been established in Theorem 27 (a). (Notice that Assumption 23 is only needed for the instability statements in Theorem 27 (a).) Similarly as in (a), we show that existence of an element $v \in \ell_+^{11}$, $v \neq 0$, $\mathcal{A}_0 v = (B_0 + C)v = 0$, leads to $\xi = -\langle \tilde{A}^{-1} z, x^* \rangle$ which is ruled out by assuming $\xi < -\langle \tilde{A}^{-1} z, x^* \rangle$. Set $x = -\tilde{A}^{-1} z$ and $w = 1$. Then $0 = \tilde{A} x + wz$ and $0 \leq -\xi w + \langle x, x^* \rangle$ which translates into $(B_0 + C)v \geq 0$ for $v = (w, x)$ by (2.55). The respective assumptions of Theorems 29, 31 and 32 are satisfied and uniform weak or uniform strong persistence follow.   □

2.12 Application to Special Metapopulation Models

In [38], we consider the following metapopulation model,

$$
\begin{cases}
w' = \displaystyle\sum_{n=1}^{\infty}(1-q_n)\beta_n x_n(t) - \left[\delta + \sum_{n=0}^{\infty}\sigma_n x_n(t)\right]w, \\[2ex]
x_0'(t) = \mu_1 x_1(t) + \displaystyle\sum_{n=1}^{\infty}\kappa_n x_n(t) - \sigma_0 w(t) x_0(t), \\[2ex]
x_n'(t) = \big[q_{n-1}\beta_{n-1} + \sigma_{n-1}w(t)\big]x_{n-1}(t) + \mu_{n+1}x_{n+1}(t) \\[0.5ex]
\qquad\quad - \big[q_n\beta_n + \sigma_n w(t) + \mu_n + \kappa_n\big]x_n(t), \\[1ex]
\qquad\quad n = 1, 2, \dots.
\end{cases}
\tag{2.62}
$$

β_n and μ_n are the birth and death rates in local populations of size n, q_n is the probability that a juvenile stays on its birth patch if the local population size is n, κ_n is the rate at which a local population of size n is completely wiped out, and σ_n the rate at which an average migrating individual settles on a patch with local population size n. Migrating individuals are assumed to not reproduce, their per capita death rate is δ.

In comparison to (2.27), we identify

$$
\begin{cases}
\alpha_{k+1,k} = q_k \beta_k, & k \in \mathbb{N}, \\
\alpha_{k-1,k} = \mu_k, & k \in \mathbb{N}, \\
\alpha_{kk} = -(q_k\beta_k + \mu_k + \kappa_k), & k \in \mathbb{N}, \\
\alpha_{0k} = \kappa_k, & k \in \mathbb{Z}_+, \\
\alpha_{k0} = 0, & k \in \mathbb{Z}_+, \\
\alpha_{jk} = 0, & |j-k| > 1,
\end{cases}
\tag{2.63}
$$

and

$$
\begin{cases}
\gamma_{k+1,k} = \sigma_k, & k \in \mathbb{Z}_+, \\
\gamma_{k,k} = -\sigma_k, & k \in \mathbb{Z}_+, \\
\gamma_{jk} = 0, & j, k \in \mathbb{Z}_+ \text{ otherwise}
\end{cases}
\tag{2.64}
$$

and $\eta_k = (1 - q_k)\beta_k$. Then

$$\sum_{j=0}^{\infty} \alpha_{jk} = 0, \qquad k = 0, 1, \ldots .$$

For $k \in \mathbb{N}$,

$$\sum_{j=1}^{\infty} j\alpha_{jk} = (k+1)q_k\beta_k + (k-1)\mu_k - k(q_k\beta_k \times \mu_k + \kappa_k)$$

$$= q_k\beta_k - \mu_k - k\kappa_k,$$

$$\eta_k + \sum_{j=1}^{\infty} j\alpha_{jk} = \beta_k - \mu_k - k\kappa_k.$$

For $k = 0$, $\sum_{j=1}^{\infty} j\alpha_{j0} = 0$. For $k \in \mathbb{Z}_+$,

$$\sum_{j=0}^{\infty} \gamma_{jk} = 0, \qquad \sum_{j=1}^{\infty} j\gamma_{jk} = \sigma_k, \qquad \sum_{j=1}^{\infty} j|\gamma_{jk}| \le 2(1+k)\sigma_k.$$

Assumption 34 (a) $\beta_n, \kappa_n \ge 0$, $\mu_n > 0$ for all $n \in \mathbb{N}$.
(b) $0 \le q_n \le 1$ for all $n \in \mathbb{N}$.
(c) $\sigma_n \ge 0$ for all $n \in \mathbb{Z}_+$, $\sup\limits_{n=0}^{\infty} \sigma_n < \infty$.

Theorem 35. *Let the Assumption 34 be satisfied. Further, if $\epsilon > 0$ is chosen small enough, let $\sup_{n=1}^{\infty} \frac{(1+\epsilon)\beta_n - \mu_n}{n} < \infty$. Then, for every $\breve{w} \ge 0$, $\breve{x} \in \ell_+^{11}$, there exists a unique integral solution of on $[0, \infty)$. Further $\|x(t)\| \le \|\breve{x}\|$ for all $t \ge 0$.*

Theorem 36. *Let the assumptions of Theorem 35 be satisfied. Further assume that there exist constants $c_4, \epsilon_4 > 0$ such that $\beta_n - \mu_n - n\kappa_n \le c_4 - \epsilon_4 n$ for all $n \in \mathbb{N}$. Then*

$$w(t) + \sum_{j=1}^{\infty} jx_j(t) \le \left(\breve{w} + \sum_{j=1}^{\infty} j\breve{x}_j \right) e^{-\epsilon_4 t} + \frac{c_4 \|\breve{x}\|}{\epsilon_4}$$

for all solutions (w, x) of (2.62) with initial data $\breve{w} \ge 0$, $\breve{x} \in \ell_+^{11}$. Further $\|x(t)\| \le \|\breve{x}\|$ for all $t \ge 0$.

We apply Theorem 21.

Theorem 37. *In addition to the Assumption 34 assume that*

$$\inf_{n=1}^{\infty} \frac{\mu_n}{n} > 0, \quad \limsup_{n \to \infty} \frac{\beta_n}{\mu_n} < 1, \quad and \quad \sup_{n=1}^{\infty} \frac{\kappa_n}{n} < \infty.$$

Then the semiflow induced by the solutions of (2.62) on $\mathbb{R}_+ \times \ell_+^{11}$ has a compact attractor for bounded sets.

2.12.1 Scenarios of Extinction

The population goes extinct without emigration from the patches or colonization of empty patches.

Theorem 38. *Let the Assumptions of Theorem 37 be satisfied. If $q_k = 1$ for all $k \in \mathbb{N}$ (i.e. there is no patch emigration) or if $\sigma_0 = 0$ (empty patches are not colonized), the total population size, $w(t) + \sum_{j=1}^{\infty} j x_j(t)$, is integrable on $[0, \infty)$ and converges to 0 as $t \to \infty$.*

Proof. This follows from Corollary 2, $\gamma_{00} = -\sigma_0$, and $\eta_k = (1 - q_k)\beta_k$. □

The population also goes extinct if on every patch the birth rate is smaller than the death rate.

Corollary 4. *Let the assumptions of Theorem 35 be satisfied. Assume that there exists some $\epsilon > 0$ such that $\beta_k - \mu_k - k\kappa_k \leq -\epsilon k$ for all $k \in \mathbb{N}$. Then the total population size, $w(t) + \sum_{j=1}^{\infty} j x_j(t)$, converges to 0 as time tends to infinity.*

Proof. The assumptions of Theorem 36 are satisfied with $c_4 = 0$. □

2.12.2 Persistence

We assume that the metapopulation is not subject to catastrophes, $\kappa_n = 0$, and introduce the following number which can be interpreted as the basic reproduction ratio of the metapopulation [38],

$$\mathcal{R}_0 = \frac{\sigma_0 N}{\sigma_0 N + \delta} \left(\sum_{j=1}^{\infty} (1 - q_j) \frac{\beta_j}{\mu_j} \prod_{k=1}^{j-1} \frac{q_k \beta_k}{\mu_k} \right). \tag{2.65}$$

Theorem 39. *Let $\sigma_0 > 0$, $\kappa_n = 0$ for all $n \in \mathbb{N}$ and $\inf_{n=1}^{\infty} \frac{\mu_n}{n} > 0$, $\limsup_{n \to \infty} \frac{\beta_n}{\mu_n} < 1$. Then the following hold:*

(a) *Let $\mathcal{R}_0 < 1$. Then the extinction equilibrium is locally asymptotically stable.*
(b) *Let $\mathcal{R}_0 > 1$. Then there exists a persistence equilibrium.*
(c) *Let $\mathcal{R}_0 > 1$ and one of the following be satisfied:*
   (c$_1$) $q_j \beta_j > 0$ *for all $j \in \mathbb{N}$ and $(1 - q_k)\beta_k > 0$ for some $k \in \mathbb{N}$,*
   or
   (c$_2$) *There exists some $k_0 \in \mathbb{N}$ such that $q_j \beta_j > 0$ for $j = 1, \ldots, k_0 - 1$, $q_j \beta_j = 0$ for all $j \geq k_0$, and that $(1 - q_j)\beta_j > 0$ for some $j \in \{1, \ldots, k_0\}$.*
   Under (c$_1$), for every $j \in \mathbb{Z}_+$, there exists some $\epsilon_j > 0$ such that

$$\liminf_{t \to \infty} w(t) \geq \epsilon_0, \qquad \liminf_{t \to \infty} x_j(t) \geq \epsilon_j \qquad \forall j \in \mathbb{N}$$

for all solutions of (2.62) with $\check{w} \geq 0$, $\check{x} \in \ell_+^{11}$, $\check{w} + \sum_{j=1}^{\infty} j \check{x}_j > 0$. Under Assumption (c$_2$), such a result holds for w and $x_1 \ldots, x_{k_0}$.

Proof. We apply Theorem 33. Let $x = -\tilde{A}^{-1}z$. Then $\sum_{k=1}^{\infty} \alpha_{jk} x_k + z_j = 0$ for $j \in \mathbb{N}$ where $x \in D(\tilde{A})$. By Remark 4, $z_j = \gamma_{j0} N$. So $z_1 = N\sigma_0$ and $z_j = 0$ for $j \geq 2$ by (31). By (2.63),

$$
\begin{aligned}
\mu_2 x_2 - q_1 \beta_1 x_1 &= \mu_1 x_1 - \sigma_0 N, \\
\mu_{j+1} x_{j+1} - q_j \beta_j x_j &= \mu_j x_j - q_{j-1} \beta_{j-1} x_{j-1}, \qquad j \geq 2.
\end{aligned}
\tag{2.66}
$$

Since $x \in D(\tilde{A})$, $\sum_{j=1}^{\infty} |\alpha_{jj}| x_j < \infty$ and (2.63) implies that the series $\sum_{j=1}^{\infty} \mu_j x_j$ and $\sum_{j=1}^{\infty} q_j \beta_j x_j$ converge. So we can add the second equality in (2.66) from j to infinity and obtain that $x_j = \frac{q_{j-1}\beta_{j-1}}{\mu_j} x_{j-1}$ for $j \geq 2$. The first equation in (2.66) implies $\mu_1 x_1 = \sigma_0 N$. This recursive equation is solved by

$$
x_j = \prod_{l=1}^{j-1} \frac{q_l \beta_l}{\mu_l} \frac{\sigma_0 N}{\mu_j}.
\tag{2.67}
$$

with the understanding that $\prod_{j=1}^{0} = 1$. By Remark 4, $\langle x, x^* \rangle = \sum_{j=1}^{\infty} \eta_j x_j$ with $\eta_j = (1 - q_j)\beta_j$, $\xi = N\sigma_0 - \delta$. This implies that $\xi + \langle \tilde{A}^{-1}z, x^* \rangle$ has the same sign as $1 - \mathcal{R}_0$. □

We refer to [38] for existence of multiple persistence equilibria, the special case of obligatory juvenile emigration, and a bang-bang principle of persistence-optimal emigration.

2.13 Special Host-Macroparasite Models and Existence of Solutions

Let x_n denote the number of hosts with n parasites and w the average number of free-living parasites,

$$
\begin{cases}
w' = \displaystyle\sum_{n=1}^{\infty} (1 - q_n)\beta_n x_n - \left[\delta + \sum_{n=0}^{\infty} \sigma_n x_n\right] w, \\[2ex]
x_0' = \displaystyle\sum_{n=0}^{\infty} \gamma_n(x) x_n + \mu_1 x_1 + \sum_{n=1}^{\infty} \kappa_n x_n - \sigma_0 w x_0 - \nu_0 x_0, \\[2ex]
x_n' = \left[q_{n-1}\beta_{n-1} + \sigma_{n-1} w\right] x_{n-1} + \mu_{n+1} x_{n+1} \\[1ex]
\qquad - \left[q_n \beta_n + \sigma_n w + \mu_n + \kappa_n + \nu_n\right] x_n, \\[1ex]
\qquad n = 1, 2, \dots.
\end{cases}
\tag{2.68}
$$

2.13.1 Explanation of Parameters

In a host with n parasites, parasites die at a rate $\mu_n \geq 0$ and are born at a rate $\beta_n \geq 0$. With probability $q_n \in [0,1]$, newborn parasites stay within the birth host.

Hosts with n parasites are found and entered by an average free-living parasite at a per capita rate σ_n. They look for treatment and are completely delivered of their parasite load at a per capita rate $\kappa_n \geq 0$. Hosts with n parasites die at a per capita rate $\nu_n \geq 0$ and give birth at a per capita rate γ_n. To be specific, we choose a Ricker type per capita reproduction function,

$$\gamma_n(x) = \tilde{\gamma}_n \exp\left(-\sum_{k=0}^{\infty} \eta_{nk} x_k\right)$$

with $\tilde{\gamma}_n, \eta_{nk} \geq 0$. Notice that no vertical transmission has been assumed, i.e., newborn hosts have no parasites.

2.13.2 Unique Existence of Solutions

To fit the host-parasite model into the general framework we identify

$$\begin{cases} \alpha_{k+1,k} = q_k \beta_k, & k \in \mathbb{N}, \\ \alpha_{k-1,k} = \mu_k, & k = 2, 3, \ldots, \\ \alpha_{01} = \mu_1 + \kappa_1, & \\ \alpha_{0k} = \kappa_k, & k = 2, 3, \ldots, \\ \alpha_{kk} = -(q_k \beta_k + \mu_k + \kappa_k + \nu_k), & k \in \mathbb{N}, \\ \alpha_{00} = -\nu_0, & \\ \alpha_{jk} = 0, & \text{otherwise}, \end{cases} \qquad (2.69)$$

$$\begin{aligned} f(w, x) &= \sum_{n=1}^{\infty} (1 - q_n) \beta_n x_n - \left[\delta + \sum_{n=0}^{\infty} \sigma_n x_n\right] w, \\ g_0(w, x) &= \sum_{n=0}^{\infty} \gamma_n(x) x_n - \sigma_0 w x_0, \\ g_j(w, x) &= w(\sigma_{j-1} x_{j-1} - \sigma_j x_j), \qquad j \in \mathbb{N}. \end{aligned} \qquad (2.70)$$

We calculate

$$\sum_{j=0}^{\infty} \alpha_{jk} = -\nu_k, \qquad k = 0, 1, \ldots.$$

For $k \in \mathbb{N}$,

$$\sum_{j=1}^{\infty} j \alpha_{jk} = (k+1) q_k \beta_k + (k-1)\mu_k - k(q_k \beta_k + \mu_k + \kappa_k + \nu_k)$$

$$= q_k \beta_k - \mu_k - k(\kappa_k + \nu_k).$$

For $k = 0$, $\sum_{j=1}^{\infty} j \alpha_{j0} = 0$. For $k \in \mathbb{Z}$,

$$\sum_{j=0}^{\infty} g_j(w, x) = \sum_{k=1}^{\infty} \gamma_k(x) x_k \quad \text{and} \quad \sum_{j=1}^{\infty} j g_j(w, x) = w \sum_{k=0}^{\infty} \sigma_k x_k.$$

Theorem 40. *Let the Assumptions 34 be satisfied and $\nu_k, \delta \geq 0$. Then, for all $\breve{w} \in \mathbb{R}_+$ and $\breve{x} \in \ell_+^{11}$, there exists a unique solution w, x on $[0, \infty)$ of (2.22).*

Per capita host mortality rates that depend on host density and parasite burden would realistically not lead to a bounded perturbation, but require a different approach.

2.14 Application to Prion Proliferation

We focus on model (2.2) and leave the models (2.4) and (2.5) for future work. We assume that the coefficients b_{jk}, σ_j, and κ_j are all non-negative and the parameters δ and Λ are positive. While the infinite matrices (α_{jk}) have been sparse (basically tri-diagonal with an additional full first row) in the special metapopulation model in Sect. 2.12 and the host-macroparasite model in Sect. 2.13, the matrix (α_{jk}) in (2.3) has a full array above the diagonal. The coefficients α_{jk} in (2.3) satisfy Assumption 1(a) (modified for the missing x_0-equation). By (2.3), for $k \geq 2$,

$$\sum_{j=1}^{\infty} \alpha_{jk} = \kappa_k + \sum_{j=1}^{k-1}(b_{jk} + b_{k-j,k}) - \kappa_k - \sum_{i=1}^{k-1} b_{ik} = \sum_{j=1}^{k-1} b_{k-j,k}.$$

We substitute $k - j = i$,

$$\sum_{j=0}^{\infty} \alpha_{jk} = \sum_{i=1}^{k-1} b_{ik}. \tag{2.71}$$

Assumption 1(b) is satisfied if we assume

$$\sup_{k \geq 1} \sum_{i=1}^{k-1} b_{ik} < \infty. \tag{2.72}$$

We cannot determine from the literature whether or not such an assumption is biologically reasonable. It seems to be mainly for mathematical reasons that the coefficients $b_{jk} = b$ are assumed to be constant in [45, App. A] because it allows a moment closure which transforms the infinite system to three ordinary differential equations which can be completely analyzed [47]. In this special case $\sum_{i=1}^{k-1} b_{ik} = b(k-1)$ and Assumption 1(b) is not satisfied. As for part (c),

$$\sum_{j=1}^{\infty} j\alpha_{jk} = \sum_{j=1}^{k-1} j(b_{jk} + b_{k-j,k}) - k\kappa_k - k\sum_{i=1}^{k-1} b_{ik}.$$

Again we substitute $i = j - k$,

$$\sum_{j=0}^{\infty} j\alpha_{jk} = -k\kappa_k + \sum_{j=1}^{k-1} jb_{jk} + \sum_{i=1}^{k-1}(k-i)b_{ik} - k\sum_{i=1}^{k-1} b_{ik} = -k\kappa_k. \tag{2.73}$$

This shows that Assumption 1(c) also follows from (2.72). Let $\tilde{\ell}_1 = \{x = (x_j)_{j=1}^\infty; \|x\|^\sim < \infty\}$ with $\|x\|^\sim = \sum_{j=1}^\infty |x_j|$. Then Theorem 2 and Lemma 2 hold mutandis mutatis under (2.72).

Since the state space of the non-linear equations involves $\tilde{\ell}_+^{11}$ rather than $\tilde{\ell}_+^1$, $\tilde{\ell}^{11} = \{x = (x_j)_{j=1}^\infty; \|x\|_1^\sim < \infty\}$ with $\|x\|_m^\sim = \sum_{j=1}^\infty j^m |x_j|$, it is sufficient, though, that the infinite matrix (α_{jk}) is associated with a positive C_0-semigroup on $\tilde{\ell}^{11}$ which follows from (2.73) by the same construction as in [39] or in [59]. In order to get a handle on the generator in a analogous fashion as in Lemma 1, we investigate

$$\sum_{j=1}^\infty j^2 \alpha_{jk} = \sum_{j=1}^{k-1} j^2 (b_{jk} + b_{k-j,k}) - k^2 \kappa_k - k^2 \sum_{i=1}^{k-1} b_{ik}.$$

With the usual substitution $j = k - i$,

$$\sum_{j=1}^\infty j^2 \alpha_{jk} = \sum_{j=1}^{k-1} j^2 b_{jk} + \sum_{i=1}^{k-1} (k-i)^2 b_{ik} - k^2 \kappa_k - k^2 \sum_{i=1}^{k-1} b_{ik}$$

$$= -2 \sum_{j=1}^{k-1} j(k-j) b_{jk} - k^2 \kappa_k.$$

If we do not want to impose (2.72), we can alternatively add the following boundedness and positivity assumptions.

Assumption 41 (a) $\sup_{k=2}^\infty \max_{j=1}^{k-1} b_{jk} < \infty$ and $\sup_{j=1}^\infty \frac{\kappa_j}{j} < \infty.$

(b) $\inf_{j=1}^\infty \kappa_j > 0$ or $\inf_{k=2}^\infty \frac{1}{k} \min_{j=1}^{k-1} b_{jk} > 0.$

It follows from these assumptions that there exist constants $c_0, c_1, \epsilon > 0$ such that

$$\sum_{j=1}^\infty j^2 \alpha_{jk} \le c_0 - \epsilon k^2 - \epsilon k |\alpha_{kk}| \qquad \forall k \in \mathbb{Z}_+.$$

The same proofs as in [39] or [59] provide the following result.

Lemma 9. *Let the Assumption 41 be satisfied. Then the operator $\breve{A}_1$ on $\tilde{\ell}^{11}$ defined by*

$$\breve{A}_1 x = \left(\sum_{k=1}^\infty \alpha_{jk} x_k \right)_{j=1}^\infty, \qquad x = (x_k)_{k=1}^\infty,$$

$$D(\breve{A}_1) = \left\{ x \in \tilde{\ell}^{11}; \sum_{k=1}^\infty k |\alpha_{kk}| \, |x_k| < \infty \right\}$$

is closable and its closure generates a positive contraction C_0-semigroup $\tilde{S}$ on $\tilde{\ell}^{11}$. $\tilde{S}$ leaves $\tilde{\ell}^{12} = \{x = (x_j); \|x\|_2^\sim < \infty\}$ invariant.

We set

$$f(t, w, x) = \Lambda - w \sum_{k=1}^{\infty} \sigma_k x_k - \delta w,$$

$$g_j(t, w, x) = w(\sigma_{j-1} x_{j-1} - \sigma_j x_j).$$

Then the Assumption 4 are satisfied. Further

$$\sum_{j=1}^{\infty} g_j(t, w, x) = 0,$$

$$\sum_{j=1}^{\infty} j g_j(t, w, x) \le w \sum_{j=1}^{\infty} \sigma_j x_j,$$

$$f(t, w, x) + \sum_{j=1}^{\infty} j g_j(t, w, x) \le \Lambda.$$

By (2.73), by similar proofs as in Theorems 5 and 7, we obtain that solutions with non-negative initial data are defined and non-negative for all $t \ge 0$ and satisfy

$$\left. \begin{aligned} w(t) &\le w(0) e^{-\delta t} + \tfrac{\Lambda}{\delta}(1 - e^{-\delta t}) \\ w(t) + \sum_{j=1}^{\infty} j x_j(t) &\le w(0) + \sum_{j=1}^{\infty} j x_j(0) + \Lambda t \end{aligned} \right\} \qquad \forall t \ge 0.$$

If $\inf_{j=1}^{\infty} \kappa_j > 0$, then

$$\limsup_{t \to \infty} \left(w(t) + \sum_{j=1}^{\infty} j x_j(t) \right) \le \frac{\Lambda}{\zeta}, \qquad \zeta = \min \left\{ \delta, \inf_{j=1}^{\infty} \kappa_j \right\} > 0.$$

If we additionally assume that the polymerization rates (σ_j) are bounded, a similar procedure as in Sect. 2.7 shows that the semiflow on $\mathbb{R}_+ \times \tilde{\ell}_+^{11}$ associated with system (2.2) has a compact attractor for bounded sets. If $\inf_{j=1}^{\infty} \kappa_j = 0$ but $\inf_{k=2}^{\infty} \frac{1}{k} \min_{j=1}^{k-1} b_{jk} > 0$, we conjecture that the semiflow has a compact attractor for bounded sets if it is restricted to the positive cone of the invariant subspace $\mathbb{R} \times \tilde{\ell}^{12}$ with the stronger norm $(w, x) = |w| + \sum_{j=1}^{\infty} j^2 |x_j|^2$.

A. Non-Differentiability of the Simple Death Process Semigroup

We prove formulas (2.19) and (2.20) which imply that the semigroups S on ℓ and S_1 on ℓ^{11} associated with the simple birth process are not differentiable at any $t \in (0, \ln 2]$. Recall that we have chosen $\bar{t} = \ln 2$ such that $e^{-\bar{t}} = 1/2$. By (2.17) and (2.18),

$$\left\| \frac{d}{dt} S(\bar{t}) e^{[2n]} \right\| = 2 \sum_{j=0}^{2n} \binom{2n}{j} 2^{-2n} |n - j|$$

$$= 2 \sum_{j=0}^{n-1} \binom{2n}{j} 2^{-2n} (n - j) + 2 \sum_{j=n+1}^{2n} \binom{2n}{j} 2^{-2n} (j - n).$$

We substitute $j = 2n - k$ in the last sum and use $\binom{2n}{k} = \binom{2n}{2n-k}$,

$$\left\| \frac{d}{dt} S(\bar{t}) e^{[2n]} \right\| = 4 \sum_{j=0}^{n-1} \binom{2n}{j} 2^{-2n} (n - j). \tag{2.74}$$

By the binomial theorem,

$$2^{2n} = \sum_{j=0}^{2n} \binom{2n}{j} = 2 \sum_{j=0}^{n-1} \binom{2n}{j} + \binom{2n}{n}. \tag{2.75}$$

By rearranging the binomial coefficients,

$$\sum_{j=0}^{n-1} \binom{2n}{j} j = 2n \sum_{j=1}^{n-1} \binom{2n-1}{j-1} = 2n \sum_{j=0}^{n-2} \binom{2n-1}{j}. \tag{2.76}$$

Again by the binomial theorem,

$$2^{2n-1} = \sum_{j=0}^{2n-1} \binom{2n-1}{j} = \sum_{j=0}^{n-2} \binom{2n-1}{j} + \sum_{j=n-1}^{2n-1} \binom{2n-1}{j}.$$

In the second sum we substitute $j = 2n - 1 - k$. Then

$$2^{2n-1} = \sum_{j=0}^{n-2} \binom{2n-1}{j} + \sum_{k=0}^{n} \binom{2n-1}{2n-1-k}$$

$$= 2 \sum_{j=0}^{n-2} \binom{2n-1}{j} + \binom{2n-1}{n} + \binom{2n-1}{n-1}.$$

We combine this formula with (2.76),

$$\sum_{j=0}^{n-1} \binom{2n}{j} j = n \left(2^{2n-1} - \binom{2n}{n} \right).$$

We combine this last formula with (2.75),

$$2\sum_{j=0}^{n-1}\binom{2n}{j}(n-j) = n\Big(2^{2n} - \binom{2n}{n}\Big) - 2n\Big(2^{2n-1} - \binom{2n}{n}\Big) = n\binom{2n}{n}.$$

By (2.74), we obtain the equation in (2.19). One checks by induction that

$$\binom{2n}{n}2^{-2n} = \frac{(1-\frac{1}{2})\cdots(n-\frac{1}{2})}{1\cdots n}.$$

(Cf. [17, II.(12.5)] and [17, II.(4.1)].) By (2.19), for $n \geq 2$,

$$\Big\|\frac{d}{dt}S(\bar{t})e^{[2n]}\Big\| = \frac{(1+\frac{1}{2})\cdots(n-1+\frac{1}{2})}{1\cdots(n-1)} = \prod_{j=1}^{n-1}\frac{j+\frac{1}{2}}{j} = \prod_{j=1}^{n-1}\Big(1+\frac{1}{2j}\Big).$$

We take the logarithm,

$$\ln\Big\|\frac{d}{dt}S(\bar{t})e^{[2n]}\Big\| = \sum_{j=1}^{n-1}\ln\Big(1+\frac{1}{2j}\Big)$$

$$\geq \int_1^{n-1}\ln\Big(1+\frac{1}{2x}\Big)dx = \frac{1}{2}\int_2^{2(n-1)}\ln\Big(1+\frac{1}{y}\Big)dy$$

$$\geq \frac{1}{2}\int_2^{2(n-1)}\Big(\frac{1}{y}-\frac{1}{2y^2}\Big)dy \geq \frac{1}{2}\big(\ln 2(n-1) - \ln 2 - 1\big)$$

$$= \frac{1}{2}\big(\ln(n-1) - 1\big).$$

We exponentiate this estimate and obtain (2.20). As for the inequality in (2.19),

$$\Big\|\frac{d}{dt}S(\bar{t})e^{[2n]}\Big\|_1 - \Big\|\frac{d}{dt}S(\bar{t})e^{[2n]}\Big\|$$

$$= \sum_{j=1}^{\infty}j\Big|\frac{d}{dt}\big[S(\bar{t})e^{[2n]}\big]_j\Big| = 2\sum_{j=1}^{2n}\binom{2n}{j}2^{-2n}j|n-j|$$

$$= 2\sum_{j=1}^{n-1}\binom{2n}{j}2^{-2n}j(n-j) + 2\sum_{j=n+1}^{2n}\binom{2n}{j}2^{-2n}j(j-n)$$

$$= 2\sum_{j=1}^{n-1}\binom{2n}{j}2^{-2n}j(n-j) + 2\sum_{k=0}^{n-1}\binom{2n}{2n-k}2^{-2n}(2n-k)(n-k)$$

$$= 2^{1-2n}2n^2 + 4n\sum_{j=1}^{n-1}\binom{2n}{j}2^{-2n}(n-j).$$

Here we have used that $\binom{2n}{2n-k} = \binom{2n}{k}$. By (2.74),

$$\Big\|\frac{d}{dt}S(\bar{t})e^{[2n]}\Big\|_1 = 2^{-2n}2n^2 + (n+1)\Big\|\frac{d}{dt}S(\bar{t})e^{[2n]}\Big\|.$$

This implies the inequality in (2.19).

Acknowledgement

This chapter has been inspired, on the biological side, by the work of Andrew Smith [10, 43, 54] and the dissertation of John Nagy [44] on *pika* and, on the mathematical side, by the papers of Francesca Arrigoni [2] and of Andrew Barbour and Andrea Pugliese [5]. H.T. thanks Doug Blount, Mats Gyllenberg, Mimmo Iannelli, Marek Kimmela, John Nagy, Andrea Pugliese, and Andrew Smith for useful discussions.

References

1. L.J.S. ALLEN, *An Introduction to Stochastic Processes with Applications to Biology*, Pearson Education, Upper Saddle River 2003
2. F. ARRIGONI, Deterministic approximation of a stochastic metapopulation model, *Adv. Appl. Prob.* **35** (2003) 691–720
3. J.M. AYERBE TOLEDANO, T. DOMINGUEZ BENAVIDES, and G. LÓPEZ ACEDO, *Measures of Noncompactness in Metric Fixed Point Theory*, Birkhäuser, Basel 1997
4. J. BANASIAK and L. ARLOTTI, *Perturbations of Positive Semigroups with Applications*, Springer, London 2006
5. A.D. BARBOUR and A. PUGLIESE, Asymptotic behaviour of a metapopulation model, *Ann. Appl. Prob.* **15** (2005) 1306–1338
6. E. BORN and K. DIETZ, Parasite population dynamics within a dynamic host population. *Prob. Theor. Rel. Fields* **83** (1989) 67–85
7. R. CASAGRANDI and M. GATTO, A persistence criterion for metapopulations, *Theor. Pop. Biol.* **61** (2002) 115–125
8. S.-N. CHOW and J.K. HALE, *Methods of Bifurcation Theory*, Springer, New York 1982
9. PH. CLÉMENT, H.J.A.M. HEIJMANS, S. ANGENENT, C.J. VAN DUIJN, and B. DE PAGTER, *One-Parameter Semigroups*, North-Holland, Amsterdam 1987
10. M.D. CLINCHY, D.T. HAYDON, and A.T. SMITH, Pattern does not equal process: what does patch occupancy really tell us about metapopulation dynamics? *Am. Nat.* **159** (2002) 351–362
11. E.A. CODDINGTON and N. LEVINSON, *Theory of Ordinary Differential Equations*, McGraw-Hill, New York 1955
12. W. DESCH and W. SCHAPPACHER, Linearized stability for nonlinear semigroups, *Differential Equations in Banach Spaces* (A. Favini, E. Obrecht, eds.) 61–73, LNiM 1223, Springer, Berlin Heidelberg New York 1986
13. K. DIETZ, Overall population patterns in the transmission cycle of infectious disease agents, *Population Biology of Infectious Diseases* (R.M. Anderson, R.M. May, eds.), 87–102, Life Sciences Report **25**, Springer, Berlin Heidelberg New York 1982
14. K.-J. ENGEL and R. NAGEL, *One-Parameter Semigroups for Linear Evolution Equations*, Springer, Berlin Heidelberg New York 2000
15. H. ENGLER, J. PRÜSS, and G.F. WEBB, Analysis of a model for the dynamics of prions II, *J. Math. Anal. Appl.* **324** (2006) 98–117

16. W. FELLER, On the integro-differential equations of purely discontinuous Markoff processes, *Trans. AMS* **48** (1940) 488–515, Errata *ibid.*, **58** (1945) 474

17. W. FELLER, *An Introduction to Probability Theory and Its Applications*, Vol. I, 3rd edition, Wiley, New York 1968

18. W. FELLER, *An Introduction to Probability Theory and Its Applications*, Vol. II, Wiley, New York 1965

19. Z. FENG, L. RONG, and R.K. SWIHART, Dynamics of an age-structured metapopulation model, *Nat. Res. Mod.* **18** (2005) 415–440

20. M. GILPIN and I. HANSKI, eds., *Metapopulation Dynamics. Empirical and Theoretical Investigations*, Academic Press, New York 1991

21. M.L. GREER, L. PUJO-MENJOUET, and G.F. WEBB, A mathematical analysis of the dynamics of prion proliferation, *J. Theor. Biol.* **242** (2006) 598–606

22. M.L. GREER, P. VAN DEN DRIESSCHE, L. WANG, and G.F. WEBB, Effects of general incidence and polymer joining on nucleated polymerization in a prion disease model, *SIAM J. Appl. Math.* **68**(1) (2007) 154–170

23. M. GYLLENBERG and I. HANSKI, Single species metapopulation dynamics: a structured model, *Theor. Pop. Biol.* **42** (1992) 35–61

24. K.P. HADELER and K. DIETZ, An integral equation for helminthic infections: global existence of solutions, *Recent Trends in Mathematics, Reinhardsbrunn 1982* (H. Kurke, J. Mecke, H. Triebel, R. Thiele, eds.) 153–163, Teubner, Leipzig 1982

25. K.P. HADELER and K. DIETZ, Population dynamics of killing parasites which reproduce in the host, *J. Math. Biol.* **21** (1984) 45–65

26. J.K. HALE, *Asymptotic Behavior of Dissipative Systems*. AMS, Providence 1988

27. J.K. HALE and P. WALTMAN, Persistence in infinite-dimensional systems, *SIAM J. Math. Anal.* **20** (1989) 388–395

28. I. HANSKI and M.E. GILPIN, eds., *Metapopulation Biology: Ecology, Genetics, and Evolution*, Academic Press, San Diego 1996

29. E. HILLE and R.S. PHILLIPS, *Functional Analysis and Semi-Groups*, AMS, Providence 1957

30. T. KATO, On the semi-groups generated by Kolmogoroff's differential equations, *J. Math. Soc. Japan* **6** (1954) 12–15

31. A.N. KOLMOGOROV, Über die analytischen Methoden der Wahrscheinlichkeitrechnung, *Math. Annalen* **104** (1931) 415–458

32. T. KATO, *Perturbation Theory for Linear Operators*, Springer, Berlin Heidelberg New York 1976

33. V.A. KOSTIZIN, Symbiose, parasitisme et évolution (étude mathématique), Herman, Paris 1934, translated in *The Golden Age of Theoretical Ecology* (F. Scudo, J. Ziegler, eds.), 369–408, Lecture Notes in Biomathematics **22**, Springer, Berlin Heidelberg New York 1978

34. M. KRETZSCHMAR, A renewal equation with a birth-death process as a model for parasitic infections, *J. Math. Biol.* **27** (1989) 191–221

35. P. LAURENÇOT and C. WALKER, Well-posedness for a model of prion proliferation dynamics, *J. Evol. Eqn.* **7** (2007) 241–264

36. R. LEVINS, Some demographic and genetic consequences of environmental heterogeneity for biological control, *Bull. Entom. Soc. Am.* **15** (1969) 237–240

37. R. LEVINS, Extinction, *Some Mathematical Questions in Biology* (M.L. Gerstenhaber, ed.) 75–107, Lectures on Mathematics in the Life Sciences 2, AMS 1970

38. M. MARTCHEVA and H.R. THIEME, A metapopulation model with discrete patch-size structure, *Nat. Res. Mod.* **18** (2005), 379–413

39. M. MARTCHEVA, H.R. THIEME, and T. DHIRASAKDANON, Kolmogorov's differential equations and positive semigroups on first moment sequence spaces, *J. Math. Biol.* **53** (2006) 642–671

40. R.H. MARTIN, *Nonlinear Operators and Differential Equations in Banach Spaces*, Wiley, New York 1976

41. J. MASEL, V.A.A. JANSEN, and M.A. NOWAK, Quantifying the kinetic parameters of prion replication, *Biophys. Chem.* **77** (1999) 139–152

42. J.A.J. METZ and M. GYLLENBERG, How should we define fitness in structured metapopulation models? Including an application to the calculation of evolutionary stable dispersal strategies, *Proc. R. Soc. Lond.* **B 268** (2001) 499–508

43. A. MOILANEN, A.T. SMITH, and I. HANSKI, Long term dynamics in a metapopulation of the American pika, *Am. Nat.* **152** (1998) 530–542

44. J. NAGY, *Evolutionary Attracting Dispersal Strategies in Vertebrate Metapopulations*, Ph.D. Thesis, Arizona State University, Tempe 1996

45. M.A. NOWAK, D.C. KRAKAUER, A. KLUG, and R.M. MAY, Prion infection dynamics, *Integr Biol* **1** (1998) 3–15

46. A. PAZY, *Semigroups of Linear Operators and Applications to Partial Differential Equations*, Springer, Berlin Heidelberg New York 1982

47. J. PRÜSS, L. PUJO-MENJOUET, G.F. WEBB, and R. ZACHER, Analysis of a model for the dynamics of prions, *Discr. Cont. Dyn. Syst.* **B 6** (2006) 215–225

48. A. PUGLIESE, Coexistence of macroparasites without direct interactions, *Theor. Pop. Biol.* **57** (2000) 145–165

49. A. PUGLIESE, Virulence evolution in macro-parasites, *Mathematical Approaches for Emerging and Reemerging Infectious Diseases* (C. Castillo-Chavez, S. Blower, P. van den Driessche, D. Kirschner, A.-A. Yakubo, eds.) 193–213, Springer, Berlin Heidelberg New York 2002

50. G.E.H. REUTER, Denumerable Markov processes and the associated contraction semigroups on ℓ, *Acta Mathematica* **97** (1957), 1–46

51. G.E.H. REUTER and W. LEDERMANN, On the differential equations for the transition probabilities of Markov processes with enumerable many states, *Proc. Camb. Phil. Soc.* **49** (1953) 247–262

52. G.R. SELL and Y. YOU, *Dynamics of Evolutionary Equations*, Springer, Berlin Heidelberg New York 2002

53. G. SIMONETT and C. WALKER, On the solvability of a mathematical model for prion proliferation, *J. Math. Anal. Appl.* **324** (2006) 580–603

54. A.T. SMITH and M.E. GILPIN: Spatially correlated dynamics in a pika metapopulation, [28], 407–428

55. H.R. THIEME, Semiflows generated by Lipschitz perturbations of non-densely defined operators, *Differential Integral Equations* **3** (1990) 1035–1066

56. H.R. THIEME, Persistence under relaxed point-dissipativity (with applications to an epidemic model), *SIAM J. Math. Anal.* **24** (1993) 407–435

57. H.R. THIEME, Remarks on resolvent positive operators and their perturbation, *Disc. Cont. Dyn. Sys.* **4** (1998) 73–90

58. H.R. THIEME, *Mathematics in Population Biology*, Princeton University Press, Princeton 2003

59. H.R. THIEME and J. VOIGT, Stochastic semigroups: their construction by perturbation and approximation, *Positivity IV – Theory and Applications* (M.R. Weber, J. Voigt, eds.), 135–146, Technical University of Dresden, Dresden 2006
60. J. VOIGT, On substochastic C_0-semigroups and their generators, *Transp. Theory Stat. Phys.* **16** (1987) 453–466
61. C. WALKER, Prion proliferation with unbounded polymerization rates, *Electron. J. Diff. Equations*, Conference **15** (2007) 387–397
62. X.-Q. ZHAO, *Dynamical Systems in Population Biology*, Springer, Berlin Heidelberg New York 2003

3

Simple Models for the Transmission of Microparasites Between Host Populations Living on Noncoincident Spatial Domains

W.-E. Fitzgibbon[1] and M. Langlais[2]

[1] Departments of Engineering Technology and Mathematics. University
of Houston, Houston, Texas 77204-3476, USA
fitz@uh.edu

[2] UMR CNRS 5466 MAB & INRIA Futurs Anubis, case 26, Université Victor
Segalen Bordeaux 2, 146 rue Léo Saignat, F 33076, Bordeaux Cedex, France
langlais@sm.u-bordeaux2.fr

Summary. The goal of this chapter is to provide a simple mathematical approach
to modeling the transmission of microparasites between two host populations living on distinct spatial domains. We shall consider two prototypical situations (1),
a vector borne disease and, (2), an environmentally transmitted disease. In our
models direct horizontal criss-cross transmission from infectious individuals of one
population to susceptibles of the other one does not occur. Instead parasite transmission takes place either through indirect criss-cross contacts between infective
vectors and susceptible individuals and vice-versa in case (1), and through indirect
contacts between susceptible hosts and the contaminated part of the environment
and vice-versa in case (2). We shall also assume the microparasite is benign in one
of the host populations, a reservoir, that is it has no impact on demography and
dispersal of individuals. Next we assume it is lethal to the second population. In
applications we have in mind the second population is human while the first one is
an animal – avian or rodent – population. Simple mathematical deterministic models with spatio-temporal heterogeneities are developed, ranging from basic systems
of ODEs for unstructured populations to Reaction–Diffusion models for spatially
structured populations to handle heterogeneous environments and populations living in distinct habitats. Besides showing the resulting mathematical problems are
well-posed we analyze the existence and stability of endemic states. Under some
circumstances, persistence thresholds are given.

3.1 Introduction

This chapter fits into the general framework of invasion and persistence of
parasites through spatially distributed host species. Within this fairly general
setting for which a large literature is already available we shall be more specifically interested into two prototypical generic situations of emerging diseases:
a vector borne disease and an environmentally transmitted disease.

In both cases a first host population, H_1, living in a spatial domain Ω_1 is a reservoir to a microparasite population. This parasite is benign in population H_1, meaning that it has no impact on fertility, mortality and dispersal of individuals. Now this parasite is fatal to a second host population, H_2, distributed over a neighboring spatial domain Ω_2 with $\Omega_1 \neq \Omega_2$. One also assumes the parasite cannot be horizontally criss-cross transmitted by direct contacts from infectives of H_1 to susceptibles of H_2, and vice-versa. A first and natural question to answer to is whether introduction of the parasite into the first host population yields its invasion and persistence in the second host population. A further question not answered to in this chapter would be assessing its impact on this second host population.

Such a dramatic epidemic pattern is quite common for vector transmitted microparasites and host populations living on either distinct spatial domains, i.e., $\Omega_1 \cap \Omega_2 = \emptyset$, or noncoincident spatial domains, i.e., $\Omega_1 \cap \Omega_2 \neq \emptyset$ and $\Omega_1 \cap \Omega_2 \neq \Omega_i$, $i = 1, 2$.

Recent outbreaks of epidemic diseases in the United States of America such as West Nile virus, cf. Rutledge et al. [55], or Saint–Louis encephalitis, cf. Shaman et al. [61], are related to the accidental introduction of an infected arthropod population or to the environmentally induced dramatic increase in numbers of permanently infective arthropod populations. These vectors were responsible for the transmission from avian species, behaving as amplifiers for the virus, to human populations of fatal arboviruses. Dengue and malaria are similarly transmitted fatal diseases mostly affecting underdeveloped countries.

Such vector transmitted epidemic diseases causing severe damages have always been a threat for humans. The spreading of the black plague through Europe in the mid fourteenth century, cf. Murray [51], killed one fourth of Europeans. The dramatic scenario depicted in Acuna-Soto et al. [2] gives a further explanation to the collapse of the Mexican population in the sixteenth century. In both cases rodents and their arthropods are thought to be responsible for the mass killing of humans.

Another situation occurs for macroparasites releasing eggs through the intestine and then feces of their hosts. Eggs and larvae are thus spatially distributed. Indirect criss-cross transmission between the two host populations occurs through the contaminated spatial subdomain $\Omega_1 \cap \Omega_2$.

A typical situation concerns the transmission of brain worm infection from white-tailed deer populations (*Odocoilus virginianus*) to moose populations (*Alces alces*) described in Schmitz and Nudds [59]. The disease is benign in the deer population but fatal to moose and other cervid. Brain worm infection was used as a weapon for competition: it has caused populations to decline in areas where deer have invaded the range of other cervid.

The case of environmentally transmitted microparasites is somewhat similar to the previous one. Under some circumstances parasites released by infectives through their excrements survive for a certain time in the environment. Then indirect criss-cross transmission between the two host populations occurs through the contaminated subdomain $\Omega_1 \cap \Omega_2$.

The domestic cat (*Felis catus*) – Feline Panleucopenia Virus (FPLV) system is a simple example of such a system with a single host population. Typically FPLV can survive up to one year in the environment, i.e., in the feces of cats, while the duration of the infective stage for exposed hosts is a few weeks. This difference in scaling is one explanation to the success of using FPLV to eradicate an invading cat population threatening local sea birds species in Marion Island, cf. Berthier et al. [12] and the bibliography therein.

A more complex system is related to the growing concern about hantavirus spreading and its transmission to humans, both in Europe and in the USA, cf. Schmaljohn and Hjelle [58]. Rodents are recognized as the more likely reservoir. In most cases hantaviruses have no impact on rodent demography and their social structures, cf. Olsson et al. [54], or dispersal patterns. Human contamination by the Puumala hantavirus occurs mainly through inhalation of contaminated dust. Bank vole (*Clethrionomys glareolus*) is the host of the Puumala virus strain causing epidemic nephropathia, a mild form of hemorrhagic fever with renal syndrome (HFRS), in human populations from Eurasia with a 0.1–20% fatal rate, cf. Sauvage et al. [56]. Deer mice (*Peromyscus maniculatus*) is the host of the American hantavirus "Sin Nombre" causing hantavirus pulmonary syndrome (HPS) in human populations from the USA with a 20–40% fatal rate.

Spatio-temporal heterogeneities are of interest when modeling the spatial propagation of infectious diseases. Spatial heterogeneities may affect local population densities and demography. Temporal heterogeneities may also affect population dynamics by producing large amplitude oscillations in carrying capacities. In both cases intraspecific and interspecific contacts between individuals may greatly differ depending on time and spatial locations.

One obvious point to be settled is whether a parasite can take advantage of these heterogeneities to invade a host population and persist, or vice-versa.

Our chapter is organized as it follows:

- In Sect. 3.2 we introduce deterministic population dynamics models – the disease free case – for mildly structured populations, this is with either time periodic or spatially dependent coefficients.
- In Sect. 3.3 we outline the derivation of basic SEIRS epidemic models for a single species host population; this yields systems of ODES for nonspatially structured populations and of PDEs for spatially structured ones.
- In Sect. 3.4 we consider unstructured host populations. We proceed to the derivation and analysis of, (1), a generic vector borne disease, referred to as MODEL H1VH2, and, (2), of an environmentally transmitted disease, referred to as MODEL H1GH2. We give stability results for the stationary states of the resulting systems of ODEs in the autonomous case. In the case of time periodic coefficients we prove the existence of periodic endemic states emerging from stable stationary states.
- In Sect. 3.5 we extend our analysis from Sect. 3.4 to the case of spatially distributed populations over distinct spatial domains. For MODEL H1VH2 this yields a system of Reaction–Diffusion equations posed on

three distinct spatial domains, while for MODEL H1GH2 this yields a system of Reaction–Diffusion equations posed on two distinct spatial domains coupled to an ODE for the environment.

3.2 Demography and Population Dynamics

In order to assess the impact of a microparasite on the dynamics of a given host species it is important to first carefully model the dynamics of the host population under consideration in the disease free case. Even for unstructured populations a scaling problem is also to be handled: comparing the average life expectancy of susceptible hosts to the duration of the infective stage for infected hosts, infected vectors and/or the contaminated part of the environment. Then various classical population dynamics models may be introduced: models ignoring demography, or models with a Malthusian growth, a logistic regulation or an Allee effect, cf. the books of Edelstein–Keshet [26], Murray [51] or Thieme [66].

Let us first consider a homogeneous and unstructured population and let $P(t) \geq 0$ be its density at time $t \geq 0$.

A simple situation is the case of a constant population density, see (3.1), used to model the outbreak of an epidemic disease on a short time range, cf. [46],

$$P(t) \equiv P(0), \ t \geq 0. \tag{3.1}$$

Malthusian growth, see (3.2), may be used to study the impact of a lethal microparasite on an exponentially growing population on a meso-scale time range,

$$P'(t) = (b - m)P(t), \ t \geq 0, \tag{3.2}$$

cf. [12]; herein, $b > 0$ and $m > 0$, are the constant birth and death rates, $b - m > 0$ being the natural growth-rate.

Simple logistic regulation models, also known as monostable dynamics, require the introduction of a carrying capacity, $K > 0$. Then starting from $P(0) > 0$ one has $P(t) \to K$ as $t \to +\infty$ with various transient behaviors depending on the size of the initial condition $P(0)$ with respect to K or $K/2$, see Fig. 3.1.

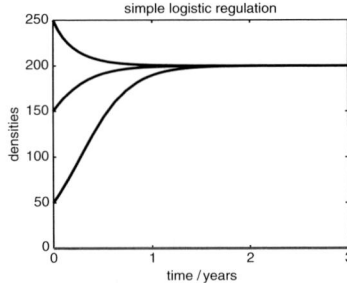

Fig. 3.1. Simple logistic regulation with $b = 5$, $m = 1$ and $k = 0.02$ or $K = 200$

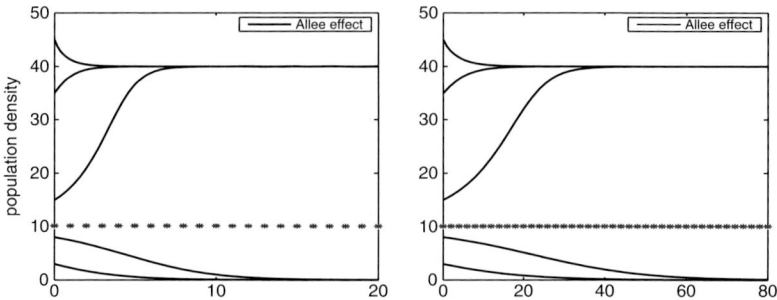

Fig. 3.2. Allee effect with $K = 40$ and $K_s = 10$, left: $a = 0.001$ and right $a = 0.0002$

A prototypical model is

$$P'(t) = (b - [m + kP(t)])P(t) = r(1 - P(t)/K)P(t), \ t \geq 0, \tag{3.3}$$

with a density-dependent effect, $k > 0$, a natural growth-rate $r = b - m > 0$ and a carrying capacity, $K = (b - m)/k$. Note that the trivial state 0 is unstable. It becomes stable as soon as $r = b - m < 0$. Generic logistic models have a density dependent growth-rate $r(p) = b(p) - m(p)$ with $p \to r(p)p$ a quadratic shaped concave function vanishing at $p = 0, K$.

To complete the description of simple population models for homogeneous and unstructured populations models with a strong Allee effect, also known as bistable models, require the introduction of both a carrying capacity, $K > 0$, and an intermediate threshold population density, $K_s > 0$ with $0 < K_s < K$. Then starting from $0 < P(0) < K_s$ one has $P(t) \to 0$ as $t \to +\infty$, while starting from $K_s < P(0)$ one has $P(t) \to K$ as $t \to +\infty$; various transient behaviors are depicted in Fig. 3.2. A prototypical model is

$$P'(t) = a(P(t) - K_s)(K - P(t))P(t), \ t \geq 0, \tag{3.4}$$

$a > 0$ being a scaling parameter related to the speed at which stabilization occurs, see Fig. 3.2; cf. the books of [15] and [26], and also [19].

Simple parametric forms for birth and death rates yielding (3.4) are $b(p) = -P^2 + (K + K_s + e)P + c$ and $m(p) = eP + KK_s + c$, cf. [41]. Generic models with a strong Allee effect have a density dependent growth-rate $r(p) = b(p) - m(p)$ with $p \to r(p)p$ a cubic shaped function vanishing at $p = 0, K_s, K$, $r(0) < 0$.

In many circumstances vital dynamics as well as carrying capacities are not constant, but offer spatio-temporal heterogeneities that may be favorable or not to the persistence of an introduced microparasite. We now consider two typical situations for an underlying logistic regulation dynamic.

3.2.1 Logistic Regulation with Temporal Variations

Let us assume that vital dynamics and carrying capacity are time dependent. The previous model in (3.3) becomes

$$P'(t) = (b(t) - [m(t) + k(t)P])P(t), \ t \geq 0. \tag{3.5}$$

A calculation, cf. [11], using $z = 1/P$ as a new state variable yields

Lemma 1. *Let b, m and k be nonnegative and time periodic functions of period $T > 0$. Assume both $k(t) \neq 0$ and $\int_0^T (b(t) - m(t))dt > 0$.*

Then (3.5) has a unique nonnegative and periodic solution, $P_{periodic}$, of period T and any solution of (3.5) starting from $P(0) > 0$ is such that $P(t) \rightarrow P_{periodic}(t)$ as $t \rightarrow +\infty$.

From both modeling and dynamical points of view two typical transient dynamical behaviors are of interest (i) populations with yearly periodical vital dynamics and, (ii) populations with oscillating to periodic carrying capacities. Figures 3.3 and 3.4 depict these two situations with data consistent with

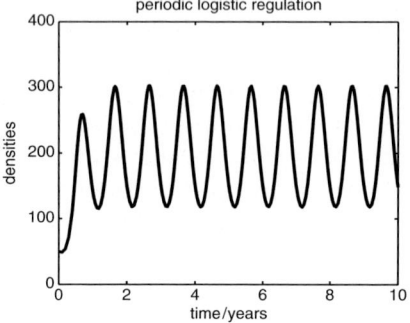

Fig. 3.3. Logistic regulation with one year periodic seasonal variations: $b = 5 + 3.5\sin(2\pi(t - 0.25))$, $m = 1$ and $k = 0.02$

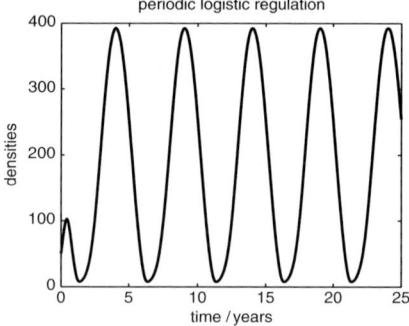

Fig. 3.4. Logistic regulation with a five years periodic carrying capacity: $b = 5$, $m = 1$ and $k = 0.02/(1.025 - \sin(2\pi t/5))$

Fig. 3.1. Oscillations in population densities driven in Fig. 3.4 by the carrying capacity are much larger than in Fig. 3.3 where they are produced by the birth-rate. The outcome of the introduction of a parasite may thus be different in the three situations exhibited in Figs. 3.1, 3.3 and 3.4.

A case study with field data from bank vole populations is found in Sauvage et al. [56, 57], Wolf et al. [70]; cf. also Yoccoz et al. [72].

3.2.2 Logistic Regulation with Spatial Heterogeneities

Let us consider a population spatially distributed over a n-dimensional domain of R^n, $n = 1, 2$ or 3. State variable $P(x, t)$ will represent the time dependent population spatial density with respect to $x \in R^n$. Reaction–Diffusion equations and multi-patch systems are the most two popular deterministic continuous models used for modeling heterogeneous spatially structured populations. In this chapter we shall only consider models built upon using Reaction–Diffusion equations.

Spatially Structured Models Using Reaction–Diffusion Equations

In that setting let Ω be a bounded domain in R^n with smooth boundary $\partial\Omega$ such that locally Ω lies on one side of $\partial\Omega$. One assumes that host individuals disperse by means of Fickian diffusion through their habitat and let $-d(x)\nabla P$ be the population flux, $d(x)$ being the positive and spatially dependent diffusivity. A simple logistic model with spatially dependent vital dynamics and carrying capacity can be expressed as a semilinear evolution equation of parabolic type,

$$\frac{\partial}{\partial t}P(x,t) - \nabla \cdot d(x)\nabla P(x,t) = (b(x) - [m(x) + k(x)P(x,t)])\, P(x,t), \quad (3.6)$$

for $x \in \Omega$ and $t > 0$. The usual assumption of an isolated population translates as a no flux boundary condition

$$d(x)\partial P/\partial\eta(x,t)) = d(x)\nabla P(x,t) \cdot \eta(x) = 0, \ x \in \partial\Omega, \ t > 0, \quad (3.7)$$

η being the unit outward normal vector to Ω along its boundary $\partial\Omega$. Last an initial population distribution is prescribed at time $t = 0$,

$$P(x,0) = P_0(x) \geq 0; \ P_0(x) \neq 0, \ x \in \Omega. \quad (3.8)$$

Various other boundary conditions can be used in the modeling process such as prescribed densities

$$P(x,t) = P_{boundary}(x,t) \geq 0, \ x \in \partial\Omega, \ t > 0,$$

$P_{boundary}(x,t) \equiv 0$ corresponding to an unfavorable or inhospitable boundary.

More general population fluxes may also be of interest, such as density dependent ones, $-d(x, P)\nabla P$, or fluxes with an additional advection term, $-d(x, P)\nabla P + V(x, P)$.

We refer to the books of Murray [51], Okubo and Levin [53], and also Cantrell and Cosner [15] or Edelstein-Keshet [26], for details and model derivation.

The Constant Coefficient Case

Let us first assume that (b, m, k) and d are positive constants. Then the large time behavior of solutions to system (3.6), (3.7) and (3.8) is identical to that of solutions to (3.3): as $t \to +\infty$ then $P(\cdot, t) \to K = (b-m)/k$ when $b - m > 0$ while $P(\cdot, t) \to 0$ when $b - m < 0$, cf. [18].

Hence for constant coefficients nothing is gained from the ODE model in (3.3).

The Genuinely Spatially Heterogeneous Case

In order to mathematically handle spatial heterogeneities and define suitable solutions to system (3.6), (3.7) and (3.8) a set of conditions has to be placed on (b, m, k) and d.

HYP 2.1 (b, m, k) and d are nonnegative elements of $L^\infty(\Omega)$; there exists positive numbers (d_{min}, d_{max}) and k_{min} such that

$$0 < d_{min} \leq d(x) \leq d_{max}, \ 0 < k_{min} \leq k(x) \ x \in \Omega.$$

HYP 2.2 There exists open subregions of Ω, $(\Omega_i^*)_{1 \leq i \leq \kappa}$, with $\overline{\Omega_i^*} \subset \Omega$ and $\overline{\Omega_i^*} \cap \overline{\Omega_j^*} = \emptyset$ for $i \neq j$ having the same smoothness properties as Ω. Set $\Omega^* = \Omega_1^* \cup \cdots \cup \Omega_\kappa^*$ and $\Omega_0^* = \Omega \setminus \overline{\Omega^*}$. See Fig. 3.5.
Then we assume $(b, m, k) \in C^{0,\alpha}(\overline{\Omega_i^*})$ and $d \in C^{2,\alpha}(\overline{\Omega_i^*})$, for $i = 0, \cdots, \kappa$.

Condition HYP 2.2 includes the case of a set of piecewise constant coefficients (b, m, k) and d. Spatially discontinuous reaction terms cannot be avoided when it comes to model interacting species living on distinct habitats, see Sect. 3.5. This will generate spatial discontinuities in derivatives of solutions.

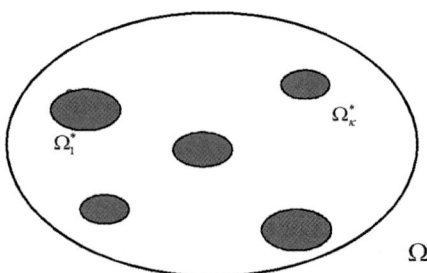

Fig. 3.5. A spatially heterogeneous domain with κ subdomains

These conditions Hyp 2.1 and Hyp 2.2 mean we consider a diffusion of compartmental or diffractive type, cf. the book Ladyzenskaja et al. [48], Seftel [60] and Stewart [63–65]. A suitable mathematical existence theory of weak or strong solutions can be developed, cf. [48], Horton [44], also Fitzgibbon and Morgan [34] and Fitzgibbon et al. [28].

In this work we are more interested in taking mathematically care of spatial heterogeneities than in looking at smoothness of weak or strong solutions. Hence we shall only consider classical solutions to system (3.6)–(3.8).

Definition 1. *A classical solution to system (3.6)–(3.8) is a continuous mapping* $P(\cdot, t) : [0, +\infty) \to C^0(\overline{\Omega})$ *with* $P(\cdot, t) \in C^{2,1}(\Omega_\ell \times (0, \infty))$ *for* $\ell = 0, \dots, \kappa$ *such that the initial condition in (3.8), the exterior boundary condition in (3.7), the compatibility conditions on the interface between* Ω_ℓ *and* Ω_0, $\ell = 1, \dots, \kappa$

$$[d\partial P/\partial \eta_\ell]_{\partial \Omega_\ell} = 0 \ on \ \partial \Omega_\ell, \ \ell = 1, \dots, \kappa,$$

where $[\xi]_{\partial \Omega_\ell}$ *is the saltus of a function* ξ *across* $\partial \Omega_\ell$, η_ℓ *is a unit normal vector to* $\partial \Omega_\ell$, *and the differential equation in (3.6) are satisfied.*

Available results can be summarized in the following statement

Theorem 1. *Let conditions Hyp 2.1 and Hyp 2.2 hold. Assume* $P_0 \in C^0(\overline{\Omega})$ *is nonnegative. Then system (3.6)–(3.8) has a unique nonnegative global classical solution with* $0 \leq P(x,t) \leq \max(\|P_0\|_{\infty,\Omega}, \|b\|_{\infty,\Omega}/k_{min})$.

Furthermore there exists a unique nonnegative classical stationary solution P_∞ *such that for any nonnegative* $P_0 \in C^0(\overline{\Omega})$, $P_0(x) \neq 0$, *one has* $P(\cdot, t) \to P_\infty(\cdot)$ *as* $t \to +\infty$ *in* $C^0(\overline{\Omega})$.

Note that P_∞ being a nonnegative stationary state solution to system (3.6)–(3.8), it is a nonnegative solution to

$$-\nabla \cdot d(x)\nabla P_\infty(x) = (b(x) - [m(x) + k(x)P_\infty(x)])P_\infty(x), \qquad (3.9)$$

for $x \in \Omega$, with a no flux boundary condition

$$d(x)\partial P_\infty/\partial \eta(x) = d(x)\nabla P_\infty(x) \cdot \eta(x) = 0, \ x \in \partial \Omega. \qquad (3.10)$$

Proof. Local existence in some maximal time interval $(0, T_{max}(P_0))$ follows from results in Horton [44]. More precisely there exists a $T_{max}(P_0)$ such that either $T_{max}(P_0) = +\infty$ yielding global existence or $T_{max}(P_0) < +\infty$ in which case introducing $\| \cdot \|_{\infty,\Omega}$ the norm in $L^\infty(\Omega)$ one has

$$\|P(\cdot, t)\|_{\infty,\Omega} \to +\infty \ as \ t \nearrow T_{max}(P_0).$$

To prove global existence one first check that local solutions remain nonnegative; this follows from a weak minimum principle, cf. [48], this is 0 is a subsolution. Next to prevent finite time blow-up we introduce a supersolution, y_+, as a solution of a basic logistic-like ODE

$$y'_+ = (\|b\|_{\infty,\Omega} - k_{min} \, y_+) \, y_+, \; y_+(0) = \|P_0\|_{\infty,\Omega}, \; t \geq 0.$$

Hence from a weak maximum principle, cf. [48], one gets

$$0 \leq P(x,t) \leq y_+(t) \leq \max(\|P_0\|_{\infty,\Omega}, \|b\|_{\infty,\Omega}/k_{min}),$$

for $x \in \Omega$ and $0 \leq t \leq T_{max}(P_0)$; this implies $T_{max}(P_0) = +\infty$.

For the large time behavior one can use Lyapunov functionals as in [16,18, 37], or a result in [49]. Using the smoothing effect of parabolic equations with time independent coefficients, cf. [48], one has $\partial P/\partial t \in L^2((\tau, \tau + T) \times \Omega)$ for $\tau > 0$ and $T > 0$, and $\nabla P(\cdot, t) \in L^2(\Omega)$ for $t > 0$. Set

$$\mathcal{F}_1(p, \cdot) = \int_0^p ((b(\cdot) - [m(\cdot) + k(\cdot)q])q dq, \; p \in R,$$

$$\mathcal{L}_1(p) = \int_\Omega \left(|\nabla p(x)|^2 - \mathcal{F}_1(p(x), x) \right) dx, \; p \in H^1(\Omega);$$

herein $H^1(\Omega)$ is the standard Sobolev space of order 1, cf. [47]. Then taking the product of (3.6) with $\partial P/\partial t$ and integrating over $(\tau, \tau + T) \times \Omega$ one gets

$$\int_{(\tau, \tau+T) \times \Omega} \left(\frac{\partial P}{\partial t} \right)^2 (x,t) dx dt = \mathcal{L}_1(P(\cdot, \tau)) - \mathcal{L}_1(P(\cdot, \tau + T)).$$

As a consequence $\mathcal{L}_1$ is a Lyapunov function and the semi-orbits $\{P(\cdot, t), \, t > 0\}$ are bounded in $H^1(\Omega) \cap L^\infty(\Omega)$. $\mathcal{L}_1$ is decreasing along the semi-orbits $\{P(\cdot, t), \, t > 0\}$, and a constant over the compact, in $L^2(\Omega)$, connected and forward invariant ω-limit set

$$\{P^\infty \in H^1(\Omega) \cap L^\infty(\Omega),$$
$$\text{there exists a sequence } (t_n)_{n \geq 0}, t_n \to +\infty \text{ as } n \to +\infty,$$
$$P(\cdot, t_n) \to P^\infty \text{ in } L^2(\Omega) \text{ as } n \to +\infty\}.$$

This implies any element in the ω-limit set is a stationary solution, cf. [37].

It remains to show the ω-limit set is made of a single element. Let μ_0 be the dominant eigenvalue of the eigenvalue problem $(\mu_j, \xi_j)_{j \geq 0}$

$$-\nabla \cdot d(x)\nabla \xi_j(x) = (b(x) - m(x))\xi_j(x) + \mu_j \xi_j(x), x \in \Omega,$$
$$d(x)\partial \xi_j/\partial \eta(x) = d(x)\nabla \xi_j(x) \cdot \eta(x) = 0, \; x \in \partial\Omega. \tag{3.11}$$

Standard sub-super solutions techniques, cf. [18], show that when $\mu_0 < 0$ there is no nontrivial nonnegative stationary solutions while when $\mu_0 > 0$ there is a unique positive one.

A concavity argument applied to the reaction term $p \to (b(\cdot) - [m(\cdot) + k(\cdot)p]p$ in (3.6) shows that $P_\infty \equiv 0$ when $\mu_0 < 0$ while P_∞ is the positive – stationary – solution of (3.9) and (3.10) when $\mu_0 > 0$, cf. [15, 18, 49]. $\square$

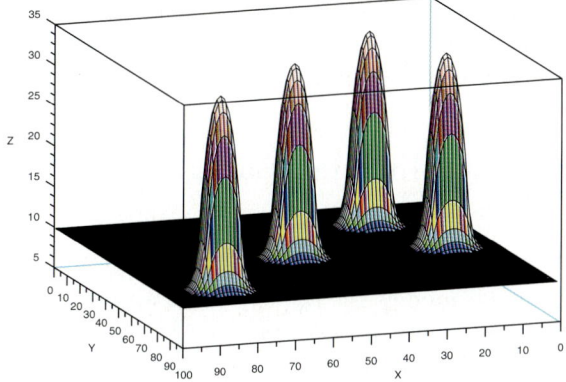

Fig. 3.6. Spatially heterogeneous stationary state solution of (3.9) and (3.10) with $\kappa = 4$ subdomains, see Fig. 3.5; see the text for the data set

In the constant coefficient case, this is (b, m, k) and d are positive constants P_∞ is a constant, the globally asymptotically stationary state of equation (3.3). As a consequence when $b \leq m$ then $P_\infty = 0$ while when $b > m$ one gets $P_\infty = K$ and, again, nothing is gained from the ODE model.

For genuinely spatially heterogeneous coefficients $P_\infty \equiv 0$ is still a stationary solution. System (3.9) and (3.10) has a unique nonconstant and stable solution when the dominant eigenvalue μ_0 is positive, cf. [15, 18].

A numerical experiment is supplied in Fig. 3.6 with $d = 1$, $\kappa = 4$ while the carrying capacity is piecewise constant: $K = 40$ on Ω_ℓ for $\ell = 1, \cdots, 4$ and $K = 16$ in Ω_0.

Satellite imagery is used in Abramson et al. [1] to derive heterogeneous carrying capacities for rodent populations, and in Tran et al. [67,68] for human populations.

Further Comments

Similar analysis can be carried out for the solution-set to models (3.5) or (3.6)–(3.8) based on (3.1), (3.2) or (3.4) with spatio-temporal heterogeneities.

Ignoring demography: For models based on (3.1) nonconstant time periodic solutions do not exist while solutions to (3.6)–(3.8) with $b = m = k = 0$ stabilize as $t \to +\infty$ toward the spatial average of their initial condition.

Malthusian growth: For models based on (3.2) nonconstant time periodic solutions do not exist unless $\int_0^T b(t)dt = \int_0^T m(t)dt$ in which case there are infinitely many periodic solutions, $P_0 \exp(\int_0^t (b(\tau) - m(\tau))d\tau)$. The large time behavior of solutions to (3.6)–(3.8) with $k = 0$ depends on the signum of the dominant eigenvalue of (3.11), μ_0; as $t \to +\infty$ nonnegative solutions may either stabilize toward 0 or become unbounded, or else converge to $c_0\xi_0$ for some nonnegative constant c_0 when $\mu_0 = 0$.

Allee effect: A more involved analysis is required for models (3.5) or (3.6)–(3.8) based on (3.4); free boundary value problems arise for the PDE problem.

3.3 Deterministic Modeling Background for the Spread of Infectious Diseases

An important literature is devoted to the derivation and analysis of deterministic mathematical models for the spread of an infectious diseases in unstructured or structured host populations, cf. the books [6, 7, 13, 14, 16, 24, 51] or the review paper in [40], and references therein. Earlier pioneering models go back to Bernoulli [11], as quoted in [7], and Kermack and McKendrick, reprinted in [46], among others.

In these models commonly referred to as – SI, SIR, SEIR, ..., models – a host population is split into several subpopulations according to the status of their individuals with respect to the infectious disease, but also to their chronological age or to any additional and relevant structuring variable, i.e, space, gender, social status.

Typically a nonstructured SEIRS model reads as follows, cf. Fig. 3.7. A susceptible class, S, is made of individuals capable of contracting the disease. After a suitably "efficient" contact with an infectious individual, newly infected individuals will enter an exposed class, E, for a latent period of duration $1/\theta$. Then they become infectious and will join an infective class, I. Depending of the infectious disease under consideration infected individuals may either remain life long permanently infectious, or die from the disease at a rate $1 - \varepsilon$, $0 \leq \varepsilon \leq 1$ being the survival rate, or else surviving individuals recover in which case they enter a resistant or immune class, R, at a rate λ, $1/\lambda$ being the duration of the infective stage. Resistant individuals do not participate to the spread of the infectious disease anymore; they may lose their immunity after a period of time $1/\gamma$ and reenter the susceptible class.

Then $P = S + E + I + R$ is the total population density from Sect. 3.2.

An important issue is horizontal transmission, that is understanding and modeling the recruitment of newly infected individuals from the susceptible

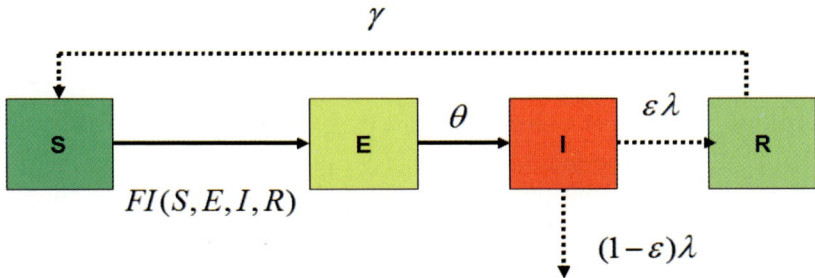

Fig. 3.7. A basic SEIRS compartmental model without demography

class by direct contacts with infectious individuals. This is the incidence term $\sigma(S, E, I, R)$ or the force of infection $FI(S, E, I, R) = \sigma(S, E, I, R)/S$. Two incidence terms are widely used in the literature:

- A density dependent or mass action form in which setting

$$\sigma(S, E, I, R) = \sigma SI;$$

  an underlying assumption requires a contact rate between individuals linearly increasing with population densities.
- A frequency dependent or proportionate mixing form in which setting

$$\sigma(S, E, I, R) = \sigma \frac{SI}{S + E + I + R};$$

  a similar underlying assumption requires a constant contact rate between individuals.

The choice of a realistic incidence term is somewhat problematic. It heavily relies on the specific host–parasite system under consideration as well as on social and spatial structures, resource availability or environmental conditions, cf. [9, 12–14, 16, 23, 24, 36, 40] and references therein.

A further issue is vertical transmission, that is transmission at birth of the disease from an infected mother to offsprings. In most models the birth-rate in the susceptible class is b, see (3.2) or (3.3), while each of the E, I and R subclasses has a specific birth-rate b_Z with $0 \le b_Z \le b$, $Z = E, I, R$. A reduced birth-rate in subclass Z, this is $0 \le b_Z < b$, indicates the parasite affects demography. Next a proportion π_Z of offsprings from the subclass Z remains in Z, $Z = E, I$ or R, the proportion $1 - \pi_Z$ being susceptible at birth. A positive π_Z indicates actual vertical transmission.

When the parasite has no impact on its host population one has $b_Z = b$ and $\pi_Z = 0$ for $Z = E, I, R$. This will be the case in Sect. 3.4 below for the reservoir populations where the parasite is benign.

An Unstructured SEIRS Model

A simple deterministic SEIRS model based on the disease free model (3.3) is a four component system of ODEs; it reads

$$\frac{d}{dt} \begin{pmatrix} S \\ E \\ I \\ R \end{pmatrix} = \Phi(S, E, I, R) = (\Phi_i(S, E, I, R))_{1 \le i \le 4}, \qquad (3.12)$$

wherein

$$\Phi_1(S, E, I, R) = bS + \sum_{Z=E,I,R} (1 - \pi_Z) b_Z Z - (m + kP)S - \sigma(S, E, I, R) + \gamma R$$
$$\Phi_2(S, E, I, R) = \qquad \pi_E b_E E - (m + kP)E + \sigma(S, E, I, R) - \theta E$$
$$\Phi_3(S, E, I, R) = \qquad \pi_I b_I I - (m + kP)I + \theta E - \lambda I$$
$$\Phi_4(S, E, I, R) = \qquad \pi_R b_R R - (m + kP)R + \varepsilon \lambda I - \gamma R,$$

together with a set of nonnegative initial data

$$(S, E, I, R)^\top (0) = (S_0, E_0, I_0, R_0)^\top \in R_+^4.$$

A $\varepsilon = 0$ corresponds to a benign parasite while a $0 \le \varepsilon < 1$ corresponds to a lethal one.

A Spatially Structured SEIRS Model

Assuming individuals in each of the subpopulations S, E, I and R disperse by means of a Fickian diffusion a spatially structured deterministic SEIRS model based on the disease free model (3.6) and on model (3.12) reads

$$\frac{d}{dt} \begin{pmatrix} S \\ E \\ I \\ R \end{pmatrix} = \mathcal{D}(S, E, I, R) + \Phi(S, E, I, R), \qquad (3.13)$$

with a diagonal diffusivity matrix

$$\mathcal{D}(S, E, I, R) = \begin{pmatrix} \nabla \cdot d_S(x)\nabla S(x,t) \\ \nabla \cdot d_E(x)\nabla E(x,t) \\ \nabla \cdot d_I(x)\nabla I(x,t) \\ \nabla \cdot 7d_R(x)\nabla R(x,t) \end{pmatrix}$$

to which must be added a suitable set of nonnegative initial data

$$(S, E, I, R)^\top (x, 0) = (S_0(x), E_0(x), I_0(x), R_0(x))^\top \in R_+^4, \ x \in \Omega,$$

and boundary conditions, such as no-flux boundary conditions,

$$d_Z(x)\nabla Z(x,t) \cdot \eta(x) = 0, \ x \in \partial\Omega, \ t \ge 0, \ Z = S, E, I, R.$$

Diffusivities may differ between subclasses, depending on the impact of the parasite on individual dispersal. In Sect. 3.5 below we shall assume identical diffusivities in reservoir populations where the parasite is benign.

R_0

A question of paramount interest when considering a host–parasite system is a question of invasion and persistence: assuming parasites are introduced in a host species will the parasite population eventually persist?

For epidemic problems this is usually reformulated into the following one: assuming a single infectious individual is introduced into a naive disease free host population at equilibrium what is the number of secondary cases caused by this infectious over its life time?

This number, labeled R_0, is referred to as the basic reproduction number, cf. the books [6, 7, 13, 14, 16, 24, 51] or the review paper in [40], and references therein. When $R_0 < 1$ the epidemic dies out while when $R_0 > 1$ one or more endemic states may exist with simple to more involved dynamics occurring, including a possible extinction of the host population.

Autonomous deterministic mathematical models such as the ones derived here can be useful in estimating R_0. A basic strategy relies on looking at the stability of the disease free equilibrium, i.e., $(K, 0, 0, 0)$ for models (3.3)–(3.12) with time independent coefficients, in which case one has to determine the sign of the real parts of the eigenvalues of the Jacobian matrix of $\Phi(S, E, I, R)$ evaluated at $(K, 0, 0, 0)$.

For highly structured models R_0 is usually expressed as the dominant eigenvalue of a suitable linear operator, cf. [22].

Further Comments

Models (3.12) and (3.13) can be adapted to disease free models devised in Sect. 3.2. For a model wherein demography is ignored one may set $b = b_z = m = k = 0$. A model with $k = 0$ corresponds to a disease free Malthusian growth. Models with an Allee effect on the host dynamics require density dependent birth rates, b_Z for $Z = S, E, I, R$, and a density dependent death rate: see the parametric forms in Sect. 3.2; cf. [25, 41] for a simple SI model.

Models with time periodic coefficient are devised accordingly.

Models (3.12) and (3.13) are obviously oversimplified albeit flexible models that can be adapted to a wide range of applications, i.e., diseases with waning immunity wherein early reinfection of temporary resistants will reinforce their immunity before they retrieve a susceptible status, cf. [35], or diseases where exposed individuals either enter the infectious stage and die afterward, this is $\varepsilon = 0$, or directly enter a recovered stage, cf. [36].

The discrete structure with respect to the status of individuals can be transformed into a continuous one upon introducing the age of the disease within an individual, cf. [27, 43, 45, 69].

Prophylaxis and control methods can also be added to these models, cf. [5].

3.4 Two Simple Basic Unstructured Epidemic Models with Two Host Populations

In this section we consider two simple models for the transmission of a microparasite between two unstructured host populations under simple circumstances (1) the case of a vector borne disease and, (2) the case of an environmentally transmitted disease.

In both cases a first host population, H_1, plays the role of a reservoir wherein the microparasite is benign while the microparasite is lethal in a second host population, H_2. An important feature concerns characteristic time scales and vital dynamics. In applications we have in mind the first host population is an animal population, i.e., birds or rodents, while the second host population is a human population.

Due to rather different time scales between life expectancies of individuals from H_1 and H_2 demography is ignored in H_2 whereas constant or time periodic logistic regulations in H_1 will have an important impact on the global dynamics of the full system.

Modified results are outlined for other constant to time periodic disease free population dynamics models in H_1 and V based on (3.1), (3.2) or (3.4)

3.4.1 A Simple Model for Vector Borne Diseases

Our model assumes two independent host populations, H_1 and H_2, and a vector population, V. Host populations are subdivided into three subclasses: susceptibles, S_i, infectives, I_i, and temporary or permanently recovered, R_i, for $i = 1, 2$. To simplify the vector population is split into susceptible, S_v and infective, I_v, subpopulations; this is equivalent to assuming infected vectors remain life long infectious. See Fig. 3.8.

The main departure from the basic SEIRS model in Fig. 3.7 arises in transmission modes. One assumes neither direct – horizontal – transmission of the disease from infective individuals I_i to susceptibles S_i in population H_i, $i = 1, 2$, nor criss-cross transmission from infective individuals I_i from population H_i to susceptibles S_j in population H_j, $i, j = 1, 2$ and $i + j = 3$.

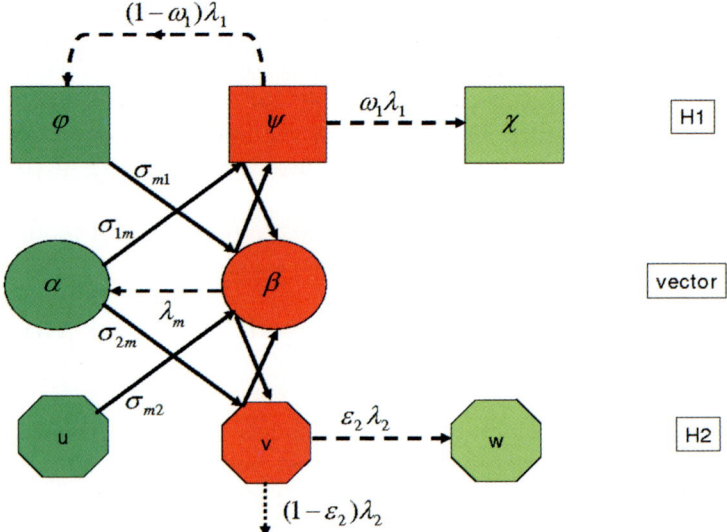

Fig. 3.8. A simple $H_1 - V - H_2$ compartmental model without demography

Infectious vectors I_v will transmit infection to susceptibles of both host populations, H_1 and H_2. Conversely susceptible vectors S_v are infected by infectious from either host population H_1 or H_2.

Concerning vital dynamics populations H_1 and V are run by a logistic growth with respective birth and death rates (b_1, m_1) and (b_m, m_m), and density dependent effects k_1 and k_m. It is assumed that the parasite has no impact on fertility: infective and recovered individuals have identical birth-rates to those of susceptibles in both H_1 and V populations; there is no vertical transmission at birth. The infectious disease is benign in populations H_1 and V, that is there is no additional mortality due to infection. Then $1/\lambda_1$ is the duration of the infective stage in H_1, a proportion ω_1 of infective becoming resistant and a proportion $1 - \omega_1$ going back to the susceptible class. Last $1/\lambda_m$ is the duration of the infective stage in the vector population, infectives retrieving a susceptible status afterwards.

Natural demography is not taken into account in population H_2. Infectives I_2 surviving the disease become permanently resistant; ε_2 is the survival rate and $1/\lambda_2$ the duration of the infective stage in H_2.

Let us now define state variables: $(\varphi, \psi, \chi)^\top$ represent the (S_1, I_1, R_1) host densities in population H_1 and $P_1 = \varphi + \psi + \chi$, $(u, v, w)^\top$ represent the (S_2, I_2, R_2) host densities in population H_2 and $P_2 = u + v + w$, while $(\alpha, \beta)^\top$ is the (S_v, I_v) vector densities and $P_3 = \alpha + \beta$.

The last point to be settled concerns the incidence terms. In this note we choose a density dependent form for each of the interactions between susceptible and infectious individuals. This is $\sigma_{m1}\varphi\beta$ is the recruitment of infectives I_1 from susceptibles S_1 in H_1 via infective vectors I_v, $\sigma_{m2}u\beta$ is the recruitment of infectives I_2 from susceptibles S_2 in H_2 via infective vectors I_v and $(\sigma_{1m}\psi + \sigma_{2m}v)\alpha$ is the recruitment of infectives I_v from susceptibles S_v in V via infective hosts I_1 and I_2.

The analysis developed below carries over to frequency dependent incidence terms. Actually given that the parasite is benign in H_1 and V it has no impact on their population densities that eventually stabilize to their carrying capacities, K_1 and K_m, see the proof of Lemma 3. Thus a frequency dependent incidence model would lead to similar results with transmission rates σ_{ij} replaced by σ_{ij}/P_i, $i, j = 1, m$ and $i \neq j$, and the persistence threshold in Lemma 8 modified accordingly, this is σ_{ij} replaced by σ_{ij}/K_i.

Putting together all of these assumptions transmission of the disease within the unstructured $H_1 - V - H_2$ model in Fig. 3.8 is modeled by a set of nonautonomous ordinary differential equations,

MODEL (H1VH2,U)

$$\varphi' = -\sigma_{m1}(t)\beta\varphi + (1 - \omega_1)\lambda_1\psi + b_1(t)P_1 - (m_1(t) + k_1(t)P_1)\varphi,$$

$$\psi' = +\sigma_{m1}(t)\beta\varphi - \lambda_1\psi - (m_1(t) + k_1(t)P_1)\psi,$$

$$\chi' = +\omega_1\lambda_1\psi - (m_1(t) + k_1(t)P_1)\chi,$$

$$u' = -\sigma_{m2}(t)\beta u,$$

$$v' = +\sigma_{m2}(t)\beta u - \lambda_2 v,$$

$$w' = +\varepsilon_2\lambda_2 v,$$

$$\alpha' = -(\sigma_{1m}(t)\psi + \sigma_{2m}(t)v)\alpha + \lambda_m\beta + b_m(t)P_3 - (m_m(t) + k_m(t)P_3)\alpha,$$

$$\beta' = +(\sigma_{1m}(t)\psi + \sigma_{2m}(t)v)\alpha - \lambda_m\beta - (m_m(t) + k_m(t)P_3)\beta,$$

supplemented with a set of nonnegative initial conditions such that each of the three populations is present at time $t = 0$ – this is $P_1(0) > 0$, $P_2(0) > 0$ and $P_3(0) > 0$ – and such that at least one infective is also present at time $t = 0$ – this is $\psi(0) + \beta(0) + v(0) > 0$.

Lemma 2. *Let the coefficients in* MODEL *(H1VH2,U) be either nonnegative constants or nonnegative smooth and bounded functions of time, $0 \le \varepsilon_2 \le 1$.*

Then for each set of nonnegative initial conditions, MODEL *(H1VH2,U) has a unique global solution $(\varphi(t), \psi(t), \chi(t), u(t), v(t), w(t), \alpha(t), \beta(t))^\top$ with nonnegative components.*

Proof. Local existence and uniqueness of a solution on some maximum time interval, $(0, T_{\max})$, depending on the initial conditions hold because the right-hand side of MODEL (H1VH2,U) is locally Lipschitz continuous. Then either $T_{\max} = +\infty$ yielding global existence or $T_{\max} < +\infty$ in which case one of the components ζ of the solution is such that $|\zeta(t)| \to +\infty$ as $t \nearrow T_{\max}$.

These local solutions have nonnegative components because the nonnegative orthant R_+^8 is forward invariant by the flow in MODEL (H1VH2,U), cf. [42].

We now prove that finite time blow-up cannot occur, yielding global existence. Adding the first three equations in MODEL (H1VH2,U) one gets $P_1'(t) \le b(t)P_1(t)$, $t \in (0, T_{\max})$; hence $0 \le P_1(t) \le P_1(0)\exp(b_{1,\max}t)$, $t \in (0, T_{\max})$ with $b_{1,\max} = \max(b_1(t), t \ge 0)$. Next along the same lines one finds $0 \le P_3(t) \le M(0)\exp(b_{m,\max}t)$, $t \in (0, T_{\max})$ with $b_{m,\max} = \max(b_m(t), t \ge 0)$. Last $0 \le P_2(t) \le P_2(0)$, $t \in (0, T_{\max})$. This shows $T_{max} = +\infty$. □

Stability Analysis for the Autonomous MODEL *(H1VH2,U)*

We look for qualitative properties of the solution set to MODEL (H1VH2,U). Let us now assume

HYP 4.1 Coefficients λ_1, ω_1, λ_m are nonnegative constants while other coefficients in MODEL (H1VH2,U) are positive constants with $b_1 - m_1 > 0$, $b_m - m_m > 0$ and $0 < \varepsilon_2 \le 1$.
$P_1(0) > 0$, $P_2(0) > 0$ and $P_3(0) > 0$.

Along the lines of [31, 33] we use the methodology of ω-limit sets for ODEs systems, see [14, 37–39, 42], to reduce the stability analysis of MODEL (H1VH2,U) in the autonomous case to the stability analysis of a much simpler system, see (3.17) below.

In the setting of Lemma 2 given a set of nonnegative initial conditions the ω-limit set for MODEL (H1VH2,U) is defined as

$$\omega_0 = \{(\varphi^\infty, \psi^\infty, \chi^\infty, u^\infty, v^\infty, w^\infty, \alpha^\infty, \beta^\infty)^\top \in R_+^8, \text{ such that}$$

there exists a sequence $(t_n)_{n \geq 0}$, $t_n \to +\infty$ as $n \to +\infty$,

$$(\varphi(t_n), \psi(t_n), \chi(t_n), u(t_n), v(t_n), w(t_n), \alpha(t_n), \beta(t_n))^\top \to$$

$$(\varphi^\infty, \psi^\infty, \chi^\infty, u^\infty, v^\infty, w^\infty, \alpha^\infty, \beta^\infty)^\top \text{ as } n \to +\infty\}.$$

Lemma 3. *Let condition* HYP *4.1 hold. Set*

$$K_1 = \frac{b_1 - m_1}{k_1}, \quad K_m = \frac{b_m - m_m}{k_m}.$$

For each set of nonnegative initial conditions there exists a compact, connected and forward invariant nonempty ω-limit set $\omega_0 = \omega(\varphi, \psi, \chi, u, v, w, \alpha, \beta)$ in R_+^8. Any element in ω_0 is of the form $(K_1 - \psi^* - \chi^*, \psi^*, \chi^*, u^*, 0, w^*, K_m - \beta^*, \beta^*)^\top$, wherein $u^* \geq 0$ and $w^* > 0$ are constants depending solely on the set of initial conditions.

Proof. State variables (φ, ψ, χ) are nonnegative. Upon adding the first three equations in MODEL (H1VH2,U) one retrieves a basic logistic equation,

$$P_1'(t) = (b_1 - [m_1 + k_1 P_1(t)]) P_1(t), \ t \geq 0.$$

Assuming both $b_1 - m_1 > 0$ and $k_1 > 0$ one may conclude that $P_1(t) \to K_1$ as $t \to +\infty$, exponentially, as soon as $P_1(0) > 0$. One also has

$$0 \leq \varphi(t), \psi(t), \chi(t) \leq P_1(t) \leq \max(P_1(0), K_1), \ t \geq 0. \qquad (3.14)$$

Along the same lines states variables (α, β) are nonnegative and P_3 is a solution of a basic logistic equation $P_3'(t) = (b_1 - [m_1 + k_1 P_3(t)]) P_3(t), \ t \geq 0$. Assuming both $b_m - m_m > 0$ and $k_m > 0$ one may conclude that $P_3(t) \to K_m$ as $t \to +\infty$, exponentially, as soon as $P_3(0) > 0$. One also has

$$0 \leq \alpha(t), \beta(t) \leq P_3(t) \leq \max(P_3(0), K_m), \ t \geq 0. \qquad (3.15)$$

Last state variables (u, v, w) are nonnegative, hence $P_2'(t) \leq 0$ because $P_2'(t) = -(1 - \varepsilon_2)\lambda_2 v(t) \leq 0, \ t \geq 0$; as a consequence

$$0 \leq u(t), v(t), w(t) \leq P_2(t) \leq P_2(0), \ t \geq 0. \qquad (3.16)$$

Conditions (3.14)–(3.16) and the resulting boundedness of the time derivatives of the state variables show the existence of an ω-limit set. It is slightly less straightforward to prove the existence of (u^*, w^*).

Lemma 4. *There exists constants (u^*, w^*), $u^* \geq 0$ and $w^* > 0$, such that $(u(t), v(t), w(t)) \to (u^*, 0, w^*)$ as $t \to +\infty$ as soon as $P_2(0) > 0$.*
Furthermore one has $u^* > 0$ if and only if $\int_0^\infty \sigma_{m2}(s)\beta(s)ds < +\infty$.

Proof. It resembles the corresponding proof for the original Kermack and McKendrick model in [46]. Integrating over time the ODE for u one gets

$$u(t) = u(0) \exp\left(-\int_0^t \sigma_{m2}(s)\beta(s)ds\right) \searrow u^* \geq 0, \text{ as } t \to +\infty.$$

Adding the equations for u and v and integrating over time one finds

$$0 \leq u(t) + v(t) + \lambda_2 \int_0^t v(s)ds = u(0) + v(0), \; t \geq 0.$$

As a consequence one has both $\int_0^{+\infty} v(s)ds < +\infty$ and $v(t) \to v^* \geq 0$ as $t \to +\infty$. This together with the boundedness of $\{v'(t), t \geq 0\}$ show $v^* = 0$.

Last integrating over time the ODE for w yields

$$w(t) = \varepsilon_2 \lambda_2 \int_0^t v(s)ds \to w^* > 0 \text{ as } t \to +\infty.$$

$\square$

This completes the proof of Lemmas 4 and 3. $\square$

The stability analysis simplifies: any element in the ω-limit set ω_0 being of the form $(K_1 - \psi^\infty - \chi^\infty, \psi^\infty, \chi^\infty, u^*, 0, w^*, K_m - \beta^\infty, \beta^\infty)^\top$ we can reduce the analysis for the autonomous MODEL (H1VH2,U) to flows on ω_0 with $(u(t), v(t), w(t)) \equiv (u^*, 0, w^*)$, this is to the reduced system

$$\psi' = +\sigma_{m1}\beta(K_1 - \psi - \chi) - (\lambda_1 + b_1)\psi,$$

$$\chi' = \omega_1\lambda_1\psi - b_1\chi, \tag{3.17}$$

$$\beta' = +\sigma_{1m}\psi(K_m - \beta) - (\lambda_m + b_m)\beta,$$

Concerning the stability analysis for (3.17) one has

Lemma 5. *Set*

$$\mathcal{T}_{0,v} = \frac{\sigma_{m1}K_1\sigma_{1m}K_m}{(\lambda_1 + b_1)(\lambda_m + b_m)}.$$

Then (3.17) has one or two stationary states with nonnegative components:

1. *When $\mathcal{T}_{0,v} < 1$ the trivial state $(0,0,0)$ is globally asymptotically stable.*
2. *When $\mathcal{T}_{0,v} > 1$ the trivial state $(0,0,0)$ is unstable and there exists a unique nontrivial stationary state $(\psi^*, \chi^*, \beta^*)$ with $\psi^* > 0$, $\chi^* \geq 0$, $0 < \psi^* + \chi^* \leq K_1$ and $0 < \beta^* \leq K_m$ that is locally asymptotically stable; furthermore $\chi^* > 0 \iff \omega_1\lambda_1 > 0$.*

Proof. Let us assume $\mathcal{T}_{0,v} < 1$. Set $\mathcal{L}_2(\psi, \chi, \beta) = (\lambda_m + b_m)\psi + (\sigma_{m1}K_1 + \theta_m)\beta + \theta_1\chi$, for positive and small enough (θ_1, θ_m). Then straightforward calculations yield $d\mathcal{L}_2/dt(t) \leq -\theta_2\mathcal{L}_2(t)$ for some small and positive θ_2. Hence $\mathcal{L}_2$ is a Lyapunov functional and $(\psi(t), \chi(t), \beta(t)) \to (0,0,0)$ as $t \to +\infty$, exponentially.

Let us now assume $\mathcal{T}_{0,v} > 1$. The Jacobian matrix of (3.17) evaluated at the trivial equilibrium, $(0,0,0)$, has a positive determinant yielding local instability of $(0,0,0)$. Let $(\psi^*, \chi^*, \beta^*)$ be a stationary state with nonnegative components. From the second and the third equations in (3.17) at equilibrium one gets

$$\chi^* = \frac{\omega_1 \lambda_1}{b_1} \psi^*, \qquad \beta^* = \frac{\sigma_{1m} K_m \psi^*}{\sigma_{1m} \psi^* + \lambda_m + b_m}.$$

Substituting this back into the first equation in (3.17) at equilibrium and simplifying by $\psi^* > 0$ leads to finding a root, $0 < \psi^* < K_1$, of an equation, $F_m(\psi^*) = b_1 + \lambda_1$, wherein

$$F_m(\psi) = \left[\frac{\sigma_{m1} \sigma_{1m} K_m}{\sigma_{1m} \psi + \lambda_m + b_m} \right] \left[K_1 - \left(1 + \frac{\omega_1 \lambda_1}{b_1} \right) \psi \right].$$

F_m is a decreasing function within the range $(0, \hat{K}_1)$, $\hat{K}_1 = (1 + \frac{\omega_1 \lambda_1}{b_1})^{-1} K_1$, satisfying $F_m(\hat{K}_1) = 0$; as a consequence there is a unique solution ψ^* in the desired range if and only if $F_m(0) > b_1 + \lambda_1$, this is $\mathcal{T}_{0,v} > 1$. Then one may check that $\psi^* > 0$, $\chi^* \geq 0$, $0 < \psi^* + \chi^* \leq K_1$ from the first equation in (3.17) at equilibrium, and $0 < \beta^* \leq K_m$ from its closed form found above.

Local stability for $(\psi^*, \chi^*, \beta^*)$ follows from the analysis of the Jacobian matrix of (3.17) evaluated at $(\psi^*, \chi^*, \beta^*)$, using the Routh–Hurwicz criterion, cf. [26, 51]. Algebraic calculations are found in Sect. 3.7 where it is shown that the eigenvalues of the Jacobian matrix are negative or have negative real parts. □

We now go back to the stability analysis of solutions to MODEL (H1VH2,U) in the logistic autonomous case.

Proposition 1. *Let condition* HYP 4.1 *hold.*
Then the solution-set $(\varphi, \psi, \chi, u, v, w, \alpha, \beta)^\top$ *for* MODEL (H1VH2,U) *satisfies:*

- *When* $\mathcal{T}_{0,v} < 1$ *as* $t \to +\infty$ $(\varphi(t), \psi(t), \chi(t), u(t), v(t), w(t), \alpha(t), \beta(t))^\top$ *goes to* $(K_1, 0, 0, u^*, 0, w^*, K_m, 0)^\top$, *for some positive* u^* *and* w^*
- *When* $\mathcal{T}_{0,v} > 1$ *let* $(\psi^*, \chi^*, \beta^*)$ *be the unique nontrivial stationary state for (3.17) with* $\psi^* > 0$, $\chi^* \geq 0$, $0 < \psi^* + \chi^* \leq K_1$ *and* $0 < \beta^* \leq K_m$; *then as* $t \to +\infty$ $(\varphi(t), \psi(t), \chi(t), u(t), v(t), w(t), \alpha(t), \beta(t))^\top$ *goes to* $(K_1 - \psi^* - \chi^*, \psi^*, \chi^*, 0, 0, w^*, K_m - \beta^*, \beta^*)^\top$, *for some positive* w^* *provided* $(\psi(0), \chi(0), \beta(0))$ *be close to* $(\psi^*, \chi^*, \beta^*)$

As a consequence $\mathcal{T}_{0,v}$ in Lemma 5 is the threshold parameter for the invasion and persistence of the parasite in the logistic autonomous MODEL (H1VH2,U).

Further Comments

Simple modifications of the logistic regulation condition HYP 4.1 lead to modified dynamics for the autonomous MODEL (H1VH2,U), in the spirit of the foregoing Proposition 1. This is derived upon adapting the proof of Lemma 3.

Ignoring demography: Assuming $b_i = m_i = k_i = 0$, $i = 1, m$ then both host population H_1 and vector population V remain constant in time. It is straightforward to derive a stability result depending on whether $\omega_1 = 0$ or $\omega_1 > 0$; cf. [31] for a simpler model without recovered class in H_1.

Malthusian growth: When $k_1 = k_m = 0$ the parasite cannot control either populations H_1 or V because there is no additional mortality due to the disease. Hence $P_i(t) \to +\infty$ as $t \to +\infty$, $i = 1, 3$.

Allee effect: The birth rate is nonlinear, see Sect. 3.2. Still assuming a parasite benign in H_1 and V the large time behavior of P_1 and P_3 is again known. For $i = 1, 3$ as $t \to +\infty$, $P_i(t) \to 0$ when $P_{0,i}$ lies below the intermediate threshold density $K_{s,i}$ while $P_i(t) \to K_i$ when $P_{0,i} > K_{s,i}$. Then the complete system can be reduced to a simpler one, see (3.17) with b_i changed into $b_i(K_i)$ when $P_{0,i} > K_{s,i}$, yielding a modified threshold for initial conditions such that $P_{0,i} > K_{s,i}$; cf. [25,41] for a single host–parasite system with an Allee effect within the host population and a lethal parasite.

MODEL (H1VH2,U) *with Time Periodic Coefficients*

We follow the techniques designed in Wolf et al. [70] to prove the existence of periodic solutions for MODEL (H1VH2,U) emerging from stable stationary states. More precisely let us assume

HYP 4.2 Coefficients $\varrho = b_j, m_j, k_j$ for $j = 1, m$, and $\varrho = \sigma_{mj}, \sigma_{jm}$ for $j = 1, 2$ in MODEL (H1VH2,U) are nonnegative time periodic functions of period T of the form

$$\varrho(t) = <\varrho> + \varepsilon\bar{\varrho}(t), \quad <\varrho> = \frac{1}{T}\int_0^T \varrho(t)dt > 0, \quad \varrho(t) \geq 0, \ t \geq 0,$$

for $\varepsilon > 0$, while $\lambda_1, \omega_1, \lambda_m$ are nonnegative constants, λ_2 is a positive constant and $0 < \varepsilon_2 \leq 1$.

Proposition 2. *Let condition* HYP 4.2 *hold. Assume* $<b_j - m_j> > 0$ *for* $j = 1, m$ *and set*

$$K_1 = \frac{<b_1 - m_1>}{<k_1>}, \quad K_m = \frac{<b_m - m_m>}{<k_m>}. \tag{3.18}$$

Set

$$T_{0,v}^\sharp = \frac{<\sigma_{m1}>K_1<\sigma_{1m}>K_m}{(\lambda_1 + <b_1>)(\lambda_m + <b_m>)}$$

When $T_{0,v}^\sharp > 1$ *for small enough* $\varepsilon > 0$ *there exists a nonnegative time periodic solution of period T to* MODEL (H1VH2,U) *with* $(u(t), v(t), w(t)) \equiv (0, 0, w^*)$ *and* $w^* > 0$.

Proof. The first part of the proof follows our proof of Proposition 1.

Let $(\varphi(t), \psi(t), \chi(t), u(t), v(t), w(t), \alpha(t), \beta(t))^\top$ be a time periodic solution with nonnegative components. Using the time periodic logistic equations satisfied by P_1 and P_3 and the result in Lemma 1 one gets the existence and uniqueness of nonnegative and time periodic functions, $(K_1^\sharp, K_m^\sharp)$, of period T such that

$$P_1(t) \equiv K_1^\sharp(t), \; P_3(t) \equiv K_m^\sharp(t), \; t \geq 0.$$

Next the semi-orbit $\{(\varphi(t), \psi(t), \chi(t), u(t), v(t), w(t), \alpha(t), \beta(t))^\top, t > 0\}$ is bounded in R_+^8, see the proof of Lemma 3. Last the conclusion of Lemma 4 still holds: assuming $P_2(0) > 0$ there exists $(u^*, w^*) \in R_+^2$

$$(u(t), v(t), w(t)) \to (u^*, 0, w^*) \text{ as } t \to +\infty,$$

and $u^* = 0 \iff \beta(t) \neq 0$. As a consequence in the time periodic model MODEL (H1VH2,U) we can reduce the search for nonnegative periodic solutions to solutions of the form $(K_1^\sharp(t) - \psi(t) - \chi(t), \psi(t), \chi(t), u^*, 0, w^*, K_m^\sharp(t) - \beta(t), \beta(t))^\top$. To prove the existence of time periodic solutions of period T one may thus as well consider the reduced system with periodic coefficients depending on $\varepsilon > 0$,

$$\psi' = +\sigma_{m1}(t)\beta(K_1^\sharp(t) - \psi - \chi) - (\lambda_1 + m_1(t) + k_1(t)K_1^\sharp(t))\psi,$$

$$\chi' = \omega_1\lambda_1\psi - (m_1(t) + k_1(t)K_1^\sharp(t))\chi, \tag{3.19}$$

$$\beta' = +\sigma_{1m}(t)\psi(K_m^\sharp(t) - \beta) - (\lambda_m + m_m(t) + k_m(t)K_m^\sharp(t))\beta.$$

Let us first assume $\varepsilon = 0$. System (3.19) reduces to system (3.17) with coefficients $<\sigma_{m1}>$, $<b_1>$, $<\sigma_{1m}>$, $<b_m>$ and carrying capacities given by (3.18). Assuming $\mathcal{T}_{0,v}^\sharp > 1$, and $\varepsilon = 0$, system (3.19) has a unique nontrivial stationary state, $(\psi^*, \chi^*, \beta^*)$ with $\psi^* > 0$, $\chi^* \geq 0$ and $\beta^* > 0$, that is locally asymptotically stable. The eigenvalues of the Jacobian matrix evaluated at $(\psi^*, \chi^*, \beta^*)$ are negative or have negative real parts, see the proof of Lemma 5 supplied in Sect. 3.7. This prevents the linearized system at $(\psi^*, \chi^*, \beta^*)$ to have nontrivial periodic solutions.

One may now use a theorem from Poincaré, cf. [39], asserting that for small enough $\varepsilon > 0$ system (3.19) possesses a unique time periodic solution of period T, $(\psi^\sharp(t, \varepsilon), \chi^\sharp(t, \varepsilon), \beta^\sharp(t, \varepsilon))$, with $(\psi^\sharp(t, 0), \chi^\sharp(t, 0), \beta^\sharp(t, 0))$ near $(\psi^*, \chi^*, \beta^*)$, the mapping $\varepsilon > 0 \to (\psi^\sharp(t, \varepsilon), \chi^\sharp(t, \varepsilon), \beta^\sharp(t, \varepsilon))^\top \in R^3$ being continuous.

When $\omega_1\lambda_1 > 0$ then $\chi^* > 0$; this together with $\psi^* > 0$ and $\beta^* > 0$ show that for possibly smaller $\varepsilon > 0$ the periodic solution $(\psi^\sharp(t, \varepsilon), \chi^\sharp(t, \varepsilon), \beta^\sharp(t, \varepsilon))$ remains nonnegative.

Now $\omega_1\lambda_1 = 0$ means the recovered class R_1 does not exist, equivalently $\chi(t) \equiv 0$, say $\chi^\sharp(t, \varepsilon) \equiv 0$, and the conclusion concerning nonnegativity follows as above.

From the proof of Lemma 4 one finds $u^* = 0$ and $w^* > 0$. □

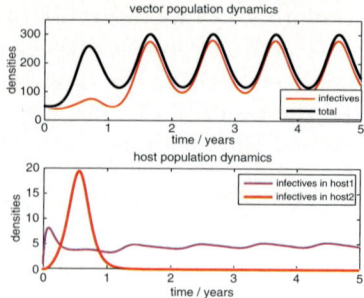

Fig. 3.9. A time periodic MODEL (H1VH2,U) numerical simulation: data from Fig. 3.3 for vectors, simple logistic regulation model for H_1 with $K_1 = 40$, $P_2(0) = 100$

Numerical simulation results are shown in Fig. 3.9. With this parameter data set large amplitude oscillations driven by one year periodic seasonal variations are observed in the vector population, data from Fig. 3.3, most vectors being infectives. It has a weak impact on the reservoir host population H_1 but a strong one in population H_2: all susceptibles become quickly infected, surviving ones enter the recovered class.

Further Comments

For nonlogistically regulated and time periodic disease free population dynamics models of Sect. 3.2 modified MODEL (H1VH2,U) can be analyzed.

3.4.2 A Simple Model for Environmentally Transmitted Diseases

Our model assumes two independent host populations, H_1 and H_2. Host populations are subdivided into three subclasses: susceptibles, S_i, infectives, I_i, and temporary or permanently recovered, R_i, for $i = 1, 2$.

The main departure from the basic SEIRS model in Fig. 3.7 and MODEL (H1VH2,U) in Fig. 3.8 arises in transmission modes. One assumes direct or horizontal transmission of the disease is feasible from infective individuals I_1 to susceptibles S_1 in population H_1. But neither horizontal transmission in population H_2, nor criss-cross transmission from infective individuals I_i from population H_i to susceptibles S_j in population H_j, $i, j = 1, 2$ and $i + j = 3$ are possible.

Instead indirect transmission through the environment drives the inter specific transmission. Infectious individuals I_i, $i = 1, 2$, will contaminate the environment through excrements and feces, cf. [12, 56]. Susceptibles of both host populations, H_1 and H_2, are contaminated by the environment. The environment has a natural decontamination rate, $\delta > 0$, $1/\delta > 0$ being the average duration of microparasite survival in the environment.

Concerning vital dynamics population H_1 is run by a logistic growth with birth and death rates (b_1, m_1) and a density-dependent effect k_1. It is assumed that the disease has no impact on fertility: infective and recovered individuals have birth-rates identical to those of susceptibles in H_1; there is no vertical transmission at birth. The infectious disease is benign in H_1, this is there is no additional mortality rate due to infection. Last $1/\lambda_1$ is the duration of the infective stage in H_1, a proportion ω_1 of infective becoming resistant and a proportion $1 - \omega_1$ going back to the susceptible class.

Natural demography is not taken into account in population H_2. Infectives I_2 surviving the disease become permanently resistant; ε_2 is the survival rate and $1/\lambda_2$ the duration of the infective stage in H_2.

The last point to be settled is the incidence term. Again we choose a density dependent form for each of the interactions between susceptible or infectious individuals and the environment.

As it is pointed out for MODEL (H1GH2,U) as long as the microparasite is benign in the reservoir population H_1 a frequency dependent incidence for horizontal transmission in H_1 would lead to minor modifications in the analysis and derivation of the threshold parameter for parasite persistence.

Let us now define state variables: (φ, ψ, χ) represent the (S_1, I_1, R_1) host densities in population H_1 and $P_1 = \varphi + \psi + \chi$, (u, v, w) represent the (S_2, I_2, R_2) host densities in population H_2 and $P_2 = u + v + w$.

A new variable, c with $0 \le c(t) \le 1$, is to be interpreted as a normalized variable representing the proportion of contaminated habitat, cf. [12, 56].

Putting together all of these assumptions transmission of the disease within the unstructured $H_1 - G - H_2$ model in Fig. 3.10 is modeled by a set of nonautonomous ordinary differential equations,

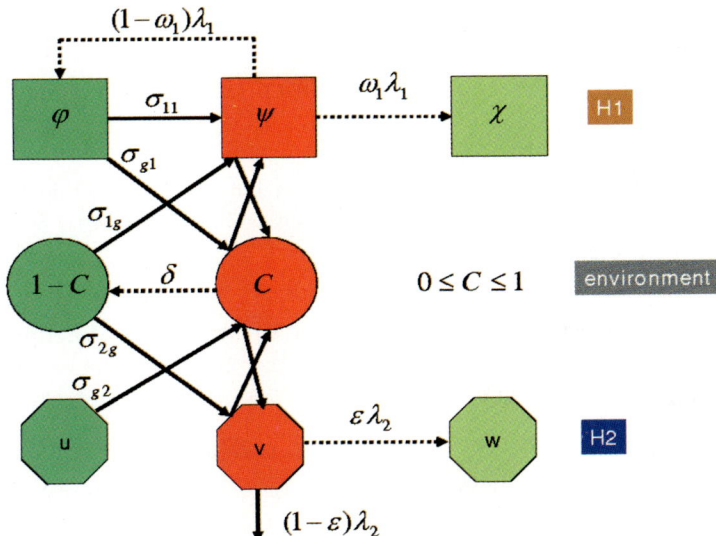

Fig. 3.10. A simple $H_1 - G - H_2$ compartmental model without demography

MODEL (H1GH2,U)

$$\varphi' = -\sigma_{11}(t)\psi\varphi - \sigma_{g1}(t)c\varphi + (1-\omega_1)\lambda_1\psi + b_1(t)P_1 - (m_1(t) + k_1(t)P_1)\varphi,$$
$$\psi' = +\sigma_{11}(t)\psi\varphi + \sigma_{g1}(t)c\varphi - \lambda_1\psi - (m_1(t) + k_1(t)P_1)\psi,$$
$$\chi' = +\omega_1\lambda_1\psi - (m_1(t) + k_1(t)P_1)\chi,$$
$$u' = -\sigma_{g2}(t)cu,$$
$$v' = +\sigma_{g2}(t)cu - \lambda_2 v,$$
$$w' = +\varepsilon_2\lambda_2 v,$$
$$c' = (\sigma_{1g}(t)\psi + \sigma_{2g}(t)v)(1-c) - \delta c.$$

supplemented with a set of nonnegative initial conditions such that each of the two populations is present at time $t = 0$ – this is $P_1(0) > 0$, $P_2(0) > 0$ – and such that infection is also present at time $t = 0$ – this is $\psi(0) + \beta(0) + c(0) > 0$.

Lemma 6. *Let the coefficients in* MODEL *(H1GH2,U) be either nonnegative constants or nonnegative smooth and bounded functions of time, $0 \le \varepsilon_2 \le 1$.*
    Then for each set of nonnegative initial conditions, $0 \le c(0) \le 1$, MODEL *(H1GH2,U) has a unique global solution $(\varphi(t), \psi(t), \chi(t), u(t), v(t), w(t), c(t))^\top$ with nonnegative components and $0 \le c(t) \le 1$.*

Proof. Compared to the proof of Lemma 2 the only difference lies in the equation for c: now the interval $0 \le c \le 1$ is forward invariant by the flow in MODEL (H1GH2,U) as soon as state variables (ψ, β) remain nonnegative. Hence $R_+^6 \times [0,1]$ is forward invariant by the flow in MODEL (H1GH2,U).

$\square$

Stability Analysis for the Autonomous MODEL *(H1GH2,U)*

Let us assume

HYP 4.3 Coefficients λ_1 and ω_1 are nonnegative constants while other coefficients in MODEL (H1GH2,U) are positive constants with $b_1 - m_1 > 0$ and $0 < \varepsilon_2 \le 1$.
    $P_1(0) > 0$ and $P_2(0) > 0$.

We still use the methodology of ω-limit sets. In the setting of Lemma 6 given a set of nonnegative initial conditions with $0 \le c(0) \le 1$ the ω-limit set for MODEL (H1GH2,U) is defined as

$$\omega_0 = \{(\varphi^\infty, \psi^\infty, \chi^\infty, u^\infty, v^\infty, w^\infty, c^\infty)^\top \in R_+^6 \times [0,1], \text{ such that}$$
$$\text{there exists a sequence } (t_n)_{n \ge 0}, \ t_n \to +\infty \text{ as } n \to +\infty,$$
$$(\varphi(t_n), \psi(t_n), \chi(t_n), u(t_n), v(t_n), w(t_n), c(t_n))^\top \to$$
$$(\varphi^\infty, \psi^\infty, \chi^\infty, u^\infty, v^\infty, w^\infty, c^\infty)^\top \text{ as } n \to +\infty\}.$$

Along the lines of the previous system MODEL (H1VH2,U) one has

Lemma 7. *Let condition* HYP 4.3 *hold. Set*

$$K_1 = \frac{b_1 - m_1}{k_1}.$$

For each set of nonnegative initial conditions with $0 \le c(0) \le 1$ there exists a compact, connected and forward invariant ω-limit set $\omega_0 = \omega(\varphi, \psi, \chi, u, v, w, c)$ in $R_+^6 \times [0,1]$.
Any element in ω_0 is of the form $(K_1 - \psi^\infty - \chi^\infty, \psi^\infty, \chi^\infty, u^*, 0, w^*, c^\infty)^\top$, u^* and w^* being nonnegative constants depending on the initial conditions.

We can reduce the stability analysis for the autonomous MODEL (H1GH2,U) to flows on ω_0 with $(u(t), v(t), w(t)) \equiv (u^*, 0, w^*)$, this is to the reduced system

$$\psi' = +(\sigma_{11}\psi + \sigma_{g1}c)(K_1 - \psi - \chi) - (\lambda_1 + b_1)\psi,$$
$$\chi' = \omega_1\lambda_1\psi - b_1\chi, \qquad\qquad\qquad\qquad (3.20)$$
$$c' = +\sigma_{1g}\psi(1 - c) - \delta c,$$

Concerning the stability analysis for (3.20) one has

Lemma 8. *Let condition* HYP 4.4 *hold. Set*

$$\mathcal{T}_{0,g} = \frac{(\sigma_{11} + \sigma_{1g}\sigma_{g1}/\delta)K_1}{\lambda_1 + b_1}.$$

Then (3.20) has one or two stationary states with nonnegative components:

1. *When $\mathcal{T}_{0,g} < 1$ the trivial state $(0,0,0)$ is globally asymptotically stable.*
2. *When $\mathcal{T}_{0,g} > 1$ the trivial state $(0,0,0)$ is unstable and there exists a unique nontrivial stationary state (ψ^*, χ^*, c^*) with $\psi^* > 0$, $\chi^* \ge 0$, $0 < \psi^* + \chi^* \le K_1$ and $0 < c^* < 1$ that is locally asymptotically stable.*

Proof. Let us assume $\mathcal{T}_{0,g} < 1$. Set $\mathcal{L}_3(\psi, \chi, c) = \delta\psi + (\sigma_{g1}K_1 + \theta_g)c + \theta_1\chi$, for positive and small enough (θ_1, θ_g). Then straightforward calculations yield $d\mathcal{L}_3/dt(t) \le -\theta_2\mathcal{L}_3(t)$ for some small and positive θ_2. Hence $\mathcal{L}_3$ is a Lyapunov functional and $(\psi(t), \chi(t), c(t)) \to (0,0,0)$ as $t \to +\infty$, exponentially.

Let us now assume $\mathcal{T}_{0,g} > 1$. The Jacobian matrix of (3.20) evaluated at the trivial equilibrium, $(0,0,0)$, has a positive determinant yielding local instability of $(0,0,0)$. Let (ψ^*, χ^*, c^*) be a nontrivial stationary state with $0 < c^* < 1$. From the second and the third equations in (3.20) at equilibrium one gets

$$\chi^* = \frac{\omega_1\lambda_1}{b_1}\psi^*, \qquad\qquad c^* = \frac{\sigma_{1g}\psi^*}{\sigma_{1g}\psi^* + \delta}.$$

Substituting this back into the first equation in (3.20) at equilibrium and simplifying by $\psi^* > 0$ leads to finding a root, $0 < \psi^* < K_1$, of an equation, $F_g(\psi^*) = b_1 + \lambda_1$, wherein

$$F_g(\psi) = \left[\sigma_{11} + \sigma_{g1}\frac{\sigma_{1g}}{\sigma_{1g}\psi + \delta}\right]\left[K_1 - \left(1 + \frac{\omega_1\lambda_1}{b_1}\right)\psi\right].$$

F_g is a decreasing function within the range $(0, \hat{K}_1)$, $\hat{K}_1 = (1 + \frac{\omega_1\lambda_1}{b_1})^{-1}K_1$, satisfying $F_g(\hat{K}_1) = 0$; as a consequence there is a unique solution ψ^* in the desired range if and only if $F_g(0) > b_1 + \lambda_1$, this is $\mathcal{T}_{0,g} > 1$. Then one can check that $\psi^* > 0$, $\chi^* \geq 0$, $0 < \psi^* + \chi^* \leq K_1$ and $0 < c^* < 1$.

Local stability for (ψ^*, χ^*, c^*) follows from the analysis of the Jacobian matrix of (3.20) evaluated at (ψ^*, χ^*, c^*), see Sect. 3.7. Eigenvalues of the Jacobian matrix are negative or have negative real parts. $\square$

We now complete the stability analysis of solutions to MODEL (H1GH2,U) in the autonomous case.

Proposition 3. *Let condition* HYP *4.3 hold.*
Then the solution-set $(\varphi, \psi, \chi, u, v, w, c)$ *to* MODEL (H1GH2,U) *satisfies:*

- *When* $\mathcal{T}_{0,g} < 1$, *as* $t \to +\infty$, $(\varphi(t), \psi(t), \chi(t), u(t), v(t), w(t), c(t))^\top$ *goes to* $(K_1, 0, 0, u^*, 0, w^*, 0)^\top$, *for some positive* u^* *and nonnegative* w^*.
- *When* $\mathcal{T}_{0,g} > 1$ *let* (ψ^*, χ^*, c^*) *be the unique nontrivial stationary state for (3.20),* $\psi^* > 0$, $\chi^* \geq 0$, $0 < \psi^* + \chi^* \leq K_1$ *and* $0 < c^* < 1$; *then,* $(\varphi(t), \psi(t), \chi(t), u(t), v(t), w(t), c(t))^\top$ *goes to* $(K_1 - \psi^* - \chi^*, \psi^*, \chi^*, 0, 0, w^*, c^*)^\top$ *as* $t \to +\infty$ *for some nonnegative* w^* *provided* $(\psi(0), \chi(0), c(0))$ *be close enough to* (ψ^*, χ^*, c^*).

As a consequence $\mathcal{T}_{0,g}$ in Lemma 8 is the threshold parameter for the invasion and persistence of the parasite in the autonomous logistic MODEL (H1GH2,U).

Further Comments

Simple modifications of condition HYP 4.3 yield modifications in dynamical behaviors for the autonomous MODEL (H1GH2,U): see the comments related to MODEL (H1VH2,U) following Proposition 1 concerning models ignoring demography, and models with Malthusian growth or Allee effect.

MODEL (H1GH2,U) *with Time Periodic Coefficients*

We still follow the methodology designed for MODEL (H1VH2,U), cf. Wolf et al. [70], to prove the existence of periodic solutions for MODEL (H1GH2,U) emerging from stable stationary states. Let us now assume

HYP 4.4 Coefficients $\varrho = b_1, m_1, k_1$ and σ_{gj}, σ_{jg} for $j = 1, 2$ in MODEL (H1GH2,U) are nonnegative time periodic functions of period T of the form

$$\varrho(t) = <\varrho> +\varepsilon\overline{\varrho}(t), \quad <\varrho> = \frac{1}{T}\int_0^T \varrho(t)dt > 0, \quad \varrho(t) \geq 0, \ t \geq 0,$$

for $\varepsilon > 0$, while λ_1, ω_1 are nonnegative constants, λ_2, δ are positive constants and $0 < \varepsilon_2 \leq 1$.

Proposition 4. *Let condition* HYP *4.4 hold. Assume* $< b_1 - m_1 >> 0$ *and set*

$$K_1 = \frac{<b_1 - m_1>}{<k_1>}.$$

Set

$$\mathcal{T}^{\sharp}_{0,g} = \frac{(< \sigma_{11} > + < \sigma_{1g} >< \sigma_{g1} > /\delta)K_1}{\lambda_1 + < b_1 >}$$

When $\mathcal{T}^{\sharp}_{0,g} > 1$ *for* $\varepsilon > 0$ *small enough there exists a nonnegative time periodic solution of period* T *to* MODEL *(H1GH2,U) with* $(u(t), v(t), w(t)) \equiv (0, 0, w^*)$ *and* $w^* > 0$.

Proof. It is quite identical to the proof of the corresponding result for MODEL (H1VH2,U), see Proposition 2, and therefore omitted. □

Numerical simulation results are shown in Figs. 3.11 and 3.12. With this parameter data set large amplitude oscillations driven by a five year periodic

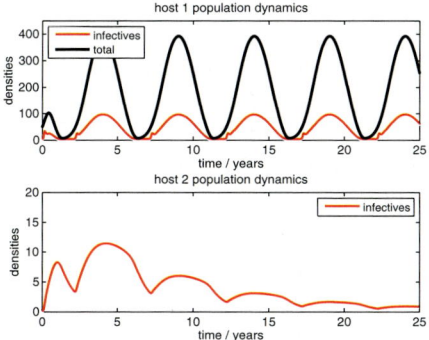

Fig. 3.11. A time periodic MODEL (H1GH2,U) numerical simulation: data from Fig. 3.4 for host population H_1 and $P_2(0) = 100$

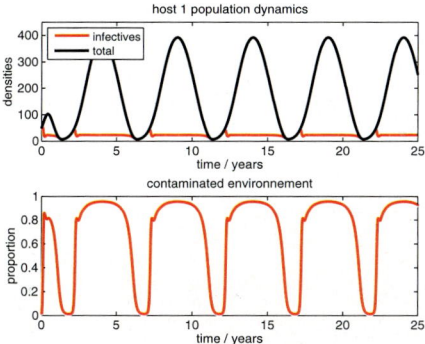

Fig. 3.12. A time periodic MODEL (H1GH2,U) numerical simulation: data from Fig. 3.4 for host population H_1 and a decontamination rate $\delta = 10$ of the environment

carrying capacity are observed in host population H_1, data from Fig. 3.4. It
has a strong impact on the contamination of the environment, see Fig. 3.12,
and also a strong impact on host population H_2, see Fig. 3.11. Note that the
periodic outbreaks of infectives in H_1 are followed by epidemic outbreaks in
population H_2 and the environment.

Much more realistic numerical simulations using field data for a bank vole–
Human–hantavirus system are found in [57], cf. [56, 70].

Further Comments

For nonlogistically regulated and time periodic disease free population dynamics models of Sect. 3.2 modified versions of MODEL (H1GH2,U) can be
devised, and suitable threshold parameters for parasite persistence exhibited.

3.5 Spatially Structured Epidemic Models on Noncoincident Spatial Domains

In this section we study systems of Reaction–Diffusion equations modeling
the transmission of a microparasite between two host populations spatially
distributed over distinct spatial habitats: a population H_1 is distributed over
a spatial domain Ω_1 while a population H_2 is distributed over a spatial domain Ω_2 with $\Omega_1 \cap \Omega_2 \neq \Omega_i$, $i = 1, 2$. Epidemiological models are those from
Sect. 3.4: a vector borne disease model with a vector population distributed
over a spatial domain Ω_m such that $\Omega_m \cap \Omega_i \neq \emptyset$, $i = 1, 2$, and an environmentally transmitted disease model in which case $\Omega_1 \cap \Omega_2 \neq \emptyset$.

3.5.1 A Simple Model for Vector Borne Diseases

Let Ω_i, $i = 1, 2, m$, be bounded domains in R^n with smooth boundary $\partial \Omega_i$,
such that locally Ω_i lies on one side of $\partial \Omega_i$. Suppose $\Omega_m \cap \Omega_i \neq \emptyset$, $i = 1, 2$.

Spatial heterogeneities arise into two manners: similarly to those from
Sect. 3.2.2, but – and mostly – because incidence terms have supports in $\overline{\Omega_m} \cap$
$\overline{\Omega_i}$, $i = 1, 2$, yielding spatial discontinuities in coefficients along the boundaries
of $\Omega_m \cap \Omega_i$ – see Fig. 3.13 and $D_{set,v}$ in HYP. 5.2 below – that prevent solutions
to be smooth there.

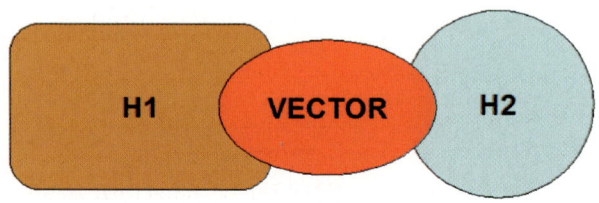

Fig. 3.13. Spatial domains for simple vector borne disease model

State variables represent time dependent spatial densities: (φ, ψ, χ) are the (S_1, I_1, R_1) densities for population H_1 in Ω_1 and $P_1 = \varphi + \psi + \chi$, (u, v, w) represent the (S_2, I_2, R_2) densities for population H_2 in Ω_2 and $P_2 = u + v + w$, while (α, β) are the (S_m, I_m) vector densities in Ω_m and $P_3 = \alpha + \beta$.

All individuals disperse by means of Fickian diffusion through their habitat. Let $-d_{ij}(x)\nabla\xi_{ij}$ be population fluxes, d_{ij} being the positive diffusivity of species i, $i = 1, 2, m$ and sub population j, $j = 1, 2, 3$, $j = 1$ corresponding to susceptibles, $j = 2$ to infectives and $j = 3$ to recovered.

A simple Reaction–Diffusion model with spatially dependent coefficients based on (3.6) can be derived from the unstructured MODEL (H1VH2,U) as a system of semilinear evolution equations of parabolic type, MODEL (H1VH2,S)

$$
\begin{aligned}
\partial\varphi/\partial t - \nabla \cdot d_{11}(x)\nabla\varphi &= -\sigma_{m1}(x)\beta\varphi \\
&\quad + (1 - \omega_1)\lambda_1\psi + b_1(x)P_1 - (m_1(x) + k_1(x)P_1)\varphi, \\
\partial\psi/\partial t - \nabla \cdot d_{12}(x)\nabla\psi &= +\sigma_{m1}(x)\beta\varphi \\
&\quad - \lambda_1\psi - (m_1(x) + k_1(x)P_1)\psi, \\
\partial\chi/\partial t - \nabla \cdot d_{13}(x)\nabla\chi &= +\omega_1\lambda_1\psi - (m_1(x) + k_1(x)P_1)\chi,
\end{aligned}
\tag{3.21}
$$

for $(x, t) \in \Omega_1 \times (0, +\infty)$,

$$
\begin{aligned}
\partial u/\partial t - \nabla \cdot d_{21}(x)\nabla u &= -\sigma_{m2}(x)\beta u, \\
\partial v/\partial t - \nabla \cdot d_{22}(x)\nabla v &= +\sigma_{m2}(x)\beta u - \lambda_2 v, \\
\partial w/\partial t - \nabla \cdot d_{23}(x)\nabla w &= +\varepsilon_2\lambda_2 v
\end{aligned}
\tag{3.22}
$$

for $(x, t) \in \Omega_2 \times (0, +\infty)$, and

$$
\begin{aligned}
\partial\alpha/\partial t - \nabla \cdot d_{m1}(x)\nabla\alpha &= -(\sigma_{1m}(x)\psi + \sigma_{2m}(x)v)\alpha \\
&\quad + \lambda_m\beta + b_m(x)P_3 - (m_m(x) + k_m(x)P_3)\alpha, \\
\partial\beta/\partial t - \nabla \cdot d_{m2}(x)\nabla\beta &= +(\sigma_{1m}(x)\psi + \sigma_{2m}(x)v)\alpha \\
&\quad - \lambda_m\beta - (m_m(x) + k_m(x)P_3)\beta,
\end{aligned}
\tag{3.23}
$$

for $(x, t) \in \Omega_m \times (0, +\infty)$.

The requirement that each population P_i remains confined to Ω_i, $i = 1, 2, m$, for all time translates as the following no flux boundary conditions, see (3.7),

$$
\begin{aligned}
d_{11}(x)\partial\varphi/\partial\eta_1 = d_{12}(x)\partial\psi/\partial\eta_1 &= d_{13}(x)\partial\chi/\partial\eta_1 = 0, \\
d_{21}(x)\partial u/\partial\eta_2 = d_{22}(x)\partial v/\partial\eta_2 &= d_{23}(x)\partial w/\partial\eta_2 = 0 \\
d_{m1}(x)\partial\alpha/\partial\eta_m &= d_{m2}(x)\partial\beta/\partial\eta_m = 0
\end{aligned}
\tag{3.24}
$$

for $(x, t) \in \partial\Omega_i \times (0, +\infty)$, η_i being the unit outward normal vector to Ω_i along its boundary $\partial\Omega_i$, $i = 1, 2, m$. Finally we specify initial conditions

$$\xi(x,0) = \xi_0(x), \ \ \xi = \varphi, \ \psi, \ \chi, \ u, \ v, \ w, \ \alpha, \ \beta \qquad (3.25)$$

that are continuous and nonnegative on their respective spatial domains of definition: Ω_1 for $(\varphi_0, \psi_0, \chi_0)$, Ω_2 for (u_0, v_0, w_0) and Ω_m for (α_0, β_0).

Generic Results

We place assumptions on the coefficients of MODEL (H1VH2,S) in the spirit of HYP. 2.1 and HYP. 2.2. See Figs. 3.5 and 3.13 for HYP. 5.2.

HYP. 5.1 (b_i, m_i, k_i) are nonnegative elements of $L^\infty(\Omega_i)$, $i = 1, m$, and d_{ij} are nonnegative elements of $L^\infty(\Omega_i)$, $i = 1, 2, m$ and $j = 1, 2, 3$; there exists positive numbers (d_{min}, d_{max}) such that

$$0 < d_{min} \le d_{ij}(x) \le d_{max}, \ x \in \Omega_i.$$

HYP. 5.2 Let $D_{set,v}$ be the set of spatial discontinuities for incidence terms, that is $D_{set,v} = (\Omega_1 \cap \partial\Omega_m) \cup (\Omega_2 \cap \partial\Omega_m) \cup (\Omega_m \cap \partial\Omega_1) \cup (\Omega_m \cap \partial\Omega_2)$. We suppose there exists open subregions of Ω_i, $(\Omega_{i\ell}^*)_{1 \le \ell \le \kappa(i)}$, with $\overline{\Omega_{i\ell}^*} \subset \Omega_i$, $\overline{\Omega_{i\ell}^*} \cap D_{set,v} = \emptyset$, $\overline{\Omega_{i\ell}^*} \cap \overline{\Omega_{ij}^*} = \emptyset$ for $\ell \neq j$ having the same smoothness properties as Ω_i, $i = 1, 2, m$.
Set $\Omega_i^* = \Omega_{i1}^* \cup \cdots \cup \Omega_{i\kappa(i)}^*$ and $\Omega_{i0}^* = \Omega_i \setminus (\overline{\Omega_i^*} \cup D_{set,v})$.
Then we assume $(b_i, m_i, k_i) \in C^{0,\alpha}(\overline{\Omega_{i\ell}^*})$ and $d_{ij} \in C^{2,\alpha}(\overline{\Omega_{i\ell}^*})$, for $\ell = 0, \cdots, \kappa(i)$, $i = 1, 2, m$ and $j = 1, 2, 3$.
HYP. 5.3 Coefficients σ_{ij} are nonnegative elements of $C^{0,\alpha}(\overline{\Omega_i \cap \Omega_j})$, $i = 1, 2, m$ and $j = 1, 2, m$, vanishing in $\Omega_j \setminus \Omega_i$.
λ_1, ω_1 and λ_m are nonnegative constants while $0 < \varepsilon_2 \le 1$.

Condition HYP. 5.2 still includes the case of piecewise constant coefficients. The set $D_{set,v}$ in HYP. 5.2 takes care of structural spatial discontinuities caused by incidence terms in MODEL (H1VH2,S), see HYP. 5.3.
We now state a generic existence/uniqueness result for classical solutions defined – componentwise – according to HYP. 5.2 and Definition 1. We refer to [29–31, 33] for a proof of related mathematical problems.

Theorem 2. *Let conditions* HYP. 5.1 *to* HYP. 5.3 *hold. Assume* $(\varphi_0, \psi_0, \chi_0)^\top$ $\in (C^0(\overline{\Omega_1}))^3$, $(u_0, v_0, w_0)^\top \in (C^0(\overline{\Omega_2}))^3$ *and* $(\alpha_0, \beta_0)^\top \in (C^0(\overline{\Omega_m}))^2$ *are componentwise nonnegative. System* MODEL (H1VH2,S) *has a unique componentwise nonnegative and global classical solution* $(\varphi, \psi, \chi, u, v, w, \alpha, \beta)^\top$.
Furthermore there exists a constant $\mathcal{K}_2$ *such that* $0 \le \|P_2(\cdot, t)\|_{\Omega_2,\infty} \le \mathcal{K}_2$, $t \ge 0$, *and a continuous positive function* $\mathcal{K} : R_+ \to R_+$ *such that*

$$0 \le \|P_1(\cdot, t)\|_{\Omega_1,\infty}, \|P_3(\cdot, t)\|_{\Omega_m,\infty} \le \mathcal{K}(t), t \ge 0.$$

Uniform bounds for solutions to system MODEL (H1VH2,S) are derived upon assuming suitable conditions on vital dynamics.

HYP. 5.4 For $i = 1, m$ there exists positive numbers r_i and $k_{i,\min}$ such that

$$0 < r_i \leq b_i(x) - m_i(x), \ 0 < k_{i,\min} \leq k_i(x), \ x \in \Omega_i.$$

Proposition 5. *Let conditions* HYP. 5.1 *to* HYP. 5.3 *hold. Assume either condition* HYP. 5.4 *hold or* $b_i(x) = m_i(x) = k_i(x) \equiv 0$ *for* $i = 1, m$. *For each set of componentwise nonnegative initial condition* $(\varphi_0, \psi_0, \chi_0)^\top \in (C^0(\overline{\Omega_1}))^3$, $(u_0, v_0, w_0)^\top \in (C^0(\overline{\Omega_2}))^3$ *and* $(\alpha_0, \beta_0)^\top \in (C^0(\overline{\Omega_m}))^2$ *there exists a positive constant* $\mathcal{K}$ *such that the solution to* MODEL (H1VH2,S) *satisfies*

$$0 \leq \|P_1(\cdot,t)\|_{\Omega_1,\infty}, \|P_2(\cdot,t)\|_{\Omega_1,\infty}, \|P_3(\cdot,t)\|_{\Omega_m,\infty} \leq \mathcal{K}, t \geq 0.$$

We refer to [29–31, 33] for a proof of the foregoing result. These uniform bounds together with regularity results for parabolic equations, cf. [48], yields compactness. Given a set of nonnegative initial conditions the ω-limit set ω_0 for MODEL (H1VH2,S) is defined as

$$\begin{aligned}
\omega_0 = \{&(\varphi^\infty, \psi^\infty, \chi^\infty, u^\infty, v^\infty, w^\infty, \alpha^\infty, \beta^\infty)^\top, \ such \ that \\
&there \ exists \ a \ sequence \ (t_n)_{n \geq 0}, \ t_n \to +\infty \ as \ n \to +\infty, \\
&(\varphi(t_n), \psi(t_n), \chi(t_n))^\top \to (\varphi^\infty, \psi^\infty, \chi^\infty)^\top \ in \ (C^0(\overline{\Omega_1}))^3, \\
&(u(t_n), v(t_n), w(t_n))^\top \to (u^\infty, v^\infty, w^\infty)^\top \ in \ (C^0(\overline{\Omega_2}))^3, \\
&(\alpha(t_n), \beta(t_n))^\top \to (\alpha^\infty, \beta^\infty)^\top \ in \ (C^0(\overline{\Omega_m}))^2, \\
&\hspace{6cm} as \ n \to +\infty\}.
\end{aligned}$$

We still refer to [29–31, 33] for a proof of the following one

Proposition 6. *Assume conditions listed in Proposition 5 are satisfied. Then trajectories* $\{(\varphi(\cdot,t), \psi(\cdot,t), \chi(\cdot,t), u(\cdot,t), v(\cdot,t), w(\cdot,t), \alpha(\cdot,t), \beta(\cdot,t))^\top, t \geq 0\}$ *are precompact in* $(C^0(\overline{\Omega_1}))^3 \times (C^0(\overline{\Omega_2}))^3 \times (C^0(\overline{\Omega_m}))^2$. *Each set of componentwise nonnegative initial condition* $(\varphi_0, \psi_0, \chi_0)^\top \in (C^0(\overline{\Omega_1}))^3$, $(u_0, v_0, w_0)^\top \in (C^0(\overline{\Omega_2}))^3$ *and* $(\alpha_0, \beta_0)^\top \in (C^0(\overline{\Omega_m}))^2$ *has a compact connected and forward invariant* ω-*limit set* ω_0.
Moreover the semidynamical system in MODEL (H1VH2,S) *has a global attractor in* $((C^0(\overline{\Omega_1}))^3 \times (C^0(\overline{\Omega_2}))^3 \times (C^0(\overline{\Omega_m}))^2)^+$.

The actual form of the ω-limit set seems out of reach of analysis in the general case, except for the second host population H_2.

Proposition 7. *Assume conditions listed in Proposition 5 are satisfied. There exists two nonnegative constants* (u^*, w^*) *such that as* $t \to +\infty$ $(u(\cdot,t), v(\cdot,t), w(\cdot,t))^\top$ *converges to* $(u^*, 0, w^*)^\top$ *in* $(C^0(\overline{\Omega_2}))^3$.

Proof. Set

$$\bar{\xi}(t) = \frac{1}{|\Omega_2|} \int_{\Omega_2} \xi(x,t) \, dx, \ \xi = u, v, w, \ t \geq 0.$$

We first show the existence of a set of nonnegative constants (u^*, v^*, w^*) such that $\bar{\xi}(t) \to \xi^*$ as $t \to +\infty$, $\xi = u, v, w$. This follows from a mere integration over Ω_2 of the equations for the nonnegative state variables (u, v, w) in (3.22). Typically from the equation for u one gets

$$\bar{u}(T) + \frac{1}{|\Omega_2|} \int_0^T \int_{\Omega_2} \sigma_{m2}(x, t)\beta(x)u(x, t)\, dxdt = \bar{u}(0).$$

This proves the existence of u^* as well as

$$\int_0^\infty \int_{\Omega_2} \sigma_{m2}(x, t)\beta(x)u(x, t)\, dxdt < +\infty.$$

From the equation for v one has

$$\bar{v}(T) + \frac{\lambda_2}{|\Omega_2|} \int_0^T \int_{\Omega_2} v(x, t)\, dxdt = \frac{1}{|\Omega_2|} \int_0^T \int_{\Omega_2} \sigma_{m2}(x, t)\beta(x)u(x, t)\, dxdt + \bar{v}(0).$$

This proves the existence of v^* as well as $v \in L^1(\Omega_2 \times (0, \infty))$ or $\bar{v} \in L^1(0, \infty)$ and the existence of w^* follows.

We now prove $v^* = 0$. Noting that

$$\frac{d}{dt}\bar{v}(t) = \frac{1}{|\Omega_2|} \int_{\Omega_2} (\sigma_{m2}\beta(x)u(x, t) - \lambda_2 v(x, t))\, dx$$

it follows the time derivative $\bar{v}'(t)$ is uniformly bounded over $(0, \infty)$. This together with $\bar{v} \in L^1(0, \infty)$ shows $v^* = 0$.

Next we strengthen these convergence results and show

Lemma 9.
$$\lim_{t \to \infty} \|\xi(\cdot, t) - \bar{\xi}(t)\|_{H^1(\Omega_2)} = 0, \ \xi = u, v, w.$$

Proof. We establish the desired result for u; the result for variables v and w follows by virtually the same arguments. Let us multiply the first equation in (3.22) through by u and integrate on Ω_2 to see that

$$\int_0^\infty \int_{\Omega_2} \|\nabla u\|^2 dxdt < +\infty. \tag{3.26}$$

Let us now multiply the first equation in (3.22) through by $\partial u/\partial t$ and integrate on Ω_2 to get

$$\int_{\Omega_2} (\partial u/\partial t)^2 dx + \frac{1}{2}d/dt \left(\int_{\Omega_2} d_{21}(x) \mid \nabla u \mid^2 dx \right) = \int_{\Omega_2} (\partial u/\partial t)\, \sigma_{m2}\beta u(x, t)dx.$$

If we apply Young's inequality we may observe that,

$$\int_{\Omega_2} (\partial u/\partial t)^2 dx + d/dt \left(\int_{\Omega_2} d_{21}(x) \mid \nabla u \mid^2 dx \right) \leq \int_{\Omega_2} (\sigma_{m2}\beta u)^2(x, t)dx.$$

As a result we may integrate on (τ, T) for any $0 < \tau < T < \infty$ to obtain,

$$\int_{\tau}^{T} \int_{\Omega_2} (\partial u/\partial t)^2 dx + d_{min} \int_{\Omega_2} \mid \nabla u(x,T) \mid^2 dx)$$
$$\leq \int_{\Omega_2} \mid \nabla u(x,\tau) \mid^2 dx + \|\sigma_{m2}\|_{\infty,\Omega_2} (\mathcal{K})^2 \int_{\tau}^{T} \int_{\Omega_2} \mid \sigma_{m2}\beta u \mid (x,t) dx.$$

$$(3.27)$$

By virtue of (3.26) above, we are assured of the existence of an increasing sequence $\{\tau_k\}_{k\geq 0}$ with $\tau_k \to \infty$ as $k \to \infty$ such that

$$\lim_{k\to\infty} \int_{\Omega_2} \mid \nabla u(x,\tau_k) \mid^2 dx = 0. \tag{3.28}$$

One also has $\int_{\tau}^{\infty} \int_{\Omega_2} \mid \sigma_{m2}\beta u(x,t) \mid dx \to 0$ as $\tau \to \infty$. Coupling this with (3.27) and (3.28) yields,

$$\lim_{t\to\infty} \int_{\Omega_2} \mid \nabla u(x,t) \mid^2 dx = 0. \tag{3.29}$$

Moreover, the Poincaré–Wirtinger Inequality, cf. [47], implies that there exists a constant $\Lambda > 0$ so that

$$\Lambda \|u(\cdot,t) - \overline{u}(t)\|_{2,\Omega_2} \leq \| \mid \nabla u(x,t) \mid \|_{2,\Omega_2}.$$

for all $t > 0$. Combining the above with (3.29) yields the desired convergence result in $H^1(\Omega_2)$ for the state variable u. $\square$

Convergence in $C^0(\overline{\Omega_2})$ follows from the $H^1(\Omega_2)$ convergence and Proposition 6. This completes the proof of Proposition 7.

A Simplified System MODEL (H1VH2,S)

This corresponds to our motivating problem wherein the microparasite has no impact on the host populations H_1 and the vector population. We now assume it has no impact on individual dispersals.

HYP. 5.5 there exists d_1 and d_m satisfying conditions HYP. 5.1 and HYP. 5.2 such that $d_{11} = d_{12} = d_{13} = d_1$ and $d_{m1} = d_{m2} = d_m$.

Upon adding the equations for (φ, ψ, χ) in MODEL (H1VH2,S) one finds a spatially heterogeneous equation for the total population P_1,

$$\partial P_1/\partial t - \nabla \cdot d_1(x)\nabla P_1 = +b_1(x)P_1 - (m_1(x) + k_1(x)P_1)P_1,$$
$$d_1(x)\partial P_1/\partial_{\eta_1}(x,t) = 0, \tag{3.30}$$
$$P_1(x,0) = \varphi_0(x) + \psi_0(x) + \chi_0(x),$$

Assuming conditions HYP. 5.1 to HYP. 5.5 hold one may use the statement in Theorem 1 to conclude to the existence of a nonnegative function $\mathcal{K}_1$ with

$$P_1(\cdot,t) \to \mathcal{K}_1(\cdot) \text{ as } t \to +\infty, \text{ in } C^0(\overline{\Omega_1}).$$

Along the same lines, from the equations for (α, β) one may conclude to the existence of a nonnegative function $\mathcal{K}_m$ such that

$$P_3(\cdot, t) \to \mathcal{K}_m(\cdot) \text{ as } t \to +\infty, \text{ in } C^0(\overline{\Omega_m}).$$

These two stabilization results and the one in Proposition 7 allows to simplify MODEL (H1VH2,S).

Proposition 8. *Assume conditions* HYP. 5.1 *to* HYP. 5.5 *hold.*
Let $(\varphi_0, \psi_0, \chi_0)^\top \in C^0(\overline{\Omega_1}))^3$, $(u_0, v_0, w_0)^\top \in C^0(\overline{\Omega_2}))^3$ and $(\alpha_0, \beta_0)^\top \in C^0(\overline{\Omega_m}))^2$ be a set of componentwise nonnegative initial conditions.
Then any element in the ω-limit set of $(\varphi_0, \psi_0, \chi_0, u_0, v_0, w_0, \alpha_0, \beta_0)^\top$ is of the form $(\mathcal{K}_1(\cdot) - \psi^* - \chi^*, \psi^*, \chi^*, u^*, 0, w^*, \mathcal{K}_m(\cdot) - \beta^*, \beta^*)^\top$.

It remains to calculate the triple $(\psi^*, \chi^*, \beta^*)$, solutions to the reduced system,

MODEL (H1VH2,S,REDUCED)

$$\partial\psi^*/\partial t - \nabla \cdot d_1(x)\nabla\psi^* = +\sigma_{m1}(x)\beta^*(\mathcal{K}_1(\cdot) - \psi^* - \chi^*)$$
$$-(\lambda_1 + m_1(x) + k_1(x)\mathcal{K}_1)\psi^*,$$

$$\partial\chi^*/\partial t - \nabla \cdot d_1(x)\nabla\chi^* = +\omega_1\lambda_1\psi^* - (m_1(x) + k_1(x)\mathcal{K}_1)\chi^*,$$

for $(x, t) \in \Omega_1 \times (0, +\infty)$,

$$\partial\beta^*/\partial t - \nabla \cdot d_m(x)\nabla\beta^* = +\sigma_{1m}(x)\psi^*(\mathcal{K}_m(x) - \beta^*)$$
$$-\lambda_m\beta^* - (m_m(x) + k_m(x)\mathcal{K}_m(x))\beta^*,$$

for $(x, t) \in \Omega_m \times (0, +\infty)$ with no flux boundary conditions:

$$d_1(x)\partial\psi^*/\partial\eta_1 = d_1(x)\partial\chi^*/\partial\eta_1 = 0, \ x \in \partial\Omega_1, \ t \geq 0,$$

$$d_m(x)\partial\beta^*/\partial\eta_m = 0, \ x \in \partial\Omega_m, \ t \geq 0,$$

and a set of componentwise nonnegative initial conditions $(\psi_0^*, \chi_0^*, \beta_0^*)$

$$(\psi_0^*, \chi_0^*, \beta_0^*)^\top \in (C^0(\overline{\Omega_1}))^2 \times C^0(\overline{\Omega_m}).$$

A comprehensive analysis of the large time behavior of solutions to MODEL (H1VH2,S, REDUCED) in the genuinely spatially heterogeneous case is difficult to derive. Numerical results from Fitzgibbon et al. [32] show that for distinct spatial domains, $\Omega_1 \neq \Omega_m$, the threshold for persistence depends on a complicated manner of $\Omega_1 \cap \Omega_m$. Hence the conjecture from [31] is not correct.

For spatially structured populations this turns out to be one of the main differences between the vector borne transmitted model MODEL H1VH2 and the environmentally transmitted model MODEL H1GH2 handled below.

For constant coefficients and identical spatial domains, $\Omega_1 = \Omega_2 = \Omega_m$, $(0, 0, 0)$ and $(\psi^*, \chi^*, \beta^*)$ from Lemma 5 are the two constant stationary states with nonnegative components. A linear stability analysis can be carried out from which it follows that $\mathcal{T}_{0,v}$ remains a threshold parameter for persistence of the microparasite; see the proofs of Lemmas 5 and 10 below.

3.5.2 A Simple Model for Environmentally Transmitted Diseases

Let Ω_i, $i = 1, 2$, be bounded domains in R^n with smooth boundary $\partial\Omega_i$ such that locally Ω_i lies on one side of $\partial\Omega_i$. Suppose $\Omega_1 \cap \Omega_2 \neq \emptyset$.

Again spatial heterogeneities arise into two manners: similarly to those from Sect. 3.2.2, but – and mostly – because incidence terms for the contamination of the environment in $\Omega_1 \cup \Omega_2$ have supports in either $\overline{\Omega_1}$ or $\overline{\Omega_2}$ yielding spatial discontinuities in coefficients along the boundaries of $\Omega_1 \cap \Omega_2$ – see Fig. 3.14 and $D_{set,g}$ in Hyp. 5.7 below – that prevent the c component of the solution-set to be smooth there.

State variables will now represent time dependent spatial densities: (φ, ψ, χ) are the (S_1, I_1, R_1) densities for population H_1 in Ω_1 and $P_1 = \varphi + \psi + \chi$, (u, v, w) represent the (S_2, I_2, R_2) densities for population H_2 in Ω_2 and $P_2 = u + v + w$, while $c(x, t)$ is the proportion of contaminated environment at location $x \in \Omega_1 \cup \Omega_2$ and time $t \geq 0$.

All host individuals disperse by means of Fickian diffusion through their habitat. Let $-d_{ij}(x)\nabla\xi_{ij}$ be the population fluxes, d_{ij} being the positive diffusivity of species i, $i = 1, 2$ and sub population j, $j = 1, 2, 3$, $j = 1$ corresponding to susceptibles, $j = 2$ to infectives and $j = 3$ to recovered.

A simple Reaction–Diffusion model with spatially dependent coefficients, based on (3.6), can be expressed as a system of semilinear evolution equations of parabolic type,

MODEL (H1GH2,S)

$$\partial\varphi/\partial t - \nabla \cdot d_{11}(x)\nabla\varphi = -\sigma_{11}(x)\psi\varphi - \sigma_{g1}(x)c\varphi$$
$$+(1 - \omega_1)\lambda_1\psi + b_1(x)P_1 - (m_1(x) + k_1(x)P_1)\varphi,$$

$$\partial\psi/\partial t - \nabla \cdot d_{12}(x)\nabla\psi = +\sigma_{11}(x)\psi\varphi + \sigma_{g1}(x)c\varphi \qquad (3.31)$$
$$-\lambda_1\psi - (m_1(x) + k_1(x)P_1)\psi,$$

$$\partial\chi/\partial t - \nabla \cdot d_{13}(x)\nabla\chi = +\omega_1\lambda_1\psi - (m_1(x) + k_1(x)P_1)\chi,$$

for $(x, t) \in \Omega_1 \times (0, +\infty)$,

$$\partial u/\partial t - \nabla \cdot d_{21}(x)\nabla u = -\sigma_{g2}(x)cu,$$

$$\partial v/\partial t - \nabla \cdot d_{22}(x)\nabla v = +\sigma_{g2}(x)cu - \lambda_2 v, \qquad (3.32)$$

$$\partial w/\partial t - \nabla \cdot d_{23}(x)\nabla w = +\varepsilon_2\lambda_2 v$$

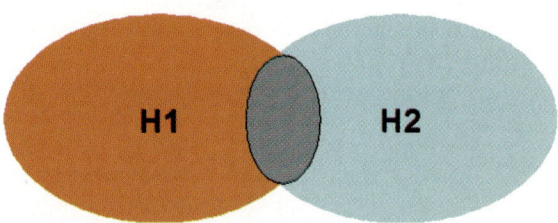

Fig. 3.14. Spatial domains for simple environmentally transmitted disease model

for $(x, t) \in \Omega_2 \times (0, +\infty)$, and

$$\partial c / \partial t = \sigma_{1g}(x)(1 - c)\psi + \sigma_{2g}(x)(1 - c)v - \delta(x)c. \tag{3.33}$$

for $(x, t) \in \Omega_1 \cup \Omega_2 \times (0, +\infty)$.

The requirement that populations H_i remain confined to Ω_i, $i = 1, 2$, for all time translates as the following no flux boundary conditions, see (3.7),

$$
\begin{aligned}
d_{11}(x)\partial\varphi/\partial\eta_1 = d_{12}(x)\partial\psi/\partial\eta_1 = d_{13}(x)\partial\chi/\partial\eta_1 = 0, \\
d_{21}(x)\partial u/\partial\eta_2 = d_{22}(x)\partial v/\partial\eta_2 = d_{23}(x)\partial w/\partial\eta_2 = 0.
\end{aligned}
\tag{3.34}
$$

η_i being the unit outward normal vector to Ω_i along its boundary $\partial\Omega_i$, $i = 1, 2$. Finally we need to specify that the initial conditions

$$\xi(x, 0) = \xi_0(x), \quad \xi = \varphi, \psi, \chi, u, v, w, c \tag{3.35}$$

are continuous and nonnegative on their respective spatial domains of definition: Ω_1 for $(\varphi_0, \psi_0, \chi_0)$, Ω_2 for (u_0, v_0, w_0) and $\Omega_1 \cup \Omega_2$ for c_0, $0 \le c_0(x) \le 1$.

Generic Results

We place assumptions on the coefficients of MODEL (H1GH2,S) in the spirit of HYP. 2.1, HYP. 2.2 and HYP. 5.2. See Figs. 3.5 and 3.14 for HYP. 5.7.

HYP. 5.6 (b_1, m_1, k_1) are nonnegative elements of $L^\infty(\Omega_1)$, and d_{ij} are nonnegative elements of $L^\infty(\Omega_i)$, $i = 1, 2$ and $j = 1, 2, 3$; there exists positive numbers (d_{min}, d_{max}) such that

$$0 < d_{min} \le d_{ij}(x) \le d_{max}, \ x \in \Omega_i.$$

HYP. 5.7 Let $D_{set,g}$ be the set of spatial discontinuities for the contamination of the environment, that is $D_{set,g} = (\Omega_1 \cap \partial\Omega_2) \cup (\Omega_2 \cap \partial\Omega_1)$.
We suppose there exists open subregions of Ω_i, $(\Omega_{i\ell}^*)_{1 \le \ell \le \kappa}$, with $\overline{\Omega_{i\ell}^*} \subset \Omega_i$, $\overline{\Omega_{i1,\ell}^*} \cap D_{set,g} = \emptyset$ and $\overline{\Omega_{i\ell}^*} \cap \overline{\Omega_{ij}^*} = \emptyset$, for $\ell \ne j$, having the same smoothness properties as Ω_i, $i = 1, 2$.
Set $\Omega_i^* = \Omega_{i1}^* \cup \cdots \cup \Omega_{i\kappa}^*$ and $\Omega_{i0}^* = \Omega_i \backslash (\overline{\Omega_i^*} \cup D_{set,g})$.
Then we assume $(b_1, m_1, k_1) \in C^{0,\alpha}(\overline{\Omega_{i\ell}^*})$ and $d_{ij} \in C^{2,\alpha}(\overline{\Omega_{i\ell}^*})$, for $\ell = 0, \cdots, \kappa$, $i = 1, 2$ and $j = 1, 2, 3$.

HYP. 5.8 Coefficients σ_{1j} are nonnegative elements of $C^{0,\alpha}(\overline{\Omega_1})$ vanishing in $\Omega_2 \backslash \Omega_1$, $j = 1, g$; coefficients σ_{2g} and σ_{g2} are nonnegative elements of $C^{0,\alpha}(\overline{\Omega_2})$ vanishing in $\Omega_1 \backslash \Omega_2$.
λ_1, ω_1 and λ_m are nonnegative constants while $\lambda_2 > 0$, $\delta > 0$ and $0 < \varepsilon_2 \le 1$ are constants.

Condition HYP. 5.7 includes the case of piecewise constant diffusivities. The set $D_{set,g}$ in HYP. 5.7 takes care of structural discontinuities caused by the incidence terms in equation (3.33), see HYP. 5.8.

We now state a generic existence/uniqueness result for classical solutions. These are defined – componentwise – according to HYP. 5.7 and Definition 1 for the $(\varphi, \psi, \chi, u, v, w)$ components. The last one, c, requires a specific treatment due to spatial discontinuities on the right-hand side of (3.33). Set

$$C(\Omega_1, \Omega_2) = \{\varrho : \Omega_1 \cup \Omega_2 \to [0,1], \quad \varrho \in C^0((\Omega_1 \cup \Omega_2) \backslash D_{set,g})\}. \quad (3.36)$$

The c component is a continuous mapping $c(\cdot, t) : [0, +\infty) \to C(\Omega_1, \Omega_2)$ with $c' \in L^\infty((0, +\infty); \Omega_1 \cup \Omega_2))$, a solution to the ordinary differential equation in (3.33) and such that the initial condition in (3.35) is satisfied.

Theorem 3. *Let conditions* HYP. 5.6 *to* HYP. 5.8 *hold. Assume* $(\varphi_0, \psi_0, \chi_0)^\top$ $\in (C^0(\overline{\Omega_1}))^3$, $(u_0, v_0, w_0)^\top \in (C^0(\overline{\Omega_2}))^3$ *and* $c_0 \in C(\Omega_1, \Omega_2)$ *are componentwise nonnegative with* $0 \le c_0(x) \le 1$. *System* MODEL (H1GH2,S) *has a unique componentwise nonnegative and global classical solution* $(\varphi, \psi, \chi, u, v, w, c)^\top$ *with* $0 \le c(x, t) \le 1$.
Furthermore there exists a constant $\mathcal{K}_2$ *such that* $0 \le \|P_2(\cdot, t)\|_{\Omega_2, \infty} \le \mathcal{K}_2$, $t \ge 0$, *and a continuous positive function* $\mathcal{K} : R_+ \to R_+$ *such that*

$$0 \le \|P_1(\cdot, t)\|_{\Omega_1, \infty}, \le \mathcal{K}(t), t \ge 0.$$

We refer to [29–31, 33] for a proof of related mathematical problems.
Uniform bounds for H_1 solution components to system MODEL (H1GH2,S) are derived upon assuming suitable conditions on vital dynamics of H_1.

HYP. 5.9 There exists positive numbers r_1 and $k_{1,\min}$ such that

$$0 < r_1 \le b_1(x) - m_1(x), 0 < k_{1,\min} \le k_1(x), \ x \in \Omega_1.$$

Proposition 9. *Let conditions* HYP. 5.6 *to* HYP. 5.8 *hold. Assume either condition* HYP. 5.9 *holds or* $b_1(x) = m_1(x) = k_1(x) \equiv 0$. *For each set of componentwise nonnegative initial condition* $(\varphi_0, \psi_0, \chi_0)^\top \in (C^0(\overline{\Omega_1}))^3$, $(u_0, v_0, w_0)^\top \in (C^0(\overline{\Omega_2}))^3$ *and* $c_0 \in C(\Omega_1, \Omega_2)$ *with* $0 \le c_0(x) \le 1$ *there exists a positive constant* $\mathcal{K}$ *such that the solution to* MODEL (H1VH2,S) *verifies*

$$0 \le \|P_1(\cdot, t)\|_{\Omega_1, \infty} \le \mathcal{K}, t \ge 0.$$

We still refer to [29–31, 33] for a proof of the foregoing result.
These uniform bounds together with regularity results for parabolic equations, cf. [48], yields compactness for H_1 and H_2 solution components.
Given a set of nonnegative initial conditions the ω-limit set ω_0 for MODEL (H1GH2,S) is now defined as

$$\omega_0 = \{(\varphi^\infty, \psi^\infty, \chi^\infty, u^\infty, v^\infty, w^\infty, c^\infty)^\top, \ such \ that$$
$$there \ exists \ a \ sequence \ (t_n)_{n \ge 0}, \ t_n \to +\infty \ as \ n \to +\infty,$$
$$(\varphi(t_n), \psi(t_n), \chi(t_n))^\top \to (\varphi^\infty, \psi^\infty, \chi^\infty)^\top \ in \ (C^0(\overline{\Omega_1}))^3,$$
$$(u(t_n), v(t_n), w(t_n))^\top \to (u^\infty, v^\infty, w^\infty)^\top \ in \ (C^0(\overline{\Omega_2}))^3,$$
$$c(t_n) \to c^\infty \ in \ C(\Omega_1, \Omega_2), \qquad as \ n \to +\infty\}.$$

We still refer to [29–31, 33] for a proof of the following one

Proposition 10. *Assume conditions of Proposition 9 are satisfied.*
Then the trajectories $\{(\varphi(\cdot,t), \psi(\cdot,t), \chi(\cdot,t), u(\cdot,t), v(\cdot,t), w(\cdot,t), c(\cdot,t))^\top, t \geq 0\}$, *are precompact in* $(C^0(\overline{\Omega_1}))^3 \times (C^0(\overline{\Omega_2}))^3 \times C(\Omega_1, \Omega_2)$.
Each set of componentwise nonnegative initial condition $(\varphi_0, \psi_0, \chi_0)^\top \in (C^0(\overline{\Omega_1}))^3$, $(u_0, v_0, w_0)^\top \in (C^0(\overline{\Omega_2}))^3$ *and* $c_0 \in C(\Omega_1, \Omega_2)$ *with* $0 \leq c_0(x) \leq 1$ *has a compact connected and forward invariant* ω-*limit set* ω_0 *with* $0 \leq c(x,t) \leq 1$.

Moreover the semidynamical system in MODEL (H1VH2,S) *has a global attractor in* $((C^0(\overline{\Omega_1}))^3 \times (C^0(\overline{\Omega_2}))^3 \times C(\Omega_1, \Omega_2)$.

The actual form of the ω-limit set seems out of reach of analysis in the general case, except for the second host population H_2 and the contaminated part of the environment in $\Omega_2 \backslash \overline{\Omega_1}$.

Proposition 11. *Assume the conditions of Proposition 9 are satisfied.*
There exists two nonnegative constants (u^*, w^*) *such that as* $t \to +\infty$
$(u(\cdot,t), v(\cdot,t), w(\cdot,t))^\top$ *converge to* $(u^*, 0, w^*)^\top$ *in* $(C^0(\overline{\Omega_2}))^3$ *and* $c(\cdot,t) \to 0$
in $C^0(\Omega_2 \backslash \overline{\Omega 1})$.

Proof. Compared to the proof of Proposition 7 the main modification lies in the behavior of $c(t)$. Point-wise integration of equation (3.33) for c yields the large time behavior of $c(t)$ over $\Omega_2 \backslash \overline{\Omega 1}$.

A Simplified System MODEL (H1GH2,S)

This corresponds to our motivating problem wherein the microparasite has no impact on the host populations H_1. We now assume it has no impact on individual dispersals.

HYP. 5.10   there exists d_1 satisfying conditions HYP. 5.1 and HYP. 5.2 such that $d_{11} = d_{12} = d_{13} = d_1$.

In that setting we can reduce our generic system MODEL (H1GH2,S) to a simpler one. Upon adding the equations for (φ, ψ, χ) in MODEL (H1GH2,S) one finds a spatially heterogeneous equation for the total population P_1, see (3.30), and conclude to the existence of a nonnegative function $\mathcal{K}_1$ with $P_1(\cdot,t) \to \mathcal{K}_1(\cdot)$ as $t \to +\infty$ in $C^0(\overline{\Omega_1})$. Thus we now have

Proposition 12. *Assume conditions* HYP. 5.6 *to* HYP. 5.10 *hold.*
Let $(\varphi_0, \psi_0, \chi_0)^\top \in (C^0(\overline{\Omega_1}))^3$, $(u_0, v_0, w_0)^\top \in (C^0(\overline{\Omega_2}))^3$, $c_0 \in C(\Omega_1, \Omega_2)$ *be a set of componentwise nonnegative initial conditions with* $0 \leq c_0(x) \leq 1$.
Then any element in the ω-*limit set of* $(\varphi_0, \psi_0, \chi_0, u_0, v_0, w_0, c_0)^\top$ *is of the form* $(\mathcal{K}_1(\cdot) - \psi^* - \chi^*, \psi^*, \chi^*, u^*, 0, w^*, c^*)^\top$.

It remains to calculate the triple (ψ^*, χ^*, c^*), solutions to the reduced system MODEL (H1GH2,S, REDUCED)

$$\partial \psi^* / \partial t - \nabla \cdot d_1(x) \nabla \psi^* = + \sigma_{11}(x) \psi^* (\mathcal{K}_1(x) - \psi^* - \chi^*)$$
$$+ \sigma_{g1}(x) c^* (\mathcal{K}_1(x) - \psi^* - \chi^*) - (\lambda_1 + m_1(x) + k_1(x) \mathcal{K}_1(x)) \psi^*,$$
$$\partial \chi^* / \partial t - \nabla \cdot d_1(x) \nabla \chi^* = + \omega_1 \lambda_1 \psi^* - (m_1(x) + k_1(x) \mathcal{K}_1) \chi^*, \qquad (3.37)$$
$$\partial c^* / \partial t = + \sigma_{1g}(x) \psi^* (1 - c^*) - \delta c^*,$$

for $(x, t) \in \Omega_1 \times (0, +\infty)$ with no flux boundary conditions:

$$d_1(x) \partial \psi^* / \partial \eta_1 = d_1(x) \partial \chi^* / \partial \eta_1 = 0, \ x \in \partial \Omega_1, \ t \geq 0, \qquad (3.38)$$

and a set of componentwise nonnegative initial conditions $(\psi_0^*, \chi_0^*, c_0^*)^\top$

$$(\psi_0^*, \chi_0^*, c_0^*)^\top \in (C^0(\overline{\Omega_1}))^3. \qquad (3.39)$$

Compared to MODEL (H1VH2,S, REDUCED) a huge simplification arises because (3.37)–(3.39) are now posed on a single spatial domain, Ω_1.

A fairly simple situation occurs when MODEL (H1GH2,S, REDUCED) has positive constant coefficients.

HYP. 5.11 Coefficients (b_1, m_1, k_1, d_1) are positive constants with $r_1 = b_1 - m_1 > 0$, $(\sigma_{11}, \sigma_{1g}, \sigma_{g1})$ and δ are positive constants while (λ_1, ω_1) are nonnegative constants.

When condition HYP. 5.11 holds one gets $\mathcal{K}_1(x) \equiv K_1$, see Sect. 3.2. Then

Lemma 10. *Assume conditions* HYP. 5.10 *and* HYP. 5.11 *hold. Set*

$$\mathcal{T}_{0,g} = \frac{(\sigma_{11} + \sigma_{1g} \sigma_{g1} / \delta) K_1}{\lambda_1 + b_1}, \quad \text{see Lemma 8.}$$

Then:

1. *When $\mathcal{T}_{0,g} < 1$ the trivial state $(0,0,0)$ is globally asymptotically stable for* MODEL (H1GH2,S, REDUCED).
2. *When $\mathcal{T}_{0,g} > 1$ the trivial state $(0,0,0)$ is unstable for* MODEL (H1GH2,S, REDUCED). *Let (ψ^*, χ^*, c^*) be the unique nontrivial stationary state for (3.20) with $\psi^* > 0$, $\chi^* \geq 0$, $0 < \psi^* + \chi^* \leq K_1$ and $0 < c^* < 1$; then (ψ^*, χ^*, c^*) is linearly stable for* MODEL (H1GH2,S, REDUCED).

Proof. Assume first $\mathcal{T}_{0,g} < 1$. For positive and small enough (θ_1, θ_g) set

$$\mathcal{L}_5(\psi, \chi, c) = \int_{\Omega_1} (\delta \psi + (\sigma_{g1} K_1 + \theta_g) c + \theta_1 \chi)(x, t) dx.$$

Then, see the proof of Lemma 8, $\mathcal{L}_5$ is a Lyapunov functional decreasing along the orbits of MODEL (H1GH2,S, REDUCED); the state variables (ψ, χ, c) experience an asymptotical exponential decay in $L^1(\Omega_1)$ and

$$0 \leq \int_0^\infty \int_{\Omega_1} (\psi + \chi + c)(x, t) \, dx dt < +\infty.$$

We now proceed as in the proof of Proposition 7. Integrating over Ω_1 yields

$$\int_0^\infty \int_{\Omega_1} (|\nabla\psi|^2 + |\nabla\chi|^2) dx dt < \infty,$$

cf. (3.26), and we conclude to the decay of ψ and χ toward 0 in $H^1(\Omega_1)$. Noting that no spatial discontinuity arises in the right-hand side of the equation for c a straightforward calculation shows that in Ω_1 one has

$$d/dt(|\nabla c|^2) \le 2\sigma_{12} |\nabla c||\nabla\psi| - 2\delta |\nabla c|^2$$

and $\||\nabla c|\|_{2,\Omega_1\times(0,+\infty)} < +\infty$; we conclude to the decay of c towards 0.

Let us now assume $\mathcal{T}_{0,g} > 1$. The trivial state $(0,0,0)$ is unstable for MODEL (H1GH2,S, REDUCED) because it is already unstable for the underling system of ODEs in (3.20).

Linear stability of (ψ^*, χ^*, c^*) for MODEL (H1GH2,S, REDUCED) is found in Sect. 3.7 below; it uses the nonnegative eigenvalues $(\mu_j)_{j\ge0}$ and eigenfunctions $(\xi_j)_{j\ge0}$ of the no flux boundary value problem

$$-d_1\Delta\phi_j = \mu_j\phi_j, \text{ in } \Omega_1, \quad d_1\partial\phi_j/\partial\eta_1 = 0, \text{ on } \partial\Omega_1. \tag{3.40}$$

$\square$

As a consequence for the environmentally transmitted disease and $\Omega_1 \ne \Omega_2$, in the constant coefficient case one may conclude that the dynamical behavior of the PDEs system is similar to the behavior of the ODEs system, a main difference from the dynamical behavior of the vector borne disease.

Proposition 13. *Let conditions* HYP. 5.5 *to* HYP. 5.11 *hold.*
Assume initial conditions $(\varphi_0, \psi_0, \chi_0)^\top \in (C^0(\overline{\Omega_1}))^3$, $(u_0, v_0, w_0)^\top \in (C^0(\overline{\Omega_2}))^3$ *and* $c_0 \in C(\Omega_1, \Omega_2)$ *are componentwise nonnegative with* $0 \le c_0(x) \le 1$, $\int_{\Omega_1} \psi(x,0)dx > 0$ *and* $\int_{\Omega_2} u(x,0)dx > 0$.
Then the solution-set $(\varphi, \psi, \chi, u, v, w, c)^\top$ *to* MODEL (H1GH2,S) *satisfies:*

- *When* $\mathcal{T}_{0,g} < 1$ *as* $t \to +\infty$ $(\varphi(\cdot,t), \psi(\cdot,t), \chi(\cdot,t))^\top$ *converges to* $(K_1, 0, 0)^\top$ *in* $(C^0(\overline{\Omega_1}))^3$, $(u(\cdot,t), v(\cdot,t), w(\cdot,t))^\top$ *converges to* $(u^*, 0, w^*)^\top$ *in* $(C^0(\overline{\Omega_2}))^3$, *for some positive* u^* *and nonnegative* w^* *constants, while* $c(\cdot,t)$ *goes to 0 in* $C(\Omega_1, \Omega_2)$.
- *When* $\mathcal{T}_{0,g} > 1$ *let* (ψ_0, χ_0, c_0) *be a set of initial data close enough to the nontrivial stationary state* (ψ^*, χ^*, c^*) *of (3.20) with* $\psi^* > 0$, $\chi^* \ge 0$, $0 < \psi^* + \chi^* \le K_1$ *and* $0 < c^* < 1$. *Then, as* $t \to +\infty$ $(\varphi(\cdot,t), \psi(\cdot,t), \chi(\cdot,t))^\top$ *goes to* $(K - \psi^* - \chi^*, \psi^*, \chi^*)^\top$ *in* $(C^0(\overline{\Omega_1}))^3$, $(u(\cdot,t), v(\cdot,t), w(\cdot,t))^\top$ *goes to* $(0, 0, w^*)^\top$ *for some nonnegative constant* w^* *in* $(C^0(\overline{\Omega_2}))^3$, *and* $c(\cdot,t)$ *goes to* c^* *in* $C^0(\overline{\Omega_1})$ *and 0 in* $C^0(\Omega_2 \backslash \overline{\Omega_1})$.

3.6 A Short Conclusion

We provided a simple deterministic mathematical approach to modeling the transmission of microparasites between host populations living on noncoincident spatial domains in two prototypical cases: a vector borne disease and an environmentally transmitted disease. Direct horizontal criss-cross transmission from infectious individuals of one population to susceptibles of the other one does not occur in our models. Instead microparasite transmission takes place via indirect criss-cross contact between infective vectors and susceptible individuals of either populations and vice-versa, or through indirect contacts between susceptible hosts and the contaminated part of the environment and vice-versa. An important assumption is the microparasite under consideration is benign in one of the host species while it is lethal to the other one. This is the case for applications to interspecific transmission of microparasites wherein the second population is Human while the first one is an avian or rodent population.

Simple models with spatio-temporal heterogeneities were devised: basic systems of ODEs for unstructured populations with time periodic vital dynamics or carrying capacities and Reaction–Diffusion models to handle spatially heterogeneous domains. The assumption concerning host species living on distinct spatial domains causes spatial discontinuities in the reaction terms, preventing solutions to be smooth. We proved these mathematical problems are well-posed in a suitable functional setting and then gave thresholds parameters for the persistence of endemic states.

Many features are not included in our models leaving room for further researches:

Chronological age and age of the disease: an important literature is devoted to age dependent models in epidemiology, cf. the books [14] and [66]; here "age" means either chronological age for any individuals or age since infection for exposed individuals. See also [43], [45], [69] or [71], and also chapters 1, 4 and 5 in this book and their references.

Homogenization problems: In many circumstances spatial heterogeneities may exhibit periodical structures, cf. [62]. This leads to complex dynamics at the local spatial scale while simpler "homogenized" dynamics may emerge at the global spatial scale, cf. [29] or [3] and references therein. See also related aggregation methods from Chapter 4 in this book.

Travelling front solutions: Invasion and persistence of parasites can be modeled using travelling waves, cf. [51] for the fox–rabies model. Extension to spatially heterogeneous environments is found in [62]. A sound mathematical analysis is devised in [10] for population dynamics models in periodically fragmented environments. See [41] and [25] for a dual – numerical simulations and mathematical analysis – approach of a simple SI model, and also Chapter 4 of this book.

L^1–*solutions*: This is a convenient and natural abstract functional setting to handle models with spatio–temporal heterogeneities; see Chapter 1 of this book and its references for a theory using semi–groups, and [8] for a direct approach using a limiting process starting from classical smooth solutions.

Discrete age or/and spatial structures: cf. the books [17] and [20], and references in the book [23]. One side advantage of these discrete models is they look simplistic to simulate from a numerical point of view. This is not totally correct many computational problems arising due to the complexity of the simulations for large size models, cf. [52] or [50]. See also chapters 1, 2, 4 and 5 in this book.

Numerical methods and simulations: Specific efficient numerical methods are to be designed for epidemic models, i.e., discretization techniques must preserve basic features of the model: nonnegativity of state variables is to be preserved, updating between two time steps must be adapted to the dynamics of both host and parasite populations, e.g., $S \to E \to I \to R \to S$, spatial heterogeneities have to be handled carefully. This is necessary to numerically estimate threshold parameters, e.g., R_0.

Stochastic models: cf. [4] or [21] and references in the book [24] for model derivation and analysis of stochastic epidemic models. This alternate approach to deterministic modeling is not considered in this Chapter, nor in this book. In some circumstances a sensitivity analysis, cf. [17], is efficiently supplemented by probabilistic methods.

3.7 Appendix: Technical Proofs

End of the Proof of Lemma 5

We prove the local stability of the unique stationary state $(\psi^*, \chi^*, \beta^*)$ of (3.17) with $\psi^* > 0$, $\chi^* \geq 0$, $0 < \psi^* + \chi^* \leq K_1$ and $0 < \beta^* < K_m$.

The Jacobian matrix, $J(\psi^*, \chi^*, \beta^*) = (\theta_{ij})_{1 \leq i,j \leq 3}$, of system (3.17) evaluated at $(\psi^*, \chi^*, \beta^*)$ reads

$$
\begin{pmatrix}
-\sigma_{m1}\beta^* - (\lambda_1 + b_1) & -\sigma_{m1}\beta^* & \sigma_{m1}(K_1 - \psi^* - \chi^*) \\
\omega_1 \lambda_1 & -b_1 & 0 \\
\sigma_{1m}(K_m - \beta^*) & 0 & -\sigma_{1m}\psi^* - (\lambda_m + b_m)
\end{pmatrix}. \qquad (3.41)
$$

Its characteristic polynomial $P(\rho)$ is a third order polynomial

$$
P(\rho) = \rho^3 + a_1 \rho^2 + a_2 \rho + a_3,
$$

wherein, using $\theta_{23} = \theta_{32} = 0$,

$$
\begin{aligned}
a_1 &= -trace(J(\psi^*, \chi^*, \beta^*)) = -\theta_{11} - \theta_{22} - \theta_{33}, \\
a_3 &= \det(J(\psi^*, \chi^*, \beta^*)) = -\theta_{31}\theta_{22}\theta_{13} + \theta_{33}[\theta_{21}\theta_{12} - \theta_{11}\theta_{22}], \\
a_2 &= -\theta_{13}\theta_{31} + \theta_{33}\theta_{11} + \theta_{33}\theta_{22} + \theta_{22}\theta_{11} - \theta_{21}\theta_{12}.
\end{aligned}
$$

Routh–Hurwicz criterion asserts the eigenvalues of $J(\psi^*, \chi^*, \beta^*)$ are negative or have negative real parts if and only if one has

$$a_1 > 0, \ a_3 > 0, \ a_1 a_2 - a_3 > 0, \tag{3.42}$$

cf. [26, 51]. First $trace(J(\psi^*, \chi^*, \beta^*))$ is obviously negative so that $a_1 > 0$. Next $\det(J(\psi^*, \chi^*, \beta^*))$ reads

$$\sigma_{1m}(K_m - \beta^*)b_1\sigma_{m1}(K_1 - \psi^* - \chi^*)$$
$$-[\sigma_{1m}\psi^* + (\lambda_m + b_m)][b_1(\sigma_{m1}\beta^* + (\lambda_1 + b_1)) + \omega_1\lambda_1\sigma_{m1}\beta^*].$$

At equilibrium stationary states of system (3.17) satisfy

$$\sigma_{1m}(K_m - \beta^*) = (\lambda_m + b_m)\frac{\beta^*}{\psi^*}, \quad \sigma_{m1}(K_1 - \psi^* - \chi^*) = (\lambda_1 + b_1)\frac{\beta^*}{\psi^*}; \tag{3.43}$$

the first term in $\det(J(\psi^*, \chi^*, \beta^*))$ simplifies into $(\lambda_m + b_m)b_1(\lambda_1 + b_1)$. Hence $\det(J(\psi^*, \chi^*, \beta^*)) < 0$, this is $a_3 > 0$. Last, one can express $a_1 a_2 - a_3$ as

$$[-\theta_{11} - \theta_{22} + \sigma_{1m}\psi^* + (\lambda_m + b_m)]$$
$$\times [-\theta_{13}\theta_{31} + \theta_{33}\theta_{11} + \theta_{33}\theta_{22} + b_1(\sigma_{m1}\beta^* + (\lambda_1 + b_1)) + \omega_1\lambda_1\sigma_{m1}\beta^*]$$
$$-\theta_{31}\theta_{22}\theta_{13} - [\sigma_{1m}\psi^* + (\lambda_m + b_m)][b_1(\sigma_{m1}\beta^* + (\lambda_1 + b_1)) + \omega_1\lambda_1\sigma_{m1}\beta^*];$$

the last term cancels out while one may check, see (3.43), that

$$-\theta_{31}\theta_{13} + \theta_{11}\theta_{33} = -(\lambda_m + b_m)(\lambda_1 + b_1) + \theta_{11}\theta_{33} > 0.$$

As a consequence $a_1 a_2 - a_3 > 0$ and local stability holds.

End of the proofs of Lemmas 8 and 10

The end of the proof of Lemma 8 is derived upon setting $\mu_k = 0$ in the algebraic calculations following (3.45) below because $J(\psi^*, \chi^*, c^*) = \Theta^0$.

Assume conditions HYP. 5.10 and HYP. 5.11 hold. Linearizing MODEL (H1GH2,S, REDUCED) about the stationary state (ψ^*, χ^*, c^*) of (3.20) one gets

$$\begin{aligned}
\partial\psi/\partial t - d_1\Delta\psi &= \sigma_{g1}(K_1 - \psi^* - \chi^*)c - (\sigma_{11}\psi^* + \sigma_{g1}c^*)\chi \\
&\quad + [\sigma_{11}(K_1 - \psi^* - \chi^*) - (\lambda_1 + b_1) - (\sigma_{11}\psi^* + \sigma_{g1}c^*)]\psi, \\
\partial\chi/\partial t - d_1\Delta\chi &= +\omega_1\lambda_1\psi - b_1\chi, \\
\partial c/\partial t &= +\sigma_{1g}(1 - c^*)\psi - (\sigma_{1g}\psi^* + \delta)c,
\end{aligned} \tag{3.44}$$

for $(x, t) \in \Omega_1 \times (0, +\infty)$ with no flux boundary conditions:

$$d_1\partial\psi/\partial\eta_1 = d_1\partial\chi/\partial\eta_1 = 0, \ x \in \partial\Omega_1, \ t \geq 0.$$

Looking for solutions to this linear system of the form

$$\psi(x, t) = \sum_{k \geq 0} \psi_k(t)\phi_k(x), \ \chi(x, t) = \sum_{k \geq 0} \chi_k(t)\phi_k(x), \ c(x, t) = \sum_{k \geq 0} c_k(t)\phi_k(x),$$

see (3.40) for ϕ_k, and substituting these into (3.44) one gets an infinite set of linear systems of ODEs,

$$
\begin{aligned}
\psi_k' &= +\left[\sigma_{11}(K_1 - \psi^* - \chi^*) - (\lambda_1 + b_1 + \mu_k) - (\sigma_{11}\psi^* + \sigma_{g1}c^*)\right]\psi_k \\
&\quad -(\sigma_{11}\psi^* + \sigma_{g1}c^*)\chi_k + \sigma_{g1}(K_1 - \psi^* - \chi^*)c_k, \\
\chi_k' &= +\omega_1\lambda_1\psi_k - (b_1 + \mu_k)\chi_k, \\
c_k' &= +\sigma_{1g}(1 - c^*)\psi_k - (\sigma_{1g}\psi^* + \delta + \mu_k)c_k,
\end{aligned}
\tag{3.45}
$$

with $\mu_k \geq 0$ by (3.40). For each $k \geq 0$ (3.45) reads

$$
(\psi_k', \chi_k', c_k')^\top = \Theta^k (\psi_k, \chi_k, c_k)^\top,
$$

wherein each 3×3 matrix $\Theta^k = \Theta^0 + \mu_k I_3$ has a structure similar to the matrix in (3.41); I_3 is the 3×3 identity matrix.

We use the Routh–Hurwitz criterion to show eigenvalues of Θ^k are negative or have negative real parts, keeping notations from (3.42). First at equilibrium stationary solutions of (3.20) satisfy

$$
\sigma_{11}(K_1 - \psi^* - \chi^*) - (\lambda_1 + b_1) = -\sigma_{g1}\frac{c^*}{\psi^*}(K_1 - \psi^* + \chi^*) < 0,
\tag{3.46}
$$

see the first equation in (3.20), so that $\sigma_{11}(K_1 - \psi^* - \chi^*) < (\lambda_1 + b_1)$. As a consequence $\theta_{11}^k < 0$ and $trace(\Theta^k) < 0$ or $a_1 > 0$. Next $\det(\Theta^k)$ reads

$$
\begin{aligned}
&\sigma_{1g}(1 - c^*)(b_1 + \mu_k)\sigma_{g1}(K_1 - \psi^* - \chi^*) \\
+&(\sigma_{1g}\psi^* + \delta + \mu_k)\{\sigma_{11}(K_1 - \psi^* - \chi^*) - (\lambda_1 + b_1 + \mu_k) - (\sigma_{11}\psi^* + \sigma_{g1}c^*)\}\{b_1 + \mu_k\} \\
-&(\sigma_{1g}\psi^* + \delta + \mu_k)\omega_1\lambda_1\{\sigma_{11}\psi^* + \sigma_{g1}c^*\};
\end{aligned}
$$

at equilibrium stationary solutions satisfy $\sigma_{1g}(1 - c^*) = \delta\frac{c^*}{\psi^*}$, see the third equation in (3.20). From this and (3.46) we see the first term in $\det(\Theta^k)$ simplifies into $-\delta(b_1 + \mu_k)[\sigma_{11}(K_1 - \psi^* - \chi^*) - (\lambda_1 + b_1)]$. Hence $\det(\Theta^k) < 0$ or $a_3 > 0$. Last, one can express $a_1a_2 - a_3$ as

$$
\begin{aligned}
&[-\theta_{11}^k - \theta_{22}^k + \sigma_{1g}\psi^* + \delta + \mu_k] \\
&\quad \times[-\theta_{13}^k\theta_{31}^k + \theta_{33}^k\theta_{11}^k + \theta_{33}^k\theta_{22}^k - (b_1 + \mu_k)\theta_{11}^k + \omega_1\lambda_1(\sigma_{11}\psi^* + \sigma_{g1}c^*)] \\
&\quad\quad -\theta_{31}^k\theta_{22}^k\theta_{13}^k - [\sigma_{1g}\psi^* + \delta + \mu_k)][-(b_1 + \mu_k)\theta_{11}^k + \omega_1\lambda_1(\sigma_{11}\psi^* + \sigma_{g1}c^*)]
\end{aligned}
$$

the last term cancels out while one has $-\theta_{31}^k\theta_{13}^k + \theta_{11}^k\theta_{33}^k > 0$, see (3.46). As a consequence $a_1a_2 - a_3 > 0$. The nontrivial stationary state (ψ^*, χ^*, c^*) is locally stable for (3.20) and linearly stable for (3.37).

Acknowledgement

This work is partially supported by a joint N.S.F.–C.N.R.S. Grant DMS 0089590 "Heterogeneous Reaction–Diffusion Advection Systems". The authors are deeply grateful to D. Pontier and J.-J. Morgan for invaluable discussions on biological considerations and mathematical analysis arguments related to the present work.

References

1. Abramson, G., Krenkre, V.M., Yates, T.L., Parmenter, R.R.: Travelling Waves of Infection in the Hantavirus Epidemics. Bull. Math. Biol., **65**, 519–534 (2003)
2. Acuna-Soto, R., Stahle, D.W., Cleaveland, M.K., Therell, M.D.: Megadrougth and Megadeath in 16th Century Mexico. Emerg. Infect. Dis., **8**, 360–362 (2002)
3. Ainseba, B., Fitzgibbon, W.E., Langlais, M., Morgan, J.J.: An Application of Homogenization Techniques to Population Dynamics Models. Commun. Pure Appl. Anal., **1**, 19–33 (2002)
4. Allen, L.J.S.: An Introduction to Stochastic Processes with Application to Biology. Prentice Hall, Upper Saddle River, N.J. (2003)
5. Anderson, R.M., Jackson, H.C., May, R.M., Smith, A.D.M.: Population Dynamics of Foxes Rabies in Europe. Nature, **289**, 765–770 (1981)
6. Anderson, R.M., May, R.M.: Population Biology of Infectious Diseases. Springer, Berlin Heidelberg New York (1982)
7. Bailey, N.T.J.: The Mathematical Theory of Infectious Diseases and its Applications, 2nd edition. Hafner Press, New York (1975)
8. Bendahmane, M., Langlais, M., Saad, M.: On Some Anisotropic Reaction–Diffusion Systems with L^1-Data Modeling the Propagation of an Epidemic Disease. Nonlinear Anal., Ser. A, Theory Methods, **54**, 617–636 (2003)
9. Begon, M., Bennett, M., Bowers, R.G., French, N.P., Hazel, S.M., Turner, J.: A Clarification of Transmission Terms in Host-Microparasite Models; Numbers, Densities and Areas. Epidemiol. Infect., **129**, 147–153 (2002)
10. Berestycki, H., Hamel, F., Roques, L.: Analysis of a Periodically Fragmented Environment Model: I. Influence of Periodic Heterogeneous Environment on Species Persistence. J. Math. Biol., **51**, 75–113 (2005)
11. Bernoulli, D.: Essai d'une nouvelle analyse de la mortalité causée par la petite vérole et des avantages de l'inoculation pour la prévenir. Mém. Math. Phys. Acad. Roy. Sci. Paris, 1–45 (1760)
12. Berthier, K., Langlais, M., Auger, P., Pontier, D.: Dynamics of Feline Virus with Two Transmission Modes Within Exponentially Growing Host Populations. Proc. Roy. Soc. Lond., B, **267**, 2049–2056 (2000)
13. Brauer, F., Castillo-Chavez, C.: Mathematical Models in Population Biology and Epidemiology. Springer, Berlin Heidelberg New York (2001)
14. Busenberg, S., Cooke, K.C.: Vertically Transmitted Diseases, Biomathematics Volume 23. Springer, Berlin Heidelberg New York (1993)
15. Cantrell, R.S., Cosner C.: Spatial Ecology Via Reaction Equations. Wiley, Chichester (2003)
16. Capasso, V.: Mathematical Structures of Epidemic Systems. Lecture Notes in Biomathematics Volume 97. Springer, Berlin Heidelberg New York (1993)
17. Caswell, H.: Matrix Population Models 2nd edition. Sinauer Associates Inc., Sunderland, Massachusetts (2001)
18. Cazenave, T., Haraux A.: An Introduction to Semilnear Evolution Equations. Oxford Lecture Series in Mathematics and its Applications, Oxford University Press, Oxford (1998)
19. Courchamp, F., Clutton-Brock, T., Grenfell, B.: Inverse Density Dependence and the Allee Effect. TREE, **14**, 405–410 (1999)
20. Cushing J.: An introduction to Structured Population Dynamics. CBMS–NSF Regional Conference Series in Applied Mathematics, SIAM, Philadelphia (1998)

21. Daley, D.J., Gani, J.: Epidemic Modelling: An Introduction. Cambridge Studies in Mathematical Biology, Cambridge University Press, Cambridge (1999)
22. Diekmann, O., Heesterbeek, J.A.P., Metz, J.A.J.: On the Definition and the Computation of the Basic Reproduction Ration R_0 in Models for Infectious Diseases in Heterogeneous Population. J. Math. Biol., **28**, 365–382 (1990)
23. Diekmann, O., De Jong, M.C.M., De Koeijer, A.A., Reijnders, P.: The Force of Infection in Populations of Varying Size: A Modeling Problem. J. Biol. Syst., **3**, 519–529 (1995)
24. Diekmann, O., Heesterbeck, J.A.P.: Mathematical Epidemiology of Infectious Diseases, Mathematical and Computational Biology. Wiley, Chichester (2000)
25. Ducrot, A., Langlais, M.: Travelling waves in invasion processes with pathogens. Mathematical Models and Methods in Applied Sciences, **18**, 1–15 (2008)
26. Edelstein-Keshet, L.: Mathematical Models In Biology. The Random House Birkhäuser Mathematical Series, New York (1988)
27. Fitzgibbon, W.E., Langlais, M.: Weakly Coupled Hyperbolic Systems Modeling the Circulation of Infectious Disease in Structured Populations. Math. Biosci., **165**, 79–95 (2000)
28. Fitzgibbon, W.E., Hollis, S., Morgan, S.: Steady State Solutions for Balanced Reaction Diffusion Systems on Heterogeneous Domains. Differ. Integral Equ., **12**, 225–241 (1999)
29. Fitzgibbon, W.E., Langlais, M., Morgan, J.J.: A Mathematical Model for the Spread of Feline Leukemia Virus (FeLV) through a Highly Heterogeneous Spatial Domain. SIAM, J. Math. Anal., **33**, 570–588 (2001)
30. Fitzgibbon, W.E., Langlais, M., Morgan, J.J.: A Reaction–Diffusion System Modeling Direct and Indirect Transmission of a Disease. DCDS B **4**, 893–910 (2004)
31. Fitzgibbon, W.E., Langlais, M., Morgan, J.J.: A Reaction Diffusion System on Non-Coincident Domains Modeling the Circulation of a Disease Between Two Host Populations. Differ. Integral Equ., **17**, 781–802 (2004)
32. Fitzgibbon, W.E., Langlais, M., Marpeau, F., Morgan, J.J.: Modeling the Circulation of a Disease Between Two Host Populations on Non Coincident Spatial Domains. Biol. Invasions, **7**, 863–875 (2005)
33. Fitzgibbon, W.E., Langlais, M., Morgan, J.J.: A Mathematical Model for Indirectly Transmitted Diseases. Math. Biosci., **206**, 233–248 (2007)
34. Fitzgibbon, W.E., Morgan, J.J.: Diffractive Diffusion Systems with Locally Defined Reactions, Evolution Equations. Ed. by Goldstein G. et al., M. Dekker, New York, 177–186 (1994)
35. Fouchet, D., Marchandeau, S., Langlais M., Pontier, D.: Waning of Maternal Immunity and the Impact of Diseases: The Example of Myxomatosis in Natural Rabbit Population. J. Theor. Biol., **242**, 81–89 (2006)
36. Fromont, E., Pontier, D., Langlais, M.: Dynamics of a Feline Retrovirus (FeLV) in Hosts Populations with Variable Structure. Proc. Roy. Soc. Lond., B, **265**, 1097–1104 (1998)
37. Hale, J.: Asymptotic Behavior of Dissiptive Systems. Mathematical Surveys and Monographs **25**, AMS Providence, RI (1988)
38. Hale, J.K., Koçak, H.: Dynamics and Bifurcations. Springer, Berlin Heidelberg New York (1991)
39. Hartman, P.: Ordinary Differential Equations. Wiley, New York (1964)
40. Hetchcote, H.W.: The Mathematics of Infectious Diseases, SIAM Rev., **42**, 599–653 (2000)

41. Hilker, F.M., Lewis, M.A., Seno, H., Langlais, M., Malchow, H.: Pathogens Can Slow Down or Reverse Invasion Fronts of their Hosts. Biol. Invasions, **7**, 817–832 (2005)
42. Hirsch, M., Smale, S.: Differential Equations, Dynamical Systems and Linear Algebra. Springer, Berlin Heidelberg New York (1974)
43. Hoppensteadt, F.C.: Mathematical Theories of Populations: Demographics, Genetics and Epidemics. CBMS, vol 20, SIAM, Philadelphia (1975)
44. Horton, P.: Global Existence of Solutions to Reaction Diffusion Systems Heterogeneous Domains, Dissertation, Texas A & M University, College Station (1998)
45. Iannelli, M.: Mathematical theory of Age-Structured Population Dynamics. Applied Mathematics Monographs no. 7, C.N.R. Pisa (1994)
46. Kermack, W.O., Mac Kendrick, A.G.: Contributions to the mathematical theory of epidemics, part I, Proc. Roy. Soc. Lond., A, **115**, 700–721 (1927). Reprinted with parts II and III in Bull. Math. Biol., **53**, 33–118 (1991)
47. Kesavan, S.: Topics in Functional Analysis and Applications, Wiley, New York (1989)
48. Ladyzenskaja, O.A., Solonnikov, V.A., Ural'ceva, N.N.: Linear and Quasilinear Equations of Parabolic Type. Translation AMS **23**, Providence, RI (1968)
49. Langlais, M., Phillips, D.: Stabilization of Solutions of Nonlinear Evolution Equations. Nonlinear Anal. T.M.A., **9**, 321–333 (1985)
50. Langlais, M., Latu, G., Roman, J., Silan, P.: Performance Analysis and qualitative Results of an Efficient Parallel Stochastic Simulator for a Marine Host–Parasite system. Concurrency Comput.: pract. exp., **15**, 1133–1150 (2003)
51. Murray, J.D.: Mathematical Biology I: An introduction, 3rd edition. Springer, Berlin Heidelberg New York (2003)
52. Naulin, J.M.: A Contribution of Sparse Matrices Tools to Matrix Population Model Analysis. Math. Biosci., **177–178**, 25–38 (2002)
53. Okubo, A, Levin, S.: Difusion and ecological problems: Modern perspectives, 2nd edn. Springer, New York (2001)
54. Olsson, G.E., White, N., Ahlm, C., Elgh, F., Verlemyr, A.C., Juto, P.: Demographic Factors Associated with Hantavirus Infection in Bank Voles (*Clethrionomys glareolus*). Emerg. Infect. Dis., **8**, 924–929 (2002)
55. Rutledge, C.R., Day, J.F., Stark, L.M., Tabachnick, W.J.: West-Nile Virus Infection Rates in Culex nigricalpus (Diptera: Culicidae) do not Reflect Transmission Rates in Florida. J. Med. Entomol., **40**, 253–258 (2003)
56. Sauvage, F., Langlais, M., Yoccoz, N.G., Pontier, D.: Modelling Hantavirus in Cyclic Bank Voles: The Role of Indirect Transmission on Virus Persistence. J. Anim. Ecol., **72**, 1–13 (2003)
57. Sauvage, F., Langlais, M., Yoccoz, N-G., Pontier, D.: Predicting the Emergence of Human Hantavirus Disease Using a Combination of Viral Dynamics and Rodent Demographic Patterns. Epidemiol. Infect., **135**, 46–56 (2007)
58. Schmaljohn, C., Hjelle, B.: Hantaviruses: A Global Disease Problem. Emerg. Infect. Dis., **3**, 95–104 (1997)
59. Schmitz, O.J., Nudds, T.D.: Parasite-Mediated Competition in Deer and Moose: How Strong is the Effect of Meningeal Worm on Moose? Ecol. Appl., **4**, 91–103 (1994)
60. Seftel, Z.: Estimates in L_q of Solutions of Elliptic Equations with Discontinuous Coefficients and Satisfying General Boundary Conditions and Conjugacy Conditions. Soviet Math. Doklady, **4**, 321–324 (1963)

61. Shaman, J., Day, J.F., Stieglitz, M.: Drought-Induced Amplification of Saint Louis Encephalitis Virus, Florida. Emerg. Infect. Dis., **8**, 575–580 (2002)
62. Shigesada, N., Kawasaki. K.: Biological Invasions: Theory and Practice, Oxford University Press, Oxford (1997)
63. Stewart, H.: Generation of Analytic Semigroups by Strongly Elliptic Operators. Trans. A.M.S., **199**, 141–162 (1974)
64. Stewart, H.: Spectral Theory of Heterogeneous Diffusion Systems. J. Math. Anal. Appl., **54**, 59–78 (1976)
65. Stewart, H.: Generation of Analytic Semigroups by Strongly El liptic Operators Under General Boundary Conditions, Trans. A.M.S., **259**, 299–310 (1980)
66. Thieme, H.R.: Mathematics in Population Biology. Princeton University Press, Princeton (2003)
67. Tran, A., Gardon, J., Weber, S., Polidori, L.: Mapping Disease Incidence in Suburban Areas Using Remotely Sensed Data. Am. J. Epidemiol, **252**, 662–668 (2004)
68. Tran, A., Deparis, X., Dussart, P., Morvan, J., Rabarison, P., Polidori, L., Gardon, J.: Dengue Spatial and Temporal Patterns, French Guiana, 2001. Emerg. Infect. Dis., **10**, 615–621 (2004)
69. Webb, G.: Theory of Nonlinear Age-Dependent Population Dynamics. Marcel Dekker, New York (1985)
70. Wolf, C., Sauvage, F., Pontier, D., Langlais, M.: A multi-Patch Model with Periodic Demography for a Bank Vole – Hantavirus System with Variable Maturation Rate. Math. Popul. Stud., **13**, 153–177 (2006)
71. Wolf, C.: Modelling and Mathematical Analysis of the Propagation of a Microparasite in a Structured Population in Heterogeneous Environment (in French). Ph.D Thesis, Bordeaux 1 University, Bordeaux (2005)
72. Yoccoz, N.G., Hansson, L., Ims, R.A.: Geographical Differences in Size, Reproduction and Behaviour of Bankvoles in Relation to Density Variations. Pol J. Ecol., **48**, 63–72 (2000)

4

Spatiotemporal Patterns of Disease Spread: Interaction of Physiological Structure, Spatial Movements, Disease Progression and Human Intervention

S.A. Gourley[1], R. Liu[2], and J. Wu[3]

[1] Department of Mathematics, University of Surrey, Guildford, Surrey,
GU2 7XH, UK
s.gourley@surrey.ac.uk
[2] Department of Mathematics, Purdue University, 150 N, University Street,
West Lafayette, IN 47907-2067, USA
rliu@math.purdue.edu
[3] Laboratory for Industrial and Applied Mathematics, Department
of Mathematics and Statistics, York University, Toronto, ON, Canada M3J 1P3
wujh@mathstat.yorku.ca

Summary. In this article we review some recent literature on the mathematical modelling of vector-borne diseases with special reference to West Nile virus and with particular focus on the role of the developmental stages of hosts in determining the transmission dynamics, the effectiveness of different approaches to controlling the vector and the spatial spread of an epidemic. A possible model incorporating the developmental stages of avian hosts is discussed which consists of equations for infective and susceptible juvenile and adult hosts and infected adult vectors. Conditions for the system to evolve to the disease free state are presented. These elucidate the role of, for example, the various death rates involved. We also review a mathematical model which incorporates culling the vector at either the larval or the adult stage and the effectiveness of the two approaches is compared. Conditions are given that are sufficient for eradication and this leads insights into the required minimum frequency of culling. Very infrequent culling is no better than no culling at all and can actually increase the time average of the number of infected vectors. We also review a reaction-diffusion extension of the model which can be used to estimate the speed at which an epidemic moves through space. Finally, we review some recent work on the use of patch models of a West Nile virus epidemic. These models are arguably easier to relate to surveillance data which is organised according to administrative regions or landscape. The patch model is used to study the situation when the dispersal of birds is not symmetric.

4.1 Introduction

There are many factors contributing to the complicated and interesting spatiotemporal spread patterns of diseases. In this article we focus on two major factors: the demographic and disease ages and the spatial movement of the disease hosts. These factors in isolation have been intensively studied using various models, but their correlation and interaction require extreme care in modeling and analysis and can lead to models with different levels of complexity.

Demographic and disease ages and spatial movement interweave in many ways, a few of which will be addressed in this article. In particular, we observe that:

(i) The spatial movement patterns of species involved in the disease transmission cycle depend on the demographic age, an example being the much higher mobility of adult birds than the nestlings and young birds in a vector-borne disease in which birds are reservoirs.

(ii) Spatial movement is influenced by the environment including the weather and therefore is not completely random but is asymmetric in terms of direction and the degree of the environmental influence on the spatial movement may vary according to the maturation status of the individuals.

(iii) Disease management measures are often age-dependent and their effectiveness may also be dependent on the mobility status of the species involved, such as larvicides and insecticide sprays for mosquito-borne diseases used to control the mosquito population at different levels of their maturity.

(iv) The spatial movement patterns of disease hosts may depend on the disease status, for example disease-induced random movement of the infected in a population in which the susceptibles are territorial.

This article aims to address some of the above issues and the relevant modeling and analysis techniques in the case of the invasion and spread of West Nile virus in North America. Models to be formulated and discussed include patch models with both long-range and short-range dispersal, delay differential systems, nonlocal delayed reaction–diffusion equations and impulsive differential equations. We shall review some of the mathematical results and numerical simulations describing the spatiotemporal patterns of disease spread whose transmission dynamics is modeled by the aforementioned nonlinear dynamical systems.

Spatial movement and reaction time lags are two important features in many ecological systems. Their interaction is one of the many factors for possible complicated spatiotemporal patterns in a single species population without an external time-dependent forcing term. Modeling this interaction is nevertheless a highly nontrivial task, and recent progress indicates diffusive (partial or lattice) systems with nonlocal and delayed reaction nonlinearities arise very naturally. Such systems were investigated in the earlier works of

Yamada [36], Pozio [19, 20] and Redlinger [21, 22]. The modeling and analysis effort in the work by Britton [3], Gourley and Britton [4] and Smith and Thieme [27] marked the beginning of the systematic study of a new class of nonlinear dynamical systems directly motivated by consideration of biological realities. See, for example, the two recent review articles [5, 6].

This new class of nonlinear dynamical systems can be derived from the classical structured population model involving maturation dependent spatial diffusion rates and nonlinear birth and natural maturation processes. More specifically, if we use $u(t, x)$ to denote the total number of matured individuals in a single species population, and if we assume the maturation time is a fixed constant τ, then we have

$$\frac{\partial}{\partial t} u(t, x) = D \frac{\partial^2}{\partial x^2} u(t, x) - du(t, x) + j(t, \tau, x), \tag{4.1}$$

where D and d are the diffusion and death rates of the adult population (assumed constants here), and $j(t, \tau, x)$ is the maturation rate at (t, x) which is basically the birth rate at the earlier time $t - \tau$, corrected for juvenile mortality, for individuals that are at position x upon maturation (these individuals having been born at various other spatial locations). This maturation rate is thus regulated by the birth process and the dynamics of the individual during the maturation phase. In the work of So, Wu and Zou [28], this is derived from the structured population model

$$\left(\frac{\partial}{\partial t} + \frac{\partial}{\partial a} \right) j(t, a, x) = D_I \frac{\partial^2}{\partial x^2} j(t, a, x) - d_I j(t, a, x) \tag{4.2}$$

for the density $j(t, a, x)$ of the immature individual with $a \in (0, \tau]$ as the variable for the demographic age, subject to some (spatial) boundary conditions (if the space is bounded) and the following boundary condition at age $a = 0$:

$$j(t, 0, x) = b(u(t, x)), \tag{4.3}$$

where $b(\cdot)$ is the birth rate function, which is assumed to depend on the total matured population at (t, x), D_I and d_I are the diffusion and death rates of the immature individual (these rates are allowed to be functions of a in [28]). The maturation rate $j(t, \tau, x)$ can be obtained by solving the linear hyperbolic–parabolic equation (4.2) subject to the boundary condition (4.3). In the case of an unbounded one-dimensional space, we have

$$j(t, \tau, x) = e^{-d_I \tau} \int_{\mathbf{R}} b(u(t - \tau, y)) f(x - y) \, dy. \tag{4.4}$$

In other words, the maturation rate at time t and spatial location x is the sum (integral) over all possible birth locations, of the birth rate at time $t - \tau$ at location y, times the probability $f(x - y)$ of an individual born at position y at time $t - \tau$ moving to position x at time t, times the probability $e^{-d_I \tau}$ of surviving the entire maturation phase. Incorporating (4.4) into (4.1) furnishes a reaction–diffusion equation with a nonlocal delayed nonlinearity, namely

$$\frac{\partial}{\partial t}u(t,x) = D\frac{\partial^2}{\partial x^2}u(t,x) - du(t,x) + e^{-d_I\tau}\int_{\mathbf{R}} b(u(t-\tau,y))f(x-y)\,dy. \quad (4.5)$$

There has been some rapid development towards a qualitative theory for the asymptotic behaviors of solutions to (4.5) with various types of assumptions on the birth functions. Notably, in comparison with analogue reaction–diffusion equations without delay, there are more prototypes than the so-called monostable and bistable cases. See [6] for more details.

The analytic form for f is given in [28]. It is possible to obtain such an analytic form here since the dynamical process during the maturation phase is governed by a linear hyperbolic–parabolic equation with time-independent constant coefficients. Such a possibility disappears in an ecological system consisting of multiple species with age or stage-dependent diffusion rates when these species interact during their maturation phases. This is also the case for the spread of a disease (even if its main carrier involves only a single species), since the model describing the infection process must involve the transfers of individuals from one compartment to another and some of these transfers such as the force of infection from the susceptible compartment to the infective compartment are nonlinear. This issue substantially complicates the modeling and may in fact make the derivation of a system of reaction–diffusion equations analogous to (4.5) impossible, although in fact an approximate system can sometimes be derived that is valid in the vicinity of an equilibrium of interest such as a disease-free equilibrium. These issues were very much apparent in the recent works of Ou and Wu [18] on rabies, and Gourley, Liu and Wu [7] on vector borne diseases such as West Nile virus.

4.2 Vector-Borne Diseases with Structured Population: Implications of Nonlinear Dynamics During the Maturation Phase

Vector borne diseases such as malaria, dengue fever and West Nile virus (WNV) are infectious diseases that are carried by insects from one host to another. In many of these diseases it is the mosquito that carries the virus but ticks, fleas and midges can also be responsible. The diseases can be spread to humans, birds and other animals.

This article will focus mainly on the mathematical modeling of West Nile virus but there are many other insect-borne diseases which currently constitute significant public health issues worldwide, and the mathematical modeling of some of these raises issues not covered in this article at all. Malaria is undoubtedly the best known vector-borne disease of all, with hundreds of millions of cases each year, and control of it has been complicated by widespread drug resistance of the parasites and insecticide resistance. Another of the more serious mosquito-borne diseases is dengue fever, epidemics of which are becoming larger and more frequent. Dengue has a global distribution comparable to that of malaria and up to 100 million cases per year, with a fatality rate in most countries of about 5% (though with proper treatment this can be reduced to below 1%). Dengue is caused by one of four closely related virus serotypes. Infection with one of these confers immunity for life but only to that particular serotype. Thus in principle a person can experience four dengue infections. Airline travel makes it possible for the viruses to disperse rapidly (in fact air travel to virtually any place in the world can take place nowadays in less time than the incubation period of most infectious diseases). In former times dengue could only be spread between continents if viruses and mosquito vectors could survive the long sea journeys, leading to long intervals between epidemics. There is currently no vaccine for dengue. Yellow fever is another example of a mosquito borne disease, endemic in South America and parts of Africa. A vaccine has been available for it for several decades.

Ticks can act as the vector for certain diseases, notably Lyme disease. Bluetongue is a midge-borne disease that can affect all ruminants with sheep being the most severely affected. Cattle act as an important reservoir species. Bluetongue appears to have originated in South Africa but has been reported in many European countries (including as far north as the Netherlands and Belgium), in the USA and in British Columbia and other places. Northern Europe is currently at the limits of the climatic conditions favorable for the spread of the disease (13–$35°C$) but climate change has the potential to change this. Since midges can be carried considerable distances by wind, it has been speculated that, in areas where bluetongue occurs only sporadically, it originates from the carriage by wind of midges from distant endemic areas.

It is widely accepted that climate change has considerable potential implications for the world wide distribution of vector-borne diseases in general, and that global warming is likely to create suitable new vector habitats. Other issues potentially contributing to the resurgence of some vector borne diseases are the excessive use of insecticide sprays to kill adult mosquitoes, the problem of insecticide and drug resistance and the general decline of the worldwide infrastructure for the surveillance, prevention and control of vector-borne diseases because the public health threat was perceived to have dropped due to previous control programs (for example, malaria had been nearly eliminated in Sri-Lanka in the 1960s).

Much has been done in terms of the mathematical modeling and analysis of the transmission dynamics and spatial spread of vector-borne diseases. However, one important biological aspect of the hosts, the stage structure, seems to have received little attention although structured population models have been intensively studied. In the context of population dynamics and spatial ecology the interaction of stage-structure with spatial dispersal has been receiving considerable attention as discussed in the last section.

The developmental stages of hosts have an important impact on the transmission dynamics of vector-borne diseases. In the case of West Nile virus the transmission cycle involves both mosquitoes and birds (the crow, jay and raven species being particularly important). Nestling crows are crows that have hatched but are helpless and stay in the nest, receiving more or less continuous care from the mother for up to two weeks and less continuous care thereafter. Fledgling crows are old enough to have left the nest (they leave it after about five weeks) but they cannot fly very well. After three or four months these fledglings will be old enough to obtain all of their food by themselves. As these facts demonstrate, the maturation stages of adult birds, fledglings and nestlings are all very different from a biological and an epidemiological perspective and a realistic model needs to take these different stages into account. For example, in comparison with grown birds, the nestlings and fledglings have much higher disease induced death rate, much poorer ability to avoid being bitten by mosquitoes, and much less spatial mobility [1,15,24].

Recently, Gourley, Liu and Wu [7] developed a model for the evolution of a general vector-borne disease with special emphasis on the transmission dynamics and spatial spread of West Nile virus. They started with the classical McKendrick von-Foerster equations for an age-structured reservoir population (birds, in the case of WNV) divided into two epidemiological compartments of susceptible and infected (and infectious), coupled with a scalar delay differential equation for the adult vector (mosquito) population under the assumption that the total vector population is maintained at a constant level. As a result, they obtained a system of delay differential equations describing the interaction of five sub-populations, namely susceptible and infected adult and juvenile reservoirs and infected adult vectors, for a vector-borne disease with particular reference to West Nile virus.

4.2.1 Model Derivation and Biological Interpretation

To discuss the model in [7] we can imagine a more general mosquito-borne disease with similar characteristics to WNV. We will also refer to the reservoir as the host, and assume that the host population is age-structured. We start with a simple division of the host population as susceptible hosts $s(t,a)$ and infected hosts $i(t,a)$ at time t and age a. These host populations are assumed to evolve according to the McKendrick von-Foerster equations for an age-structured population:

$$\frac{\partial s}{\partial t} + \frac{\partial s}{\partial a} = -d_s(a)s(t,a) - \beta(a)s(t,a)m_i(t), \qquad (4.6)$$

and

$$\frac{\partial i}{\partial t} + \frac{\partial i}{\partial a} = -d_i(a)i(t,a) + \beta(a)s(t,a)m_i(t), \qquad (4.7)$$

where $m_i(t)$ is the number of infected adult mosquitoes, the functions $d_s(a)$ and $d_i(a)$ are the age-dependent death rates of susceptible and infected hosts.

A host becomes infected when it is bitten by an infected mosquito. Here and in what follows, we use mass action though other incidences can be used.

The host population is split into juveniles and adults, defined respectively as those of age less than some number τ, and those of age greater than τ. We assume:

$$d_s(a) = \begin{cases} d_{sj} \ a < \tau \\ d_{sa} \ a > \tau, \end{cases} \qquad d_i(a) = \begin{cases} d_{ij} \ a < \tau \\ d_{ia} \ a > \tau, \end{cases} \tag{4.8}$$

and

$$\beta(a) = \begin{cases} \beta_j \ a < \tau \\ \beta_a \ a > \tau. \end{cases} \tag{4.9}$$

The subscripts in these quantities refer to disease and juvenile/adult status; thus for example the per capita death rates for susceptible juveniles and infected adults would be d_{sj} and d_{ia} respectively. The above choices enable us to formulate a closed system of delay differential equations involving only the total numbers of hosts classified as adult susceptibles, adult infected, juvenile susceptibles and juvenile infected. These total numbers are given respectively, using self explanatory notation, by

$$A_s(t) = \int_\tau^\infty s(t,a)\, da, \qquad A_i(t) = \int_\tau^\infty i(t,a)\, da, \qquad J_s(t) = \int_0^\tau s(t,a)\, da,$$

$$J_i(t) = \int_0^\tau i(t,a)\, da. \tag{4.10}$$

We consider the situation when the number of offsprings produced by infected adult hosts can be ignored, and the offsprings are therefore always susceptible. We also assume:

$$s(t,0) = b(A_s(t)), \qquad i(t,0) = 0, \tag{4.11}$$

where $b(\cdot)$ is the birth rate function for hosts.

Differentiating the expression for $A_s(t)$ and using integration along characteristics of equation (4.6) to find $s(t,\tau)$, we find that

$$\frac{dA_s}{dt} = b(A_s(t-\tau)) \exp\left(-\int_{t-\tau}^t (d_{sj} + \beta_j m_i(u))\, du\right) - d_{sa} A_s(t) - \beta_a m_i(t) A_s(t). \tag{4.12}$$

In a similar way,

$$\frac{dJ_s}{dt} = b(A_s(t)) - b(A_s(t-\tau)) \exp\left(-\int_{t-\tau}^t (d_{sj} + \beta_j m_i(u))\, du\right)$$
$$- d_{sj} J_s(t) - \beta_j m_i(t) J_s(t), \tag{4.13}$$

$$\frac{dA_i(t)}{dt} = -d_{ia} A_i(t) + \beta_a m_i(t) A_s(t)$$
$$+ \beta_j b(A_s(t-\tau)) \int_{t-\tau}^t m_i(\xi) e^{-d_{ij}(t-\xi)} \exp\left(-\int_{t-\tau}^\xi (d_{sj} + \beta_j m_i(v))\, dv\right) d\xi, \tag{4.14}$$

$$\frac{dJ_i(t)}{dt} = -d_{ij}J_i(t) + \beta_j m_i(t)J_s(t)$$

$$- \beta_j b(A_s(t-\tau)) \int_{t-\tau}^{t} m_i(\xi) e^{-d_{ij}(t-\xi)} \exp\left(-\int_{t-\tau}^{\xi} (d_{sj} + \beta_j m_i(v))\, dv\right) d\xi.$$

$$(4.15)$$

Let $m_T(t)$ be the total number of (adult) mosquitoes, divided into infected mosquitoes $m_i(t)$ and susceptible mosquitoes $m_T(t) - m_i(t)$. Death and reproductive activity for mosquitoes is assumed not to depend on whether they are carrying the disease or not, and so the total number of adult mosquitoes is assumed to obey

$$\frac{dm_T(t)}{dt} = e^{-d_l \sigma} B(m_T(t-\sigma)) - d_m m_T(t), \qquad (4.16)$$

where d_l and d_m denote the death rates of larval and adult mosquitoes respectively and σ is the length of the larval phase from egg to adult. It is possible but unnecessary to write down a differential equation for larval mosquitoes. Infected adult mosquitoes $m_i(t)$ are assumed to obey

$$\frac{dm_i(t)}{dt} = -d_m m_i(t) + \beta_m(m_T(t) - m_i(t))(J_i(t) + \alpha A_i(t)). \qquad (4.17)$$

Thus, the rate at which mosquitoes become infected is given by mass action as the product of susceptible mosquitoes $m_T(t) - m_i(t)$ and infected hosts (birds) which may be either juvenile or adult. The factor α accounts for the possibility that juvenile and adult birds might not be equally vulnerable to being bitten.

Despite the complications of equations (4.12)–(4.14), the epidemiology and ecological interpretation of the integral term appearing in (4.14) and (4.14) seems to be straightforward. The last term in (4.14) tells us the rate at which infected immatures become infected adults having contracted the disease while immature. This term is the rate at which infected individuals pass through age τ. An individual that is of age τ at time t was born at time $t - \tau$. However, all individuals are born as susceptibles, which is why the birth rate $b(A_s(t-\tau))$ is involved. The individuals we are presently discussing have each acquired the infection at some stage during childhood, so assume a particular individual acquires it at a time $\xi \in (t - \tau, t)$. This particular individual remained susceptible from its birth at time $t - \tau$ until time ξ, and the probability of this happening is

$$\exp\left(-\int_{t-\tau}^{\xi} (d_{sj} + \beta_j m_i(v))\, dv\right).$$

The other exponential term, namely $e^{-d_{ij}(t-\xi)}$, in (4.14) is the probability that the individual will survive from becoming infected at time ξ until becoming an adult at time t. The product $\beta_j m_i(\xi)$ is the per capita conversion rate of susceptible juveniles to infected juveniles at time ξ, and ξ running from $t - \tau$ to t totals up the contributions from all possible times at which infected individuals passing into adulthood might have acquired the infection.

4.2.2 Disease Free Equilibrium

The following assumptions on the birth function $B(\cdot)$ for mosquitoes are minimal in the sense that they ensure that the mosquito population $m_T(t)$ does not tend to zero even in the absence of the disease:

$$
\left.
\begin{array}{l}
B(0) = 0, \ B(\cdot) \text{ is strictly monotonically increasing, there exists } m_T^* > 0 \\
\text{such that } e^{-d_l \sigma} B(m) > d_m m \text{ when } m < m_T^* \text{ and } e^{-d_l \sigma} B(m) < d_m m \text{ when} \\
m > m_T^*.
\end{array}
\right\}
$$
(4.18)

The quantity $m_T^* > 0$ in (4.18) is an equilibrium of (4.16), and $m_T(t) \to m_T^*$ as $t \to \infty$ provided $m_T(\theta) \geq 0$ and $m_T(\theta) \not\equiv 0$ on $\theta \in [-\sigma, 0]$. Accordingly, (4.17) is asymptotically autonomous and we may replace $m_T(t)$ by m_T^* in (4.17), thereby lowering the order of the system to be studied, which now consists of (4.12)–(4.14) together with

$$
\frac{dm_i(t)}{dt} = -d_m m_i(t) + \beta_m(m_T^* - m_i(t))(J_i(t) + \alpha A_i(t)).
$$
(4.19)

This system does not explicitly involve the delay σ, but this delay is still involved via the quantity m_T^*. The appropriate initial data for the abovementioned system is

$$
\begin{aligned}
& A_s(\theta) = A_s^0(\theta) \geq 0, \qquad \theta \in [-\tau, 0], \\
& m_i(\theta) = m_i^0(\theta) \in [0, m_T^*], \qquad \theta \in [-\tau, 0], \\
& A_i(0) = A_i^0(0) \geq 0, \\
& J_s(0) = \int_{-\tau}^0 b(A_s^0(\xi)) \exp\left(-\int_\xi^0 [d_{sj} + \beta_j m_i^0(u)] \, du \right) d\xi, \\
& J_i(0) = \int_{-\tau}^0 b(A_s^0(\xi)) \left\{ \int_\xi^0 \beta_j m_i^0(\eta) e^{d_{ij}\eta} e^{-\int_\xi^\eta [d_{sj} + \beta_j m_i^0(v)] \, dv} \, d\eta \right\} d\xi,
\end{aligned}
$$
(4.20)

where $A_s^0(\theta)$, $m_i^0(\theta)$ and $A_i^0(0)$ are prescribed. The last two conditions in (4.20) are compatibility conditions on the initial data. It is common to impose such conditions in stage structured population modeling. Ecologically, the need for such conditions is obvious. If the number of adult hosts on the initial interval $[-\tau, 0]$ is known, then so is the number of juvenile hosts at time $t = 0$, being given by the formulae above for $J_s(0)$ and $J_i(0)$. For example, $J_s(0)$ is the number of juvenile susceptibles at time $t = 0$. Each of these susceptibles was born at some time $\xi \in [-\tau, 0]$, hence the presence of the birth rate $b(A_s^0(\xi))$, and each has to have survived and remained susceptible until time 0, hence the exponential term which represents the probability of this actually happening. The interpretation of the expression for $J_i(0)$ is similar but more complicated. Of the infected juveniles $J_i(0)$ at time 0, each one was born at some time $\xi \in [-\tau, 0]$ as a susceptible, and each of these newborns at time ξ then became infected at some subsequent time $\eta \in [\xi, 0]$.

From a mathematical point of view, the compatibility conditions in (4.20) make it possible to show that solutions remain nonnegative. A theorem to this effect was proved in [7].

Notice that J_s, J_i can be solved as:

$$J_s(t) = \int_{t-\tau}^{t} b(A_s(\xi)) \exp\left(-\int_{\xi}^{t} [d_{sj} + \beta_j m_i(u)] \, du\right) d\xi, \qquad (4.21)$$

$$J_i(t) = \int_{t-\tau}^{t} b(A_s(\xi)) \left\{ \int_{\xi}^{t} \beta_j m_i(\eta) e^{-d_{ij}(t-\eta)} e^{-\int_{\xi}^{\eta} [d_{sj} + \beta_j m_i(v)] \, dv} \, d\eta \right\} d\xi, \qquad (4.22)$$

therefore, by (4.19), m_i satisfies the following equation:

$$\frac{dm_i(t)}{dt} = -d_m m_i(t) + \beta_m (m_T^* - m_i(t))$$

$$\times \left(\int_{t-\tau}^{t} b(A_s(\xi)) \left\{ \int_{\xi}^{t} \beta_j m_i(\eta) e^{-d_{ij}(t-\eta)} e^{-\int_{\xi}^{\eta} [d_{sj} + \beta_j m_i(v)] \, dv} \, d\eta \right\} d\xi + \alpha A_i(t) \right). \qquad (4.23)$$

In [7], sufficient conditions were obtained for the system to evolve to the disease free state (i.e. conditions that ensure A_i, J_i and m_i go to zero as $t \to \infty$). Since the differential equations (4.13) and (4.14) can be solved to give (4.21) and (4.22) respectively, it is sufficient to study the system consisting of equations (4.12), (4.14) and (4.23), with initial data taken from (4.20). These equations form a closed system for $A_s(t)$, $A_i(t)$ and $m_i(t)$. The key is to establish using these three equations a differential inequality for the variable $m_i(t)$ only, and to use this to find conditions which ensure that $m_i(t) \to 0$ as $t \to \infty$. Note that if $m_i(t) \to 0$ then, from (4.22) it follows immediately that $J_i(t) \to 0$ and, furthermore, (4.14) then becomes an asymptotically autonomous ODE from which we find that $A_i(t)$ tends to zero.

In [7] the following reasonable assumptions were made concerning the birth rate function b:

$$\left. \begin{array}{l} b(0) = 0, \ b(A) > 0 \text{ when } A > 0, \ b_{\sup} := \sup_{A \geq 0} b(A) < \infty, \text{ there exists} \\ A_s^* > 0 \text{ such that } e^{-d_{sj}\tau} b(A) > d_{sa} A \text{ when } A < A_s^* \text{ and } e^{-d_{sj}\tau} b(A) < \\ d_{sa} A \text{ when } A > A_s^*. \end{array} \right\} \qquad (4.24)$$

The quantity A_s^* in (4.24) is a nonzero equilibrium value for $A_s(t)$ in the case when the disease is absent. Assumptions (4.24) ensure that the population $A_s(t)$ of adult susceptible hosts will not go to zero even without the disease, otherwise the model is not interesting.

Under further conditions involving the following functions:

$$a_1(\epsilon) = d_m d_{ia} + d_m d_{ij} + d_{ia} d_{ij}$$
$$- \frac{\beta_m m_T^* b_{\sup} \beta_j}{d_{sj}} - \beta_m m_T^* \alpha \beta_a \left(\frac{b_{\sup} e^{-d_{sj}\tau}}{d_{sa}} + \epsilon \right)$$
$$- e^{-d_{sj}\tau} \left(\frac{1 - e^{-\tau(d_{ij} - d_m - d_{sj})}}{d_{ij} - d_m - d_{sj}} \right) \beta_m m_T^* \alpha \beta_j b_{\sup} \tag{4.25}$$

and

$$a_0(\epsilon) = d_m d_{ia} d_{ij} - \frac{d_{ia} \beta_m m_T^* b_{\sup} \beta_j}{d_{sj}} - d_{ij} \beta_m m_T^* \alpha \beta_a \left(\frac{b_{\sup} e^{-d_{sj}\tau}}{d_{sa}} + \epsilon \right)$$
$$- d_{ij} e^{-d_{sj}\tau} \left(\frac{1 - e^{-\tau(d_{ij} - d_m - d_{sj})}}{d_{ij} - d_m - d_{sj}} \right) \beta_m m_T^* \alpha \beta_j b_{\sup}, \tag{4.26}$$

with $b_{\sup} = \sup_{A \geq 0} b(A)$, the following theorem on disease eradication was proved in [7]:

Theorem 1 *Let (4.18) and (4.24) hold, and let $A_s(t)$, $A_i(t)$ and $m_i(t)$ satisfy (4.12), (4.14) and (4.23), with initial data taken from (4.20). Assume further that*

$$a_1(0) > 0, \quad a_0(0) > 0 \quad and \quad (d_m + d_{ia} + d_{ij})a_1(0) > a_0(0), \tag{4.27}$$

where the functions a_1, a_0 are defined by (4.25) and (4.26). Then $(A_i(t), m_i(t)) \to (0,0)$ as $t \to \infty$.

If (4.24) holds then the model (4.12)–(4.14) and (4.19) has a disease-free equilibrium (DFE), obtained by substituting $J_i = 0$, $A_i = 0$ and $m_i = 0$ into the right hand sides of those equations and setting them to zero, given by

$$E_0 = (A_s^*, J_s^*, 0, 0, 0), \tag{4.28}$$

where $A_s^* > 0$ and $J_s^* > 0$ are given by

$$\begin{cases} b(A_s^*)e^{-d_{sj}\tau} - d_{sa}A_s^* = 0, \\ J_s^* = \dfrac{b(A_s^*)}{d_{sj}}(1 - e^{-d_{sj}\tau}). \end{cases} \tag{4.29}$$

Using specific features of the spectral theory of positive semiflows, [7] obtained the following theorem which gives a sufficient condition for linear stability of the disease free state, that turns out also to be the necessary condition.

Theorem 2 *Let (4.18) and (4.24) hold and assume that $d_{sa} > |b'(A_s^*)|e^{-d_{sj}\tau}$ and that*

$$d_m > \beta_m m_T^* \left\{ \frac{b(A_s^*)\beta_j}{d_{ij} - d_{sj}} \left[\frac{1 - e^{-d_{sj}\tau}}{d_{sj}} - \frac{(1 - e^{-d_{ij}\tau})}{d_{ij}} \right] \right.$$
$$\left. + \frac{\alpha}{d_{ia}} \left[\beta_a A_s^* + \beta_j b(A_s^*)e^{-d_{sj}\tau} \frac{(1 - e^{-(d_{ij} - d_{sj})\tau})}{d_{ij} - d_{sj}} \right] \right\}. \tag{4.30}$$

Then the disease free equilibrium E_0 given by (4.28) is linearly asymptotically stable as a solution of the full model (4.12)–(4.14), (4.19).

4.2.3 Numerical Simulations

If we introduce the new variable W_1 defined by

$$W_1(t) = \int_{t-\tau}^t m_i(\xi) e^{-d_{ij}(t-\xi)} \exp\left(-\int_{t-\tau}^\xi (d_{sj} + \beta_j m_i(v))\, dv\right) d\xi,$$

then the model (4.12)–(4.14) and (4.19) can be recast in the form:

$$\frac{dJ_s(t)}{dt} = b(A_s(t)) - b(A_s(t-\tau)) e^{-d_{sj}\tau} e^{-\int_{t-\tau}^t \beta_j m_i(v) dv}$$
$$- d_{sj} J_s(t) - \beta_j m_i(t) J_s(t),$$

$$\frac{dA_s(t)}{dt} = b(A_s(t-\tau)) e^{-d_{sj}\tau} e^{-\int_{t-\tau}^t \beta_j m_i(v) dv} - d_{sa} A_s(t) - \beta_a m_i(t) A_s(t),$$

$$\frac{dJ_i(t)}{dt} = -d_{ij} J_i(t) + \beta_j m_i(t) J_s(t) - \beta_j b(A_s(t-\tau)) W_1(t),$$

$$\frac{dA_i(t)}{dt} = -d_{ia} A_i(t) + \beta_a m_i(t) A_s(t) + \beta_j b(A_s(t-\tau)) W_1(t),$$

$$\frac{dm_i(t)}{dt} = -d_m m_i(t) + (m_T(t) - m_i(t)) \beta_m (J_i(t) + \alpha A_i(t)),$$

$$\frac{dW_1(t)}{dt} = W_1(t)(d_{sj} - d_{ij} + \beta_j m_i(t-\tau)) + m_i(t) e^{-d_{sj}\tau} e^{-\int_{t-\tau}^t \beta_j m_i(v) dv}$$
$$- e^{-d_{ij}\tau} m_i(t-\tau). \tag{4.31}$$

In the simulations reported below, we take the birth function of mosquitoes and that of birds as

$$B(m_T) = b_m m_T e^{-a_m m_T}, \qquad b(A_s) = b_b A_s e^{-a_b A_s}, \tag{4.32}$$

respectively. These forms for the birth function have been used, for example, in the well-studied Nicholson's blowflies equation [9].

Various parameter values are given in Table 4.1, taken from [2, 14, 15, 35] with reference to West Nile virus. We took the initial conditions to be

$$A_s(t) = 700, \qquad M_I(t) = 0,$$

for $t \in [-\tau, 0]$ and $A_i(0) = 2$. This, together with the matching condition (4.20), gives $J_s(0) = 5470$ and $J_i(0) = 0$.

In Fig. 4.1, parameter values are $\beta_j = 4.7015 \times 10^{-6}$, $\beta_a = 2.3705 \times 10^{-6}$, $\beta_m = 1.1853 \times 10^{-6}$, $\alpha\beta_m = 4.3657 \times 10^{-7}$ and other parameters have the values shown in Table 4.1. In this case d_m is larger than the right hand side of (4.30) which equals 0.0382. One can check the condition (4.30) is satisfied and the infected populations go to zero. However, as we increase the contact rates, i.e., parameter values are $\beta_j = 6.7021 \times 10^{-6}$, $\beta_a = 3.3792 \times 10^{-6}$, $\beta_m =$

Table 4.1. Meaning of parameters of model (4.31)

Para.	Meaning of the parameter	Value
b_b	Maximum per capita daily birds production rate	0.5
$1/a_b$	Size of birds population at which the number of new born birds is maximized	1,000
b_m	Maximum per capita daily mosquito egg production rate	5
$1/a_m$	Size of mosquito population at which egg laying is maximized	10,000
d_{sj}	Mortality rate of uninfected juveniles (per day)	0.005
d_{ij}	Mortality rate of infected juveniles (per day)	0.05
d_{sa}	Mortality rate of uninfected adults (per day)	0.0025
d_{ia}	Mortality rate of infected adults (per day)	0.015
d_m	Mortality rate of mosquito (per day)	0.05
β_j	Contact rate between uninfected juvenile and infected mosquito	Variable
β_a	Contact rate between uninfected adult and infected mosquito	Variable
β_m	Contact rate between uninfected mosquito and infected juvenile	Variable
$\alpha\beta_m$	Contact rate between uninfected mosquito and infected juvenile	Variable
τ	Duration of more vulnerable period of bird (day)	160
σ	Maturation time of mosquito (day)	10
d_l	Mortality rate of larva mosquito (per day)	0.1

1.6896×10^{-6}, $\alpha\beta_m = 6.2234 \times 10^{-7}$ and other parameters have the values shown in Table 4.1, d_m is larger than the right hand side of (4.30) which equals 0.0777. In this case, the condition (4.30) fails and the disease sustains in the bird and mosquito population as shown in Fig. 4.2. If we continue to increase the contact rates: parameter values are $\beta_j = 1.8 \times 10^{-5}$, $\beta_a = 9.0756 \times 10^{-6}$, $\beta_m = 4.5378 \times 10^{-6}$, $\alpha\beta_m = 1.6714 \times 10^{-5}$ and other parameters have the values shown in Table 4.1, d_m is less than the right hand side of (4.30) which equals 0.5605. We eventually find oscillatory behaviors as shown in Fig. 4.3 suggesting the possibility of a Hopf bifurcation to periodic solutions.

4.3 Age-Structured Control Measure: Interaction of Delay, Impulse and Nonlinearity

Culling has been a common method for pest control and ecosystem management. Despite different formats such as shooting, trapping and crop spraying, culling often takes place at certain particular times only. These culling times are regulated by many factors including the maturation status of individuals of the species involved. For example, crop spraying may be exercised at certain times coinciding with critical stages in the insects' development.

Culling has also been a widely adopted tool to control vector-borne diseases in the hope that culling the vector at carefully chosen times may intervene the transmission cycle and reduce the infection. A specific example is larvicides

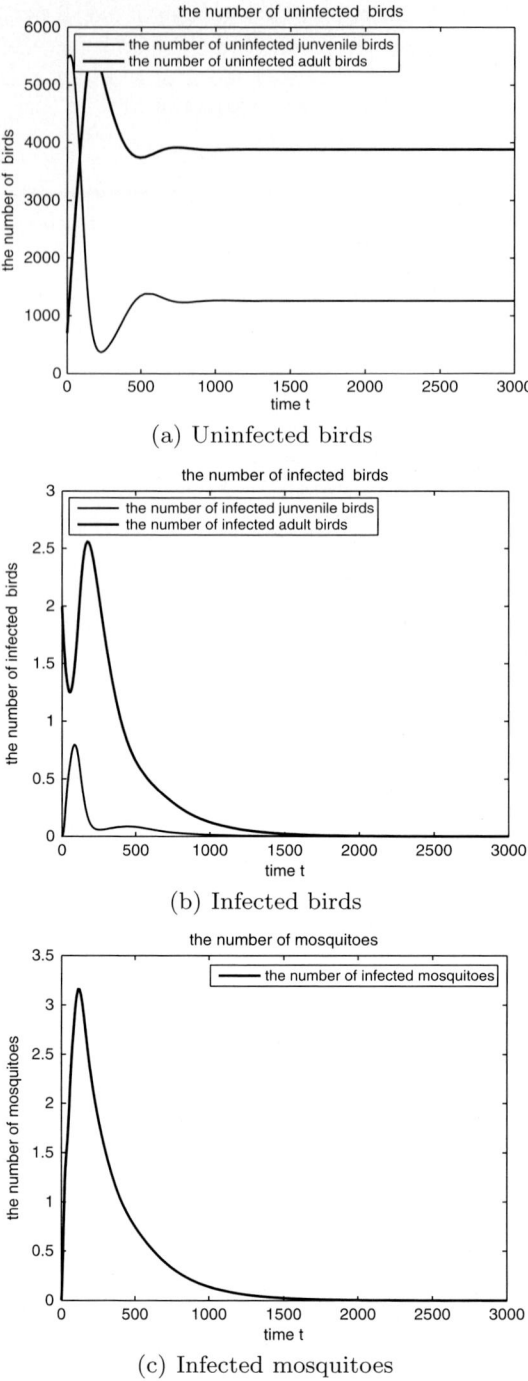

(a) Uninfected birds

(b) Infected birds

(c) Infected mosquitoes

Fig. 4.1. The DFE is stable

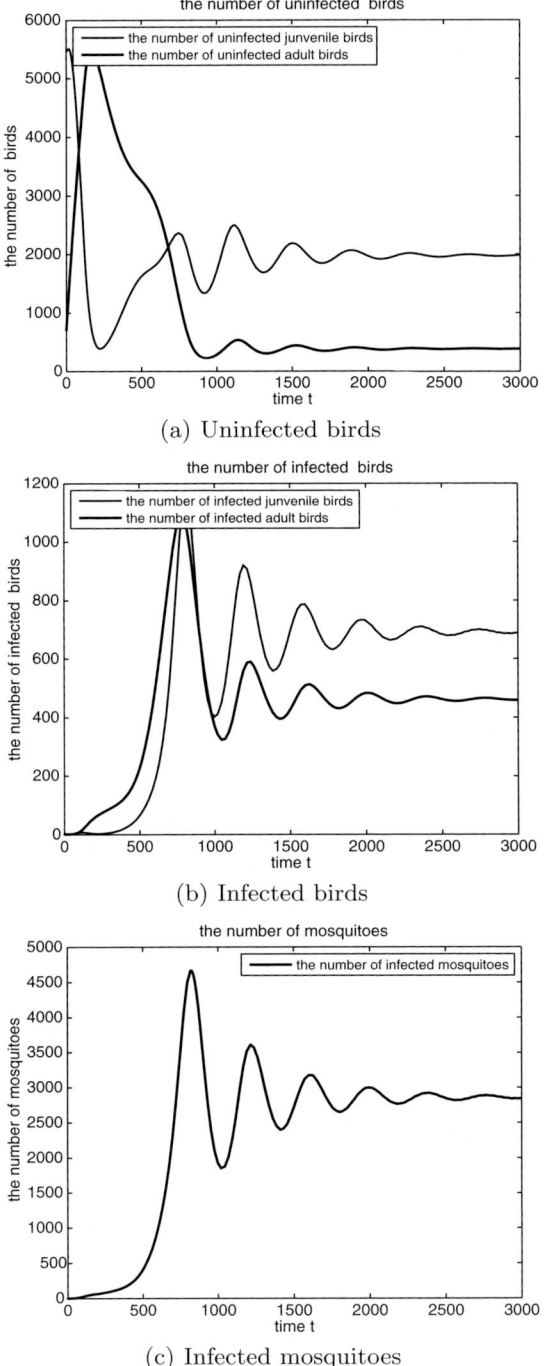

(a) Uninfected birds

(b) Infected birds

(c) Infected mosquitoes

Fig. 4.2. The DFE is unstable and the solution evolves to an endemic equilibrium

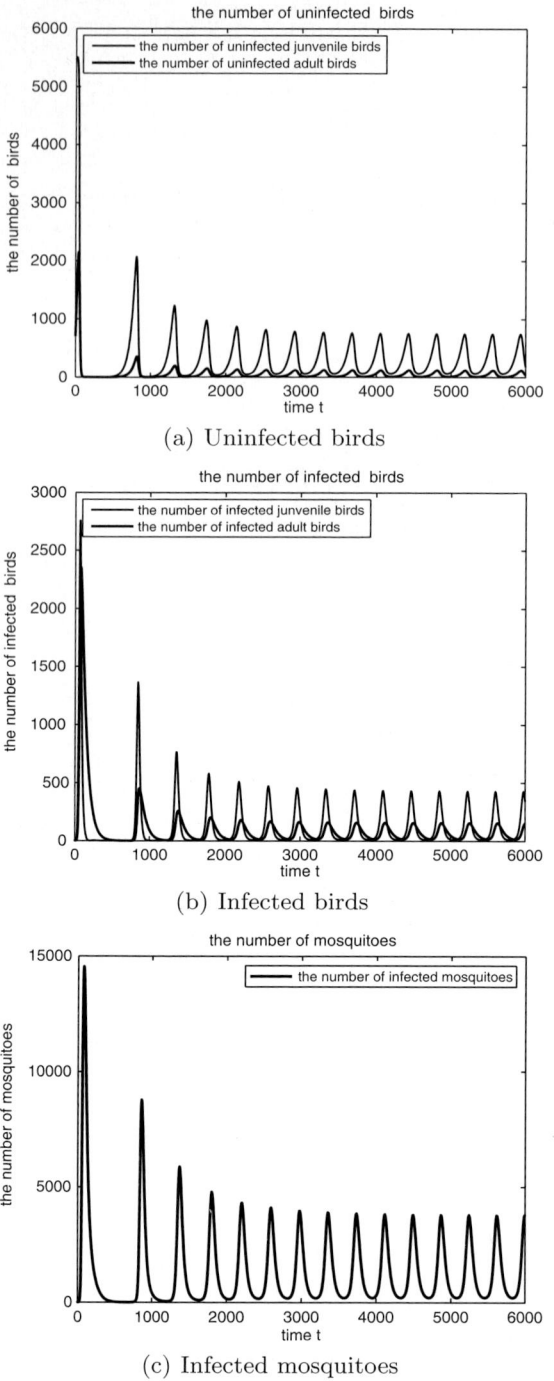

(a) Uninfected birds

(b) Infected birds

(c) Infected mosquitoes

Fig. 4.3. The DFE is unstable and the solution is oscillating

and insecticide sprays as techniques in the war against mosquitoes, the vector for many mosquito-born diseases including West Nile virus and dengue fever.

Larvicides are substances that destroy mosquito larva, the pre-adult insects that breed and mature in standing water. Biological larvicides are made from the bacteria *Bacillus thuringiensis israaelensis* or *Bacillus sphaericus*. The bacteria produce a crystal which is poisonous to mosquito larvae but virtually harmless to other forms of aquatic life. Unfortunately, as with many insect species, the larva can be difficult to find. An alternative to larvicides is insecticide sprays to kill adult mosquitoes. Called adulticides and used in many residential areas, some kill both mosquitoes and "good insects" that eat mosquitoes. Another drawback, in addition to limited effect, is the perceived public health implication. Large numbers of mosquitoes survive adulticide sprays by flying away or finding shelter in gutter downspouts and eaves, under foliage, and other protected areas. Therefore, it is highly desirable to develop appropriate models so that we can qualitatively examine the effectiveness of larvicides and insecticide sprays.

In Gourley, Liu and Wu [8], appropriate mathematical models were derived and utilized to assess the effectiveness of culling as a tool to eradicate vector-borne diseases. The model, focused on the culling strategies determined by the stages during the development of the vector, becomes either a system of autonomous delay differential equations with impulses (in the case where the adult vector is subject to culling) or a system of nonautonomous delay differential equations where the time-varying coefficients are determined by the culling times and rates (in the case where only the immature vector is subject to culling). Sufficient conditions were derived to ensure eradication of the disease, and simulations provided to compare the effectiveness of larvicides and insecticide sprays for the control of West Nile virus. Their results show that eradication of vector-borne diseases is possible by culling the vector at either the immature or the mature phase, even though the size of the vector is oscillating and above a certain level. The work [8] is based on the study of Simons and Gourley [26] of a time dependent stage structured population in which the adults (but not juveniles) are subject to culling or trapping which occurs only at certain particular times $t_1, t_2, t_3, \ldots$. Their model equation is

$$u'_m(t) = e^{-\mu\tau} b(u_m(t-\tau)) - d(u_m(t)) - \sum_{j=1}^{\infty} b_j u_m(t_j^-) \delta(t-t_j), \quad t > 0 \quad (4.33)$$

where $\mu > 0$ represents juvenile mortality, $u_m(t)$ is the total number of adults at time t, $u_m(t_j^-)$ is the population just before the impulsive cull at time t_j, τ is the maturation time, b_j is the proportion of the mature species trapped or culled at time t_j and δ denotes the Dirac delta function. In (4.33), $b(u_m(t))$ is a function representing the birth rate of the immature species while $d(u_m(t))$ is the natural death rate of the mature species.

4.3.1 Impulsive Systems and Analytic Results

The idea in Simons and Gourley [26] was developed further in [8] with the development of a stage-structured model for a single species population in which only the immatures are culled. This strategy can result in eradication as long as the unculled adults have some intrinsic death rate. To describe the work [8], we follow Simons and Gourley [26] and assume culling occurs only at certain discrete times t_j. At this stage, we assume that the immatures are culled. Let $u(t,a)$ be the density of individuals of age a at time t. Then

$$\frac{\partial u}{\partial t} + \frac{\partial u}{\partial a} = -\mu(a)u - \sum_{j=1}^{\infty} b_j(a)u(t_j^-,a)\delta(t - t_j), \quad 0 < a < \tau \quad (4.34)$$

where τ is the age at which an individual becomes a mature reproducing adult, $\mu(a)$ is the natural death rate for immatures, $b_j(a) \in [0,1]$ is the fraction of individuals of age a that are removed at the cull at time t_j and δ is the Dirac delta function. The superscript $-$ on the variable t_j in $u(t_j^-,a)$ denotes the limit of $u(t,a)$ as t approaches t_j from below (in other words, the population just prior to the cull at time t_j). We shall frequently also need the right limit, denoted using a superscript $+$, to refer to the situation immediately after a cull. We assume that

$$u(t,0) = b(u_m(t)), \quad (4.35)$$

where $b(\cdot)$ is the birth function and $u_m(t)$ is the total number of adults, given by

$$u_m(t) = \int_{\tau}^{\infty} u(t,a)\, da. \quad (4.36)$$

The solution of (4.34) will be continuous in time except for discontinuous jumps at the particular times t_j when culls occur. To see that $b_j(a)$ does indeed have the interpretation of being the fraction of age a removed at time t_j, integrate (4.34) from time t_j^- to t_j^+ to obtain

$$u(t_j^+,a) = (1 - b_j(a))u(t_j^-,a). \quad (4.37)$$

We assume (for now) that the adults are not subject to culling and also that their intrinsic death rate is a constant, μ_m. Thus

$$\frac{\partial u}{\partial t} + \frac{\partial u}{\partial a} = -\mu_m u \quad \text{for } a > \tau, \quad (4.38)$$

where μ_m is some constant. Recall that $u_m(t)$ is defined by (4.36). Differentiating this expression and assuming that $u(t,\infty) = 0$ gives

$$\frac{du_m(t)}{dt} = \int_{\tau}^{\infty} \left(-\frac{\partial u}{\partial a} - \mu_m u(t,a) \right) da = u(t,\tau) - \mu_m u_m(t) \quad (4.39)$$

and so we need $u(t, \tau)$ in terms of the function u_m. This is a little tricky in that the culls do not have to be equally spaced in time, and so as time progresses the issue is mainly one of keeping track of how many culls have occurred in the previous τ units of time.

For a general time t let

$$i(t) = \max\{i : t_i \leq t\} \tag{4.40}$$

and

$$k(t) = \min\{i : t_i > t - \tau\}. \tag{4.41}$$

Then, for a given t, relevant culls are those at the times t_j with j between $k(t)$ and $i(t)$ inclusive. The expression for $u(t, \tau)$ is shown in [8] to be the following, in which the exponential term is the probability of not dying a natural death during the maturation phase from age 0 to τ:

$$u(t, \tau) = b(u_m(t - \tau)) \exp\left(-\int_0^\tau \mu(s)\, ds\right) \prod_{j=k(t)}^{i(t)} (1 - b_j(\tau - (t - t_j))). \tag{4.42}$$

Therefore the delay differential equation (4.39) for the total number of adults $u_m(t)$ becomes

$$\frac{du_m(t)}{dt} = S(t) \exp\left(-\int_0^\tau \mu(s)\, ds\right) b(u_m(t - \tau)) - \mu_m u_m(t), \tag{4.43}$$

where

$$S(t) = \prod_{j=k(t)}^{i(t)} (1 - b_j(\tau - (t - t_j))), \tag{4.44}$$

with $i(t)$ and $k(t)$ given by (4.40) and (4.41). Each term in this product represents the probability of surviving a particular cull. All information relating to culling is contained in the function $S(t)$ and features nowhere else.

With the above preparation, we can now formulate a mathematical model for the situation when only immature (larval) mosquitoes are culled. By solving the von Foerster equation for the larval mosquitoes, we can formulate a model involving only three state variables: $M_S(t)$, $M_I(t)$ and $B_I(t)$ which denote respectively the total numbers of susceptible adult mosquitoes, infected adult mosquitoes and infected birds.

Larval mosquitoes, whose densities are denoted by $l(t, a)$, are assumed not to interact with the adults or the birds. The larval stage (considered as the only stage prior to adulthood) is of duration τ. Larvae are culled and so, following the modeling described above, their evolution equation is taken to be of the form

$$\frac{\partial l}{\partial t} + \frac{\partial l}{\partial a} = -\mu(a)l - \sum_{j=1}^{\infty} b_j(a)l(t_j^-, a)\delta(t - t_j) \qquad 0 < a < \tau, \tag{4.45}$$

where the t_j are the times at which culls happen. Both susceptible and infected mosquitoes may lay eggs but the virus is not passed on to offspring. The birth rate $l(t,0)$ of mosquitoes is therefore assumed to be a function of the total number of adult mosquitoes $M_S(t) + M_I(t)$, so that

$$l(t,0) = b(M_S(t) + M_I(t)), \tag{4.46}$$

where $b(\cdot)$ is the birth rate function. Susceptible adult mosquitoes are assumed to satisfy an equation of the form

$$\frac{dM_S}{dt} = l(t,\tau) - \gamma B_I M_S - d_S M_S,$$

where $l(t,\tau)$ is the rate at which mosquitoes become mature. In this equation, $\gamma B_I M_S$ is the rate at which susceptible mosquitoes become infected mosquitoes (a mosquito becomes infected when it bites an infected bird) and $d_S M_S$ is the death rate for susceptible mosquitoes. By analogy with the earlier analysis for a single species, we may derive an equation for the susceptible mosquitoes $M_S(t)$ of the form

$$\frac{dM_S}{dt} = S(t) \exp\left(-\int_0^\tau \mu(s)\, ds\right) b(M_S(t-\tau) + M_I(t-\tau)) - \gamma B_I M_S - d_S M_S, \tag{4.47}$$

with $S(t)$ again defined by (4.44). This equation is then coupled with the following equations for the infected mosquitoes and the infected birds:

$$\frac{dM_I}{dt} = \gamma B_I M_S - d_I M_I, \tag{4.48}$$

$$\frac{dB_I}{dt} = \beta(N_B - B_I)M_I - d_B B_I. \tag{4.49}$$

The meaning of the terms in (4.48) is obvious. As regards (4.49), we are assuming that the total number of birds is some constant $N_B > 0$, so that $N_B - B_I$ is the number of susceptible birds. Thus $\beta(N_B - B_I)M_I$ is the rate at which susceptible birds become infected birds, assumed to be given by the law of mass action. A bird becomes infected when it is bitten by an infected mosquito. Systems (4.47)–(4.49) for $t > 0$ is supplemented with the initial data

$$M_S(\theta) = M_S^0(\theta) \geq 0, \qquad \theta \in [-\tau, 0],$$

$$M_I(\theta) = M_I^0(\theta) \geq 0, \qquad \theta \in [-\tau, 0], \tag{4.50}$$

$$B_I(0) = B_I^0 \in [0, N_B]$$

with $M_S^0(\theta)$, $M_I^0(\theta)$ and B_I^0 prescribed.

It is shown in [8] that if the birth function $b(\cdot)$ satisfies $b(0) = 0$ and $b(M) > 0$ for all $M > 0$, then the solution of systems (4.47)–(4.49) for $t > 0$, subject to (4.50), satisfies $M_S(t) \geq 0$, $M_I(t) \geq 0$, $B_I(t) \in [0, N_B]$ for all $t > 0$.

The following theorem, proved in [8], provides conditions sufficient for the eradication of the disease.

Theorem 3 *Consider system (4.47)–(4.49) for $t > 0$, subject to (4.50). Suppose the birth function $b(\cdot)$ satisfies $b(0) = 0$ and $b(M) > 0$ for all $M > 0$, and let $S(t)$ be defined by (4.44). Let $S^\infty = \limsup_{t\to\infty} S(t)$. Assume that either*

$$\min(d_I, d_S) > S^\infty b'_{max} \exp\left(-\int_0^\tau \mu(s)\,ds\right), \qquad (4.51)$$

or

$$d_I d_B > \frac{\gamma\beta N_B b_{\max} S^\infty \exp\left(-\int_0^\tau \mu(s)\,ds\right)}{\min(d_I, d_S)}, \qquad (4.52)$$

where $b_{\max} = \sup_{m\geq 0} b(m)$ and $b'_{\max} = \sup_{m\geq 0} b'(m)$. Then $B_I(t) \to 0$ and $M_I(t) \to 0$ as $t \to \infty$.

Note that S^∞ describes the accumulated effect of culling. Since $0 < b_j < 1$, it becomes evident that the more frequent culling occurs, the smaller S^∞ is; and the higher the culling rate, the smaller S^∞. Theorem 3 shows that high culling frequency or rate can both eradicate the disease.

Let us now discuss the issue of culling of mature mosquitoes. We continue to assume that the total number of birds in an area is some constant N_B. If birds are divided into two classes: uninfected B_S and infected B_I, then $B_S = N_B - B_I$. Then the change rate of infected birds is increased through infection of uninfected birds when they are bitten by infected mosquitoes and reduced by the natural death and disease-induced death (at a rate d_B). Thus,

$$\frac{dB_I}{dt} = \beta(N_B - B_I)M_I - d_B B_I, \qquad (4.53)$$

where β is the contact rate between infected mosquitoes and uninfected birds.

As far as mosquitoes are concerned, we assume now that only the adults are subject to culling. Adult mosquitoes are divided into two classes: uninfected M_S and infected M_I. Since it would be difficult in practice to cull only infected ones, culling will be applied equally to both classes. The total number of adult mosquitoes will be denoted $M_T = M_S + M_I$. Culling occurs only at the particular prescribed times t_j, $j = 1, 2, 3, \ldots$, satisfying the assumptions below. At the cull which occurs at time t_j a proportion c_j of the adult mosquito population is culled, causing a sharp decrease in the population and consequently a discontinuity in the evolution of $M_S(t)$ and $M_I(t)$ at each time t_j. The following assumption is made:

$$0 < t_1 < t_2 < \cdots < t_j < \cdots \text{ with } t_j \to \infty \text{ as } j \to \infty,$$

$$\inf_{j\geq 1} \delta_j > 0, \text{ where } \delta_j = t_j - t_{j-1}, \qquad (4.54)$$

$$c_j \in (0, 1] \text{ for each } j = 1, 2, 3, \ldots.$$

Note that no c_j is allowed to be zero (we can of course eliminate any "null culls" by relabelling the sequence t_j to include only "genuine" culls with $c_j > 0$, and we are assuming that this has been done). The evolution of $B_I(t)$ (infected birds) will remain continuous in time, but its derivative will have discontinuities at the times t_j.

Let $l(t, a)$ be the density of larval mosquitoes at time t of age a, and assume that a mosquito becomes mature on reaching the age τ. Since immature mosquitoes are not subject to culling but only to natural death, we have

$$\frac{\partial l}{\partial t} + \frac{\partial l}{\partial a} = -d_L l, \ t > 0, \ 0 < a < \tau \tag{4.55}$$

with $d_L > 0$ constant. The birth rate $l(t, 0)$ is a function of the total number of adult mosquitoes, so that

$$l(t, 0) = b(M_T(t)). \tag{4.56}$$

We assume there is no vertical transmission between mosquitoes, so the uninfected mosquito population is increased via the maturation rate $l(t, \tau)$. It is diminished by infection, which may be acquired when uninfected mosquitoes feed from the blood of infected birds, by natural death at a rate d_M and by culling at the times t_j, $j = 1, 2, 3, \ldots$. Thus

$$\frac{dM_S}{dt} = l(t, \tau) - \gamma M_S B_I - d_M M_S - \sum_{j=1}^{\infty} c_j M_S(t_j^-)\delta(t - t_j), \tag{4.57}$$

where γ is the contact rate between uninfected mosquitoes and infected birds.

The infected mosquito population is generated via the infection of uninfected mosquitoes by infected birds and diminished by natural death at a rate d_M and culling at the times t_j, $j = 1, 2, 3, \ldots$. Thus,

$$\frac{dM_I}{dt} = \gamma M_S B_I - d_M M_I - \sum_{j=1}^{\infty} c_j M_I(t_j^-)\delta(t - t_j). \tag{4.58}$$

It is assumed that the uninfected mosquitoes and infected mosquitoes are equally mixed, so that at each cull the proportions of each class removed are the same.

From (4.55) and (4.56),

$$l(t, \tau) = b(M_T(t - \tau))e^{-d_L \tau}.$$

So the model equations assume the form

$$\begin{cases} \dfrac{dB_I}{dt} = \beta(N_B - B_I)M_I - d_B B_I, \\[2mm] \dfrac{dM_S}{dt} = b(M_T(t - \tau))e^{-d_L \tau} - \gamma M_S B_I - d_M M_S - \displaystyle\sum_{j=1}^{\infty} c_j M_S(t_j^-)\delta(t - t_j), \\[2mm] \dfrac{dM_I}{dt} = \gamma M_S B_I - d_M M_I - \displaystyle\sum_{j=1}^{\infty} c_j M_I(t_j^-)\delta(t - t_j) \end{cases}$$

$$\tag{4.59}$$

for $t > 0$ subject to initial conditions of the form (4.50). The model may also be written

$$
\begin{cases}
\dfrac{dB_I}{dt} = \beta(N_B - B_I)M_I - d_B B_I, \\[2mm]
\dfrac{dM_S}{dt} = b(M_T(t - \tau))e^{-d_L \tau} - \gamma M_S B_I - d_M M_S, \quad t \neq t_j, \\[2mm]
M_S(t_j^+) = (1 - c_j)M_S(t_j^-), \\[2mm]
\dfrac{dM_I}{dt} = \gamma M_S B_I - d_M M_I, \quad t \neq t_j, \\[2mm]
M_I(t_j^+) = (1 - c_j)M_I(t_j^-)
\end{cases}
\tag{4.60}
$$

again subject to (4.50).

It was shown in [8] that if we assume (4.54) holds and if the birth function $b(\cdot)$ satisfies $b(0) = 0$ and $b(M) > 0$ for all $M > 0$, then the solution of system (4.60) for $t > 0$, subject to (4.50), satisfies $M_S(t) \geq 0$, $M_I(t) \geq 0$, $B_I(t) \in [0, N_B]$ for all $t > 0$.

Henceforth we assume the birth function satisfies

$$
\left.
\begin{array}{l}
b(0) = 0, \ b(\cdot) \text{ is strictly monotonically increasing, there exists } M_T^* > 0 \text{ such} \\
\text{that } e^{-d_L \tau} b(M) > d_M M \text{ when } M < M_T^* \text{ and } e^{-d_L \tau} b(M) < d_M M \text{ when} \\
M > M_T^*.
\end{array}
\right\}
\tag{4.61}
$$

From system (4.60), note that the total number $M_T(t)$ of adult mosquitoes obeys

$$
\frac{dM_T}{dt} = b(M_T(t - \tau))e^{-d_L \tau} - d_M M_T(t),
$$
$$
M_T(t_j^+) = (1 - c_j)M_T(t_j^-).
\tag{4.62}
$$

In the absence of culling, the quantity $M_T^* > 0$ referred to in (4.61) is an equilibrium of (4.62) and $M_T(t) \to M_T^*$ as $t \to \infty$ (see Kuang [11]). Due to the assumptions in (4.61) which imply that the differential equation in (4.62) has the properties of a monotone system, solutions of (4.62) with culling are bounded above by the corresponding solutions without culling. It is therefore easy to appreciate that, with culling, there exists a finite time beyond which $M_T(t) \leq M_T^*$, and hence also $M_S(t) \leq M_T^*$. In [8], it was shown that this estimate can be improved to involve c_j and an upper bound $\delta_{\sup}$ on the amount of time that elapses between two successive culls. More precisely, they proved that

Theorem 4 *Assume (4.54) and (4.61) hold, and let*

$$
c_{\inf} = \inf_{j \geq 1} c_j, \qquad \delta_{\sup} = \sup_{j \geq 1}(t_{j+1} - t_j)
$$

and assume $c_{\inf} > 0$ and $\delta_{\sup} < \infty$. Then solutions $M_T(t)$ of (4.62) satisfy

$$
M_T(t) \leq M^{**} := M_T^*(1 - c_{\inf} e^{-d_M \delta_{\sup}})
\tag{4.63}
$$

for all t sufficiently large. Consequently, $M_S(t) \leq M^{**}$ for t sufficiently large.

This result then yields some useful sufficient conditions for disease eradication.

Theorem 5 *Suppose (4.54) and (4.61) hold, and that*

$$d_B d_M > \gamma \beta N_B M^{**}, \tag{4.64}$$

where M^{**} is defined by (4.63). Then $B_I(t) \to 0$ and $M_I(t) \to 0$ as $t \to \infty$, where $B_I(t)$ and $M_I(t)$ satisfy (4.60) subject to (4.50).

Condition (4.64) predicts disease eradication when death rates are high, contact rates are low, the total number of birds is low, and the quantity M^{**} is low. Recall that this latter quantity, being defined by (4.63), involves information about the culling and is low when large fractions are removed at each cull and the culls are frequent. The following theorem shows that even if condition (4.64) is violated, eradication of the disease is still possible. Its proof can be found in [8].

Theorem 6 *Suppose (4.54) and (4.61) hold, that $\delta_{\sup} = \sup_{j \geq 1}(t_{j+1} - t_j) < \infty$ and that $c_{\inf} > 0$ where $c_{\inf} = \inf_{j \geq 1} c_j$. Suppose also that $d_B d_M \leq \gamma \beta N_B M^{**}$ and*

$$\sup_{j \geq 1} \left[(\lambda_2(1 - c_j) - \lambda_1)e^{\lambda_1 \delta_j} + (\lambda_2 + (c_j - 1)\lambda_1)e^{\lambda_2 \delta_j} + c_j d_M (e^{\lambda_2 \delta_j} - e^{\lambda_1 \delta_j}) \right.$$
$$\left. - (\lambda_2 - \lambda_1)(1 - c_j)e^{-(d_B + d_M)\delta_j} \right] < \lambda_2 - \lambda_1, \tag{4.65}$$

where $\lambda_1 < 0$ and $\lambda_2 \geq 0$ satisfy

$$\lambda^2 + (d_B + d_M)\lambda + (d_B d_M - \gamma \beta N_B M^{**}) = 0. \tag{4.66}$$

Then $B_I(t) \to 0$ and $M_I(t) \to 0$ as $t \to \infty$, where $B_I(t)$ and $M_I(t)$ satisfy (4.60) subject to (4.50).

4.3.2 Simulations and Discussion

In this section we present the results of some numerical simulations to compare the effectiveness of larval culling versus adult culling. Larval culling is described by system (4.47)–(4.49), and let us recall that for larval culling all information about the culling is embodied in the function $S(t)$ defined by (4.44). Culling of adults is described by system (4.60).

In the simulations we take the birth function of mosquitoes as

$$b(M) = bMe^{-aM} \tag{4.67}$$

which we feel to be an ecologically reasonable choice, being linear in M only for small densities M, levelling off as a consequence of intraspecific competition working to reduce per capita fecundity, and then actually dropping at very large densities M due to the available resources in these circumstances being utilised by the adults only for their own physiological maintenance and not reproduction.

Figures 4.4–4.6 are intended to compare larval culling with adult culling in a variety of culling regimes. Each figure contains nine plots in all; these being the variables $M_S(t)$, $M_I(t)$ and $B_I(t)$ in each of the situations of no culling, culling of adult mosquitoes and culling of larval mosquitoes. Where a simulation is of a variable with adult mosquito culling, the simulation is of model (4.60) with the c_j value shown in the caption. Where larval culling is mentioned the simulation is of system (4.47)–(4.49) with the b_j given in the caption. In all simulations the culls are at equally spaced times, although we do examine the effect of different spacings, i.e. different frequencies of culling. The interval between two consecutive culls we shall denote as Δt. The cull times are given by $t_j = t_0 + j\Delta t$, $j = 1, 2, 3, \ldots$ with $t_0 = 4$. The initial conditions were taken to be

$$M_S(t) = 5,000, \qquad M_I(t) = 600, \qquad B_I(t) = 100$$

for $t \in [-\tau, 0]$. Table 4.2 gives the meanings and the values used for the various parameters. Note that the parameter b is that which appears in our choice for the birth rate function (4.67) and equals the maximum daily egg production per adult mosquito.

We made the following general observations which are based on the results of numerous simulations:

• Under the same culling rates (i.e. if $b_j = c_j$) and frequencies, adulticide seems to be more effective than larvicide. However, for the reasons given below, adulticide is more difficult in practice. Larvicide alone is perfectly capable of eradicating the disease. If the culling frequency is such that a typical larval cohort is likely to experience only one cull, then a large fraction of the larvae have to be killed at each cull.

• If we increase the culling frequency (i.e. decrease Δt), the effect of both larvicide and adulticide increases. If the culling frequency is high enough the disease dies out.

• In the cases of both larvicide and adulticide, the disease dies out for sufficiently large proportions b_j and c_j respectively.

• When the infected mosquitoes disappear under some culling regime, whether the susceptible mosquitoes die out depends on the death rate of infected birds. If the death rate of infected birds is large enough, the disease will die out in both mosquito and bird population. While if the death rate of infected birds is small (i.e., the disease has no impaction on birds), the WNv will sustain in the mosquito and bird population except all mosquitoes die out.

The purpose of Figs. 4.5 and 4.6 is to illustrate what happens if the interval between successive culls is larger than the maturation delay $\tau = 10$, a

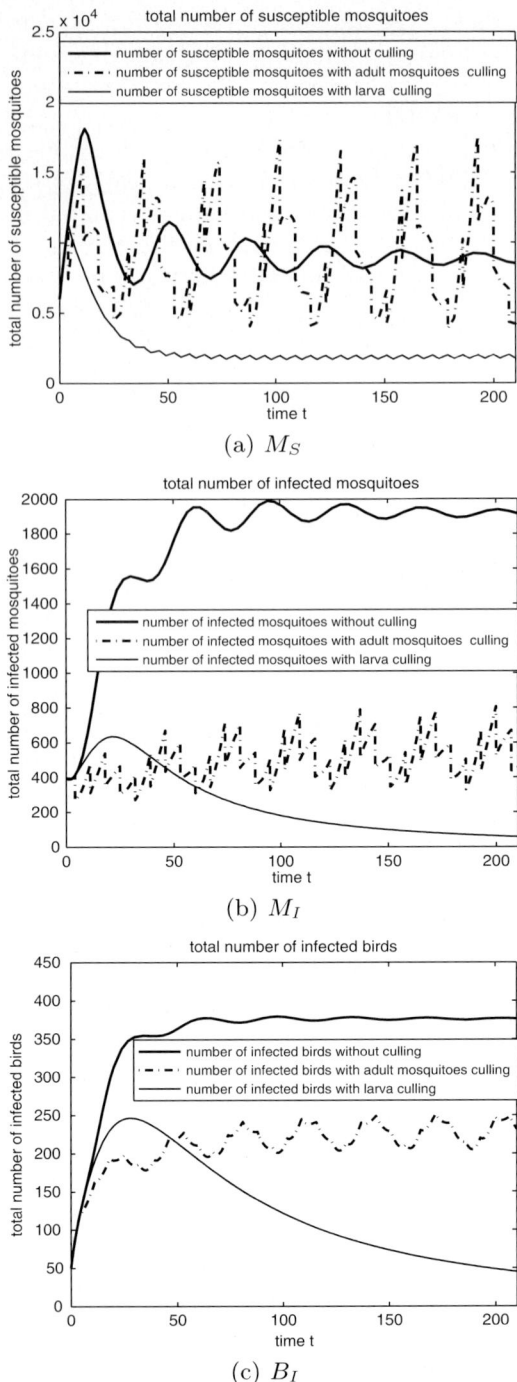

(a) M_S

(b) M_I

(c) B_I

Fig. 4.4. Parameter values are $b_j = 0.95$, $c_j = 0.35$, $\Delta t = 7$ and other parameters have the values shown in Table 4.2

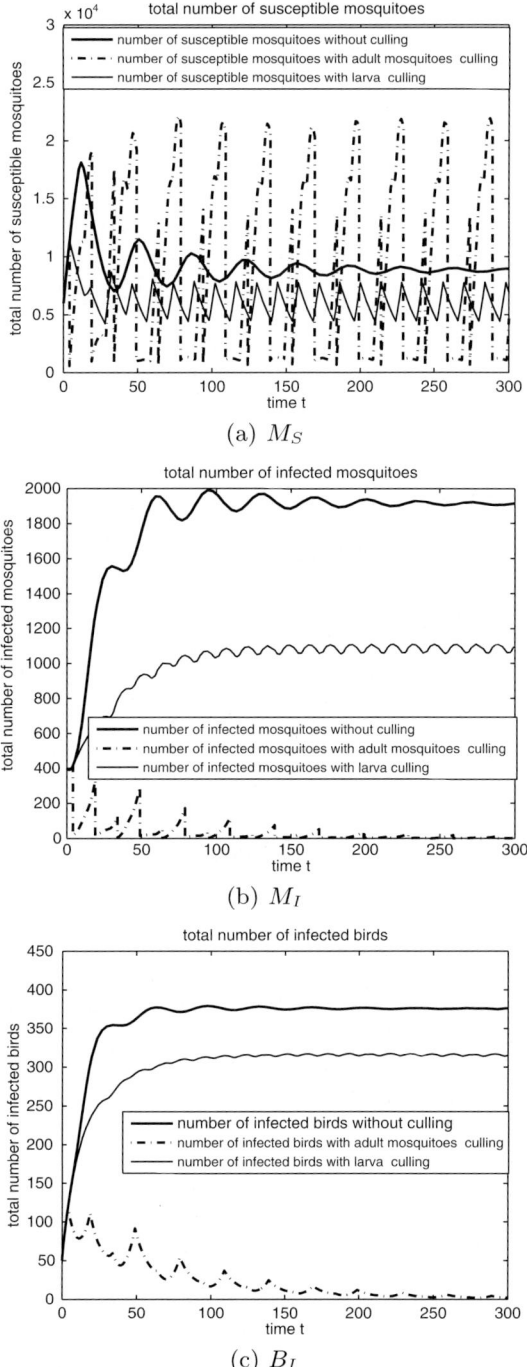

(a) M_S

(b) M_I

(c) B_I

Fig. 4.5. Parameter values are $b_j = 0.99$, $c_j = 0.95$, $\Delta t = 15$ and other parameters have the values shown in Table 4.2

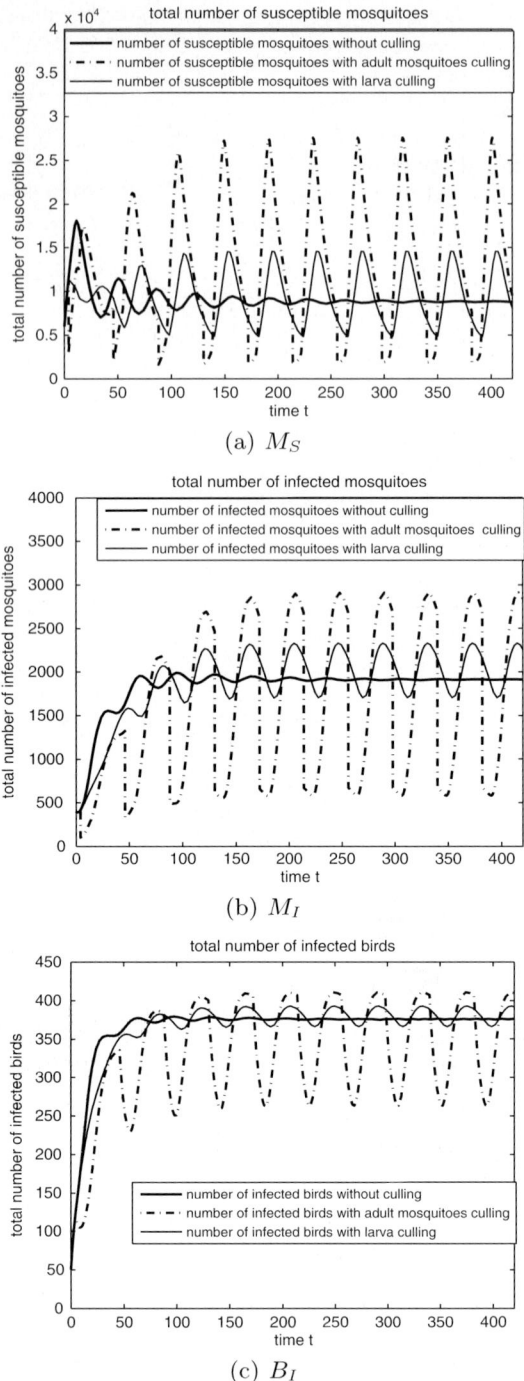

(a) M_S

(b) M_I

(c) B_I

Fig. 4.6. Parameter values are $b_j = 0.75$, $c_j = 0.75$, $\Delta t = 42$ and other parameters have the values shown in Table 4.2

Table 4.2. Parameter values used for the simulations. Those that vary from simulation to simulation are shown in the figure captions. Literature used [2, 14, 35]

Para.	Meaning	Value
d_L	Per capita death rate of mosquito larva (per day)	0.1
τ	Maturation time of mosquito larva (days)	10
d_M	Per capita natural death rate of mosquito (per day)	0.05
d_B	Disease-induced death rate of infected bird (per day)	0.1
N_B	Total number of birds	500
β	Contact rate between infected mosquito and susceptible birds	$0.0144/N_B$
γ	Contact rate between susceptible mosquito and infected birds	$0.0792/N_B$
b	Maximum per capita daily egg production rate	10
$1/a$	Size of mosquito population at which egg laying is maximized	2,500
b_j	Fraction of larva removed at the cull at time t_j	Variable
c_j	Fraction of adult mosquito removed at the cull at time t_j	Variable
$1/\Delta t$	Culling frequency	Variable

situation that is not really covered by the analytical results. If this happens, certain larva cohorts may completely escape a cull. Figure 4.5 shows that, as a consequence of this, the disease can persist even when larva culling is maximized ($b_j = 0.99$). The disease can still be eradicated via adult culling but only with a very high proportion $c_j = 0.95$ removed each time. The number of susceptible mosquitoes oscillates wildly. Figure 4.6 illustrates that if the interval between culls is very high indeed compared to the maturation delay (we have used $\Delta t = 50$ with $\tau = 10$), then the culling is not having any useful effect at all. Indeed, with culling of adults, the number of infected mosquitoes appears to oscillate with an even higher mean than the oscillation with no culling at all.

Let us remark that condition (4.52), though only a sufficient condition for disease eradication, seems close in at least some parameter regimes to being a necessary condition as well. If we take $b = 10$, $\Delta t = 7$ and other parameter values given in Table 4.2, then condition (4.52) predicts disease eradication if $b_j > 0.9676$ for each j. Condition (4.52) is a sufficient condition. Trial and error numerical simulation indicates that a necessary and sufficient condition for eradication is approximately $b_j > 0.949$ so that the critical b_j is close to the analytical estimate of 0.9676. Figure 4.4 gives the results when $b_j = 0.95$, showing that the disease slowly disappears in this case. This further emphasizes our point that if the disease is to be eradicated via larval culling only, then very large fractions of larvae have to be destroyed at each cull if we cull at a frequency of once every 7 days ($\Delta t = 7$). We chose this frequency to ensure that every larval cohort (we have taken the larval stage as lasting 10 days) is subject to at least one cull with some cohorts experiencing two (the function $S(t)$ defined by (4.44) takes care of this automatically). However, one could of course increase the culling frequency. The function $S(t)$, and therefore the number S^∞ in (4.52), goes down quickly as the culling becomes more frequent, leading to vastly less stringent conditions on the b_j.

Mosquito control programs often emphasize larval control, possibly due to the greater difficulty in organising spraying of adults. One knows where to look for mosquito breeding activity (a pool cover, ornamental pool, bird bath, gutter or even an old tire is all they need). Larvicides can provide up to a month of control (adulticides only a few hours) and larvicides can be applied in such a way that there is less human exposure. In WNv endemic areas of the US the use of mosquito adulticides is in fact usually a measure of last resort because of health risks associated from exposure to the insecticide, which is released into the atmosphere in the form of very fine droplets. People need to be advised in advance and to be given precautions such as remaining indoors during spraying and to take other precautions. To justify the use of adulticides public health officials have to have reached the view that the risks from WNv are higher than those associated with exposure to the insecticide, and they need to inform the public and advise on precautions.

4.4 Spatial Spread: Interaction of Individual Movements and Physiological Status

We now address the issue of spatial spread of vector-borne diseases involving age-structure. In this section we summarise the work by Gourley, Liu and Wu [7]. Adding spatial diffusion using Fick's law into equations (4.6) and (4.7) gives

$$\frac{\partial s}{\partial t} + \frac{\partial s}{\partial a} = D_s(a)\frac{\partial^2 s}{\partial x^2} - d_s(a)s(t,a,x) - \beta(a)s(t,a,x)m_i(t,x) \qquad (4.68)$$

and

$$\frac{\partial i}{\partial t} + \frac{\partial i}{\partial a} = D_i(a)\frac{\partial^2 i}{\partial x^2} - d_i(a)i(t,a,x) + \beta(a)s(t,a,x)m_i(t,x) \qquad (4.69)$$

on a one-dimensional spatial domain $x \in (-\infty, \infty)$, where $m_i(t,x)$ is the number of infected adult mosquitoes at (t,x) satisfying a reaction–diffusion equation mentioned below. We assume that the age-dependent diffusivities $D_s(a)$, $D_i(a)$ have the special form

$$D_s(a) = \begin{cases} D_{sj} & a < \tau \\ D_{sa} & a > \tau, \end{cases} \qquad D_i(a) = \begin{cases} D_{ij} & a < \tau \\ D_{ia} & a > \tau. \end{cases} \qquad (4.70)$$

If the diffusivities are chosen as above, we may derive a system of reaction–diffusion equations, valid near the disease-free region $x \approx -\infty$ only, for the quantities

$$A_s(t,x) = \int_\tau^\infty s(t,a,x)\,da, \qquad A_i(t,x) = \int_\tau^\infty i(t,a,x)\,da,$$

$$J_s(t,x) = \int_0^\tau s(t,a,x)\,da, \qquad J_i(t,x) = \int_0^\tau i(t,a,x)\,da. \qquad (4.71)$$

The derivation of such an equation for A_s involves computing the quantity $s(t, \tau, x)$, which in turn involves solving the following equation for the function $s_\xi(a, x) := s(\xi + a, a, x)$:

$$\frac{\partial s_\xi}{\partial a} = D_s(a) \frac{\partial^2 s_\xi}{\partial x^2} - d_s(a) s_\xi(a, x) - \beta(a) s_\xi(a, x) m_i(\xi + a, x). \quad (4.72)$$

Unfortunately (4.72) cannot be solved explicitly for $s_\xi(a, x)$ because of the presence of the variable m_i which satisfies a separate nonlinear partial differential equation (see below). However, the analysis to be summarised here concerns the spatial spread of the disease in the form of a traveling wave solution which moves leftwards through the spatial domain $x \in (-\infty, \infty)$ and which constitutes a connection between the disease free state and an endemic state. It is possible to derive a system of partial differential equations that are valid in the spatial region of interest, i.e. the region far ahead of the advancing epidemic ($x \to -\infty$). We assume that the linearised equations in this region determine the speed of the epidemic wave. In the disease free region $x \approx -\infty$, the variables $A_i(t, x)$, $J_i(t, x)$ and $m_i(t, x)$ are all close to zero. Equation (4.72) is therefore solved in the case when m_i is zero. The solution subject to the first condition appearing below:

$$s(t, 0, x) = b(A_s(t, x)), \qquad i(t, 0, x) = 0 \quad (4.73)$$

is, for $a \leq \tau$ and $\xi \geq 0$,

$$s_\xi(a, x) = s(\xi + a, a, x) = \int_{-\infty}^{\infty} \Gamma(D_{sj}a, x - y)b(A_s(\xi, y))e^{-d_{sj}\tau} \, dy \quad (4.74)$$

where

$$\Gamma(t, x) = \frac{1}{\sqrt{4\pi t}} e^{-x^2/4t}. \quad (4.75)$$

From (4.74) an expression for $s(t, \tau, x)$ can be found and we deduce that for $t \geq \tau$ the partial differential equation for $A_s(t, x)$:

$$\frac{\partial A_s}{\partial t} = \int_{-\infty}^{\infty} \Gamma(D_{sj}\tau, x - y)b(A_s(t - \tau, y))e^{-d_{sj}\tau} \, dy$$

$$+ D_{sa} \frac{\partial^2 A_s}{\partial x^2} - d_{sa} A_s(t, x) - \beta_a m_i(t, x) A_s(t, x) \quad (4.76)$$

is valid in the far left of the spatial domain $x \in (-\infty, \infty)$. Similarly, we find an approximate equation for $J_s(t, x)$, also valid only in the far field $x \to -\infty$:

$$\frac{\partial J_s}{\partial t} = b(A_s(t, x)) - \int_{-\infty}^{\infty} \Gamma(D_{sj}\tau, x - y)b(A_s(t - \tau, y))e^{-d_{sj}\tau} \, dy$$

$$+ D_{sj} \frac{\partial^2 J_s}{\partial x^2} - d_{sj} J_s(t, x) - \beta_j m_i(t, x) J_s(t, x). \quad (4.77)$$

The derivation of a partial differential equation for $A_i(t,x)$ involves computing $i(t,\tau,x)$, which involves solving the following equation for the function $i_\xi(a,x) := i(\xi + a, a, x)$:

$$\frac{\partial i_\xi}{\partial a} = D_{ij}\frac{\partial^2 i_\xi}{\partial x^2} - d_{ij}i_\xi(a,x) + \beta_j m_i(\xi + a, x)s(\xi + a, a, x).$$

The solution of this equation satisfying the second condition in (4.73) is

$$i_\xi(a,x) = \beta_j \int_0^a e^{-d_{ij}(a-\zeta)} \int_{-\infty}^{\infty} \Gamma(D_{ij}(a-\zeta), x-y)m_i(\xi+\zeta,y)s(\xi+\zeta,\zeta,y)\,dy\,d\zeta$$

where Γ is given by (4.75). For $s(\xi + \zeta, \zeta, y)$ we use expression (4.74). Then, setting $a = \tau$ and $\xi = t - \tau$ in the above expression gives us $i(t,\tau,x)$ and we find that the variable $A_i(t,x)$ representing the number of adult infected hosts satisfies, for $t \geq \tau$,

$$\frac{\partial A_i}{\partial t} = D_{ia}\frac{\partial^2 A_i}{\partial x^2} - d_{ia}A_i(t,x) + \beta_a m_i(t,x)A_s(t,x)$$

$$+ \beta_j \int_0^\tau e^{-d_{ij}(\tau-\zeta)} \int_{-\infty}^{\infty} \Gamma(D_{ij}(\tau - \zeta), x - y)m_i(t - \tau + \zeta, y) \quad (4.78)$$

$$\times \int_{-\infty}^{\infty} \Gamma(D_{sj}\zeta, y - \eta)b(A_s(t - \tau, \eta))e^{-d_{sj}\zeta}\,d\eta\,dy\,d\zeta$$

which is again valid only in the far field $x \to -\infty$, since we have used expression (4.74). The last term in the right hand side of (4.78) is the rate at which infected immatures become infected adults and has a similar interpretation to a term in the right hand side of (4.14). The term involves additional integrals because of diffusion, but in some respects is a little simpler than one might expect. This is because of the approximations that have been made due to restricting to the $x \approx -\infty$ zone. The interpretation of the term we are discussing is as follows. Each individual that reaches adulthood at point x at time t as an infected individual was born as a susceptible at time $t - \tau$ at some other point η. For an amount of time ζ that individual drifted around as a susceptible individual with diffusivity D_{sj} until reaching a point y where it became infected at time $t - \tau + \zeta$. For an amount of time $\tau - \zeta$, constituting the remainder of its childhood, it drifted around as an infected individual with diffusivity D_{ij} to reach point x at time t where it becomes an adult. The two exponential factors represent the probability of surviving the susceptible and infected portions of childhood.

The partial differential equation for $J_i(t,x)$ is

$$\frac{\partial J_i}{\partial t} = D_{ij}\frac{\partial^2 J_i}{\partial x^2} - d_{ij}J_i(t,x) + \beta_j m_i(t,x)J_s(t,x)$$

$$- \beta_j \int_0^\tau e^{-d_{ij}(\tau-\zeta)} \int_{-\infty}^{\infty} \Gamma(D_{ij}(\tau - \zeta), x - y)m_i(t - \tau + \zeta, y) \quad (4.79)$$

$$\times \int_{-\infty}^{\infty} \Gamma(D_{sj}\zeta, y - \eta)b(A_s(t - \tau, \eta))e^{-d_{sj}\zeta}\,d\eta\,dy\,d\zeta.$$

It is assumed that the infected adult mosquitoes $m_i(t, x)$ satisfy

$$\frac{\partial m_i}{\partial t} = D_m \frac{\partial^2 m_i}{\partial x^2} - d_m m_i(t, x) + \beta_m(m_T^* - m_i(t, x))(J_i(t, x) + \alpha A_i(t, x)).$$
(4.80)

The complete system to be solved consists of equations (4.76)–(4.79) and (4.80), and in [7] solutions were considered which have the form of a leftward moving traveling wave-front, constituting invasion into what was formerly a disease-free zone. So, as $x \to -\infty$, the variables tend to the disease-free values in which A_i, J_i and m_i are zero while $A_s^* > 0$ and $J_s^* > 0$ are given by (4.29), assuming that (4.24) holds (if (4.24) does not hold then the host population is eradicated even in the absence of the disease).

In fact, [7] looked at wave-front solutions that constitute a transition from the disease free state to an endemic steady state. The endemic state cannot be found explicitly but a condition for its existence is known. This condition is the opposite of (4.30), so it is assumed in the subsequent discussion that

$$d_m < \beta_m m_T^* \left\{ \frac{b(A_s^*)\beta_j}{d_{ij} - d_{sj}} \left[\frac{1 - e^{-d_{sj}\tau}}{d_{sj}} - \frac{(1 - e^{-d_{ij}\tau})}{d_{ij}} \right] \right.$$
$$\left. + \frac{\alpha}{d_{ia}} \left[\beta_a A_s^* + \beta_j b(A_s^*) e^{-d_{sj}\tau} \frac{(1 - e^{-(d_{ij}-d_{sj})\tau})}{d_{ij} - d_{sj}} \right] \right\}.$$
(4.81)

If the equations for A_i, J_i and m_i ((4.78)–(4.80)) are converted into traveling wave form, with $z = x + ct$ (and $c \geq 0$) as the independent variable, and linearised in the region $x \to -\infty$ where $A_s \to A_s^*$, $J_s \to J_s^*$ and the other variables approach zero, nontrivial solutions of the traveling wave equations of the form $(A_i, J_i, m_i) = (q_1, q_2, q_3) \exp(\lambda z)$ can be sought. The characteristic equation for λ is rather complicated.

An epidemiologically feasible wave-front is one in which all the variables remain nonnegative as $x \to -\infty$ (i.e. as $z \to -\infty$ in the traveling wave variable formulation). The decay of A_i, J_i and m_i to zero as $z \to -\infty$ must not be oscillatory. It is therefore necessary that there should exist at least one strictly positive real root λ of the characteristic equation with the property that the corresponding eigenvector (q_1, q_2, q_3) points into the positive octant in $\mathbf{R}^3$. This happens only for c above some minimum value $c_{\min} > 0$. The need to examine carefully the eigenvectors as well as the eigenvalues makes the problem especially nontrivial, and the reader is referred to [7] for all the details.

For the case when the birth functions $b(\cdot)$ and $B(\cdot)$ are chosen as in (4.32) the minimum speed of spread, according to the predictions of the linearised analysis, was computed in [7] to be about 2.62 km per day, i.e. about 956 km per year. This is roughly consistent with the speed at which West Nile virus has spread across the USA. The disease first emerged in New York in 1999 and had reached the West coast five years later. There is, however, great uncertainty about some of the parameter values, especially the diffusivities. It seems to be difficult to find good data on diffusion coefficients for bird species generally. Both fledgling and adult crows are involved in the model being summarised here, and good data on their respective diffusivities is not really available. It was nevertheless noted in [7] that the speed of spread is not particularly sensitive to the values of certain particular parameters (e.g. the diffusivity of mosquitoes) but very sensitive to others (particularly the contact rates).

There also remains the issue of whether it is correct to compute the minimum speed according to the predictions of the linearised analysis, and then declare that solutions starting from realistic initial data will evolve to that speed. The mathematical theory of the speed of spread in reaction–diffusion equations with functional terms is still far from complete, especially for coupled systems such as those being described here. Considerable progress in this area has, however, taken place recently for scalar equations (see Thieme and Zhao [30]).

4.5 Spatial Spread in Patches: Asymptotic Dispersal

To tie the model simulations to the surveillance data, it is sometimes desirable to develop patchy models, since a public surveillance system is normally organized by medical and administrative regions and landscape. Also, the surveillance seems to indicate the jump or discontinuous spatial spread patterns in the establishment phase of WNv, as shown in the 2000–2003 Health Canada map of dead birds submitted for WNv diagnosis by health region. This discontinuous spatial spread seems to be the consequence of the combination of the local interaction and spatial diffusion of birds and mosquitoes and long-range dispersal of birds, and this also motivated the use of patchy models instead of the reaction–diffusion model.

In [13], a patchy model for the spatial spread of West Nile virus was formulated and analyzed, with a goal to see how the interaction of the ecology of birds and mosquitoes, the epidemiology of bird–mosquito cycles, and the diffusion and immigration patterns of birds affect the long-term and transient transmission of the diseases within the whole region consisting of multiple patches. This was partially achieved by calculating the basic reproduction number of the region as a function of the basic reproduction number of each patch, the spatial dispersal rates and patterns of birds, and the spatial scale of the birds' flying range in comparison with the mosquitoes' flying range.

The work focused on the one-dimensional patch model, which can only be regarded as a theoretical approximation of the West Nile virus landscape in a given region and better understanding of the West Nile virus spread in a real medical landscape can be achieved only by extending this work to a two-dimensional model and by incorporating more spatiotemporal heterogeneities.

To formulate the patchy model in [13] for the spatial spread of the West Nile virus, we assume that there are N patches under consideration where, depending on the purpose of modeling, availability of data, and implementation of surveillance, control, and prevention measures, the partition of the whole region into nonoverlapping patches changes. In the modeling and simulations in [13], the average distance a female mosquito can fly during its lifetime was used as a measuring unit for the partition. Therefore, if we assume the region is one-dimensional and we use $1, \dots, N$ to denote the corresponding patches, then mosquitoes belonging to the ith patch can fly only to their nearest neighboring patches $i - 1$ and $i + 1$, while birds belonging to the i-th patch fly to their mth neighbor patches $i - m, \dots, i - 1, i + 1, \dots, i + m$, with $m \geq 1$.

We also make the following assumptions (i) the virus does not have any adverse effect on mosquitoes and vertical transmission in mosquitoes can be ignored; (ii) most birds will recover from the virus and become immune to further infection and new-born birds have no immunity; and (iii) birds and mosquitoes have fixed recruitment rates in each patch.

We denote the number of individuals of birds and mosquitoes on the ith patch at time t respectively by

$$B_{Si} \; : \; \text{the susceptible birds in patch } i$$
$$B_{Ii} \; : \; \text{the infectious birds in patch } i$$
$$M_{Si} \; : \; \text{the susceptible mosquitoes in patch } i$$
$$M_{Ii} \; : \; \text{the infectious mosquitoes in patch } i$$

The model formulated in [13], based on the model set up in [2] for the dynamics between birds and mosquitoes within a patch and linear spatial dispersal among patches, and takes the form

$$
\begin{cases}
\dfrac{dB_{Si}}{dt} = b_i - d_{bi}B_{Si} + \displaystyle\sum_{\substack{(j=-m+i \\ j \neq i)}}^{m+i} D_{bji}B_{Sj} - \displaystyle\sum_{\substack{(j=-m+i \\ j \neq i)}}^{m+i} D_{bij}B_{Si} - \dfrac{C_{mbi}M_{Ii}B_{Si}}{N_{Bi}}, \\[2em]
\dfrac{dB_{Ii}}{dt} = -d_{b2i}B_{Ii} + \dfrac{C_{mbi}M_{Ii}B_{Si}}{N_{Bi}} + \displaystyle\sum_{\substack{(j=-m+i \\ j \neq i)}}^{m+i} D_{bji}B_{Ij} - \displaystyle\sum_{\substack{(j=-m+i \\ j \neq i)}}^{m+i} D_{bij}B_{Ii}, \\[2em]
\dfrac{dM_{Si}}{dt} = m_i - d_{mi}M_{Si} + \displaystyle\sum_{|k-i|=1} D_{mki}M_{Sk} - \displaystyle\sum_{|k-i|=1} D_{mik}M_{Si} - \dfrac{C_{bmi}M_{Si}B_{Ii}}{N_{Bi}}, \\[2em]
\dfrac{dM_{Ii}}{dt} = -d_{mi}M_{Ii} + \displaystyle\sum_{|k-i|=1} D_{mki}M_{Ik} - \displaystyle\sum_{|k-i|=1} D_{mik}M_{Ii} + \dfrac{C_{bmi}M_{Si}B_{Ii}}{N_{Bi}},
\end{cases}
$$

$$(4.82)$$

with $1 \leq i \leq N$. The total number of birds in patch i is $N_{Bi} = B_{Si} + B_{Ii}$. All parameters are defined in Table 4.3. Based on the biological fact that the death rate of infected birds is greater than that of susceptible birds, we assume $d_{b2i} \geq d_{bi}$ for all i.

We assume the dispersion rates of birds depend on the distance from the starting patch to their destination, but these rates may depend on the direction, thus

$$
\begin{cases}
D_{bij} = 0, D_{bji} = 0, \text{ if } \; |i - j| > m, \\
D_{bji} = g_b(i - j), \quad \text{if } 0 < |i - j| \leq m,
\end{cases}
$$

$$(4.83)$$

Table 4.3. Definitions for parameters in the model (4.82)

Parameter	Meaning
b_i	Recruitment rate of birds in patch i
d_{bi}	Death rate of birds in patch i
C_{mbi}	Effective contact rate between susceptible birds and infectious mosquitoes in patch i
d_{b2i}	Death rate of infectious birds in patch i
m_i	Recruitment rate of mosquitoes in patch i
C_{bmi}	Contact rate between susceptible mosquitoes and infectious birds in patch i
d_{mi}	Death rate of mosquitoes in patch i
D_{bij}	Diffusion rate of birds from the ith patch to the jth patch
D_{mij}	Diffusion rate of mosquitoes from the ith patch to the jth patch

where $g_b : \{-m, \dots, -1, 1, \dots, m\} \longrightarrow [0, \infty)$ is the dispersion function. We assume Neumann boundary conditions; namely, if $j < 0$ or $j > N$, then $D_{bji} = D_{bij} = 0$.

The dispersal rates of mosquitoes are given by

$$\begin{cases} D_{mik} = 0, D_{mki} = 0, & \text{if } |k - i| \neq 1, \\ D_{mki} = d_{m12}, & \text{if } |k - i| = 1, \end{cases} \tag{4.84}$$

where $d_{m12} > 0$ is a constant. Again, we assume that if $k < 0$ or $k > N$, then $D_{mki} = D_{mik} = 0$.

The disease-free equilibrium (DFE) is given by solving the vector equation

$$\begin{cases} \mathbf{B_b}\overrightarrow{\mathbf{B_S}} = \overrightarrow{\mathbf{b}}, \\ \mathbf{M_m}\overrightarrow{\mathbf{M_S}} = \overrightarrow{\mathbf{m}}, \end{cases} \tag{4.85}$$

where $\overrightarrow{\mathbf{B_S}} = (B_{S1}, \dots, B_{SN})^T$, $\overrightarrow{\mathbf{b}} = (b_1, \dots, b_N)^T$; $\mathbf{B_b}$ is a $N \times N$ matrix with

$$\mathbf{B}_{bii} = d_{bi} + \sum_{j=-m+i, j \neq i}^{m+i} D_{bij},$$

and $\mathbf{B}_{bij} = -D_{bji}$ for $0 < |i - j| \leq m$ and $1 \leq j \leq N$, otherwise $\mathbf{B}_{bij} = 0$; $\overrightarrow{\mathbf{M_S}} = (M_{S1}, \dots, M_{SN})^T$, $\overrightarrow{\mathbf{m}} = (m_1, \dots, m_N)^T$; $\mathbf{M_m}$ is a tridiagonal matrix with

$$\mathbf{M}_{mii} = d_{mi} + \sum_{|k-i|=1} D_{mik},$$

and if $|i - j| = 1$ and $1 \leq j \leq N$, $\mathbf{M}_{mij} = -D_{mji}$, otherwise $\mathbf{M}_{mij} = 0$.

Simple matrix analysis shows that (4.85) has exactly one positive solution, denoted by

$$\overrightarrow{\mathbf{Bs}} = (B_{S1}^*, \ldots, B_{SN}^*)^T,$$

$$\overrightarrow{\mathbf{Ms}} = (M_{S1}^*, \ldots, M_{SN}^*)^T. \tag{4.86}$$

The basic reproduction number, denoted by $\mathcal{R}_0$, is "the expected number of secondary cases produced, in a completely susceptible population, by one typical infectious individual". If $\mathcal{R}_0 < 1$, then on average an infected individual produces less than one new infected individual over the course of its infectious period, and the infection cannot grow. Conversely, if $\mathcal{R}_0 > 1$, then each infected individual produces, on average, more than one new infection, and the disease can invade and spread in the population. In [13], the formula in [33] was used to calculate the reproduction number $\mathcal{R}_0$. Namely, $\mathcal{R}_0 = \rho(FV^{-1})$, where

$$\mathcal{F} = \begin{pmatrix} F_1 \\ & F_2 \end{pmatrix} \quad \text{and} \quad \mathcal{V} = \begin{pmatrix} B \\ & M \end{pmatrix},$$

with an empty element or block in a matrix meaning zero (number or matrix), and

$$F_1 = \mathrm{diag}(C_{mb1}, \ldots, C_{mbN}),$$

$$F_2 = \mathrm{diag}(C_{bm1}M_{S1}^*/B_{S1}^*, \ldots, C_{bmN}M_{SN}^*/B_{SN}^*),$$

$$B = \begin{pmatrix} B_{11} & \cdots & B_{1,m+1} & & & \\ & \ddots & & \ddots & & \\ B_{m+1,1} & \ddots & & & \ddots & \\ & & \ddots & & & B_{N,N-m} \\ & & & \ddots & & \ddots \\ & & B_{N-m,N} & & & B_{NN} \end{pmatrix},$$

with

$$B_{ii} = d_{b2i} + \sum_{\substack{j=-m+i \\ (j \neq i, 1 \leq j \leq N)}}^{m+i} D_{bij},$$

$$B_{ij} = -D_{bji}, \text{ if } 0 < |i - j| \leq m \text{ and } 1 \leq j \leq N,$$

and

$$M = \begin{pmatrix} M_{11} & M_{12} & & & \\ M_{21} & \ddots & & \ddots & \\ & \ddots & & \ddots & M_{N,N-1} \\ & & M_{N-1,N} & & M_{NN} \end{pmatrix},$$

with

$$M_{11} = d_{m1} + d_{m12},$$
$$M_{ii} = d_{mi} + 2d_{m12}, \ i = 2, \ldots, N - 1,$$
$$M_{NN} = d_{mN} + d_{m12},$$
$$M_{ik} = -d_{m12}, \qquad \text{if } |i - k| = 1 \text{ and } 1 \leq i, k \leq N.$$

Unfortunately, it is a nontrivial task to find the explicit form of $\mathcal{R}_0$ in the general case. [13] considered the special case where all patches are identical from the aspect of ecology and epidemiology:

$$b_i = b, \quad d_{bi} = d_b, \quad d_{b2i} = d_{b2}, \quad d_{mi} = d_m,$$
$$m_i = \tilde{m}, \ C_{mbi} = C_{mb}, \ C_{bmi} = C_{bm} \ D_{bij} = D_{bji}.$$

In this case, a straightforward calculation gives the coordinates for the DFE as

$$\overrightarrow{\mathbf{Bs}} = (B_{S1}^*, \ldots, B_{SN}^*)^T = \left(\frac{b}{d_b}, \ldots, \frac{b}{d_b} \right)^T,$$

$$\overrightarrow{\mathbf{Ms}} = (M_{S1}^*, \ldots, M_{SN}^*)^T = \left(\frac{\tilde{m}}{d_m}, \ldots, \frac{\tilde{m}}{d_m} \right)^T. \tag{4.87}$$

An explicit expression for the reproduction number $\mathcal{R}_0$ is given by

$$\mathcal{R}_0 = \sqrt{\frac{C_{bm} C_{mb} M_S^*}{d_{b2} d_m B_S^*}}.$$

This shows that a region consisting of identical patches coupled by symmetric dispersal of birds has the same reproduction number as if each patch is isolated from the others.

This conclusion is not true anymore, however, if the dispersal of birds is not symmetric, as shown next. [13] considered the case where the dispersal rates of birds depend on the direction, using the perturbation theory to calculate the basic reproduction number of model (4.82) in the special case of three identical patches.

To describe the results in [13], we denote the diffusion rate of birds to the left by D_{bl} and to the right by D_{br}. To address the impact of the direction-selective dispersal of birds on $\mathcal{R}_0$, we write $D_{br} = D_{bl} + \epsilon$, where ϵ is a small positive number. Note that $\epsilon = 0$ implies the symmetric dispersal of birds. Let

$$p_2 = \frac{8d_{b2} B_S^* (d_{b2} - d_b + 2D_{bl})}{9(d_b + 3D_{bl})^2 D_{bl}}. \tag{4.88}$$

Then, [13] obtained

$$\mathcal{R}_0 = \frac{1}{\sqrt{\dfrac{d_{b2} d_m B_S^*}{C_{bm} C_{mb} M_S^*} + \epsilon^2 p_2 + O(\epsilon^3)}}. \tag{4.89}$$

Note that if we fix $D_{bl} > 0$, p_2 is always positive. Therefore the breaking of symmetry in spatial dispersal of birds always decreases $\mathcal{R}_0$.

Reference [13] also reported some numerical simulation results to demonstrate the effect of different dispersal patterns of birds on the spatial spread of WNv, and to illustrate possible discrepancy between surveillance data and the model-based simulations in different time scales. Their focus is on the time when a particular patch has recorded WNv activities, namely, when at least one bird has died of WNv infection. This allows them to compare the simulation results with surveillance data in Canada, since dead birds with WNv was used as an indicator for determining whether a region has WNv activities.

The ranges of parameters involved were obtained in biological literatures [2, 15, 23, 31, 32, 35]. It was also assumed that all patches are identical from the aspect of ecology and epidemiology as discussed before, and the dispersal rates of birds are a decreasing function of the distance from the origin, but the spatial dispersal may be nonsymmetric in terms of spatial direction (left vs. right, in the case of one-dimensional space). For the sake of simplicity, the dispersal rate of birds $g_b(k)$, with $k = i - j$, from patch i to j, is a piecewise linear function, was given by

$$g_b(k) = \begin{cases} \frac{h_1}{m}(m - |k|), & \text{if } 0 < k \le m, \\ \frac{h_2}{m}(m - |k|), & \text{if } -m \le k < 0, \end{cases} \tag{4.90}$$

where $m \pm i$ are the furthest patches that a bird can fly during the average life span of female mosquitoes and h_1 measures the diffusivity rate of birds to the left, while h_2 measures the diffusivity rate of birds to the right. The net rate at which a bird flies out of a given patch should be less than 1; therefore, $0 \le (h_1 + h_2)m/2 < 1$. Notice that $h_1 = h_2$ corresponds to the bidirectional dispersal symmetric with respect to the spatial direction, while $h_1 \ne h_2$ corresponds to the nonsymmetric spatial direction selective dispersal that seems to be closer to the ecological reality of birds in Canada within the time scale under consideration.

On average, birds can fly 13.4 km per day or 1,000 km per year [17]. During the average life span of 30 days, most female mosquitoes remain within 1.6 km of their breeding site. A few species may range up to 10 km or more. Thus in the average lifespan of female mosquitoes, the flying range of a bird is about 40 times that of mosquitoes. Hence, $m = 40$ is assumed.

The distance from British Columbia to South Ontario is about 3,000 km, and hence, the total number of patches is assumed $N = 300$. In the simulation of [13], the time unit is one day. Since WNv is a seasonal disease, we consider the period from late April to early October to be a total of about 180 days.

The simulations in [13] with $h_1 = h_2 = 0.005$ give the spread speed of WNv about 1,000 km per year, which coincides with the observed spread rate in North America [12]. Increasing h_2 slightly while keeping h_1 unchanged yielded nonsymmetric dispersal of birds, but this minor breaking of symmetry has limited impact on the number of infected birds and their spatial spread.

However, if h_2 is further increased to $h_2 = 0.01$, the spread speed of the disease is much faster and the magnitude of outbreaks is higher compared to the cases with symmetry or with minor symmetric breaking. Naturally, we notice that the spatial spread is continuous in the sense that there is no patch i escaping from WNv if patch $j > i$ has WNv activities (i.e., had infected birds). This is due to the continuous spread of birds.

In reality, birds may skip some patches during their long-range dispersal. To model this special dispersal pattern, a dispersal function of the following form considered was

$$
g_b(k) = \begin{cases} \frac{h_1}{m}(m - |k|), & \text{if } |k| \bmod 4 = 0 \text{ and } 0 < k \le 40, \\ \frac{h_2}{m}(m - |k|), & \text{if } |k| \bmod 4 = 0 \text{ and } -40 \le k < 0, \\ 0, & \text{otherwise.} \end{cases} \tag{4.91}
$$

In other words, the birds in patch i jump to patches $i \pm 4, \ldots, i \pm 4J$, where J is the integer part of $m/4$.

Note that the model (4.82) is still a continuous model even though the dispersal function of birds is not continuous. Simulation results in [13] using the above jump dispersal function and in the case $h_1 = h_2 = 0.01$ show obvious jumps in the transmission of WNv and the disease spread speed is about 1,000 km per year. In this case, some patches avoid the disease because of the discontinuous dispersal of birds.

4.6 Discussion

We will briefly describe some aspects of mosquito and bird behavior, and control, that might benefit from further mathematical work.

Much work remains to be done in the modeling of mosquito dispersal behavior, which varies from one species to another. Although mosquitoes are capable of flying significant distances, they usually only go far enough to find a blood meal. Wind and air turbulence play an important role in mosquito movement. For example, species that are active during the day may be more likely to be carried into the upper air, by turbulence and convection, and conveyed long distances than species that are active at night (see Service [25]). Large scale weather events can be associated with large scale migration of insects in general and mosquitoes in particular. Wind can have other more indirect implications for the mosquito life cycle too, since the larvae require water that is relatively unaffected by wave action (they need slow moving or stagnant water). Indeed, a reason for the preference for water containing vegetation is that this reduces wave action. It has been suggested that mosquito memory may limit oviposition in unsuitable habitats (McCall and Kelly [16]).

There is a need for models that incorporate aspects of bird dispersal behavior not considered in the works we have summarised here. It is really only in the breeding season that crows, once paired, seek to establish individual territories to raise their broods. In the nonbreeding season crow activities tend to be centered around large communal roosts, which may contain tens of thousands of birds, to which they return in the evenings after searching for food during the day. Recent work by Ward et al. [34] suggests that roosting behavior may be an important component in regulating West Nile virus transmission because of the nocturnal feeding behavior of *Culex* mosquitoes. Moreover, large roosts are often in mosquito-friendly habitats: areas with large trees protected by wetlands. An extension of the model we have described involving simple Fickian diffusion might be used to incorporate the fact that in the final days of a crow's life after contracting the virus the crow is effectively a sitting duck for feeding mosquitoes. An additional compartment of nondiffusing crows which are in the final stages of disease might constitute an appropriate extension of the modeling described here. It should be noted that although crows are particularly susceptible to West Nile virus, there are numerous other bird species including House Sparrows and Cardinals which are vulnerable to the disease but have lower disease induced mortality rates. Some bird species do not become ill from the virus but harbor high levels of it in their blood and therefore serve as reservoirs. Although birds seem to be the primary hosts for West Nile, the virus has been known to infect horses, cats, dogs, squirrels and rabbits.

Biological control methods provide an alternative to the use of larvicides to kill mosquito larvae. For example, the mosquitofish *Gambusia affinis* is a small surface feeding minnow that can be stocked seasonally in water sites where mosquitoes are known to be active. These fish can eat over 100 larvae per day, have short gestation periods and their young can begin eating larvae immediately. However, there can be problems associated with the introduction of fish to areas where they would not normally occur. For example, the fish may have an undesirable impact on native species that already exist, and once introduced can be hard to remove. In parts of the southwestern United States the use of another predator of mosquito larvae and pupae, the arroyo chub *Gila orcutti*, is presently being investigated.

Biological control of adult mosquitoes seems to be more difficult than biological control of larvae. The use of birds and bats is sometimes suggested, but some bird species can make matters worse by eating dragonflies, which are important predators of both larval and adult mosquitoes (dragonflies can also be killed by the adulticide sprays that are aimed at killing adult mosquitoes).

In the case of West Nile virus the transmission dynamics is determined mainly by the behavior of mosquitoes and birds. Humans are considered as dead-end hosts, in the sense that they are incapable of continuing the virus transmission cycle. However, based on a study of the *Aedes aegypti* mosquito in Puerto Rico and Thailand, Harrington et al. [10] have suggested that people rather than mosquitoes are the primary mode of dengue virus dissemination.

Finally, there is undoubtedly a need for more work on the spread of mosquito borne diseases between countries and continents via air and sea travel. Recently, Tatem et al. [29] used a database of international ship and aircraft traffic movements in a study of the spread of the mosquito *Aedes albopictus*, which is known to be a vector of numerous arboviruses including dengue, yellow and West Nile fever.

Acknowledgment

This work was partially supported by the Public Health Agency of Canada, Ontario Ministry of Health and Long-Term Care, Natural Sciences and Engineering Research Council of Canada, Canada Research Chairs Program and Mathematics for Information Technology and Complex Systems.

References

1. J. S. Blackmore and R. P. Dow, *Differential feeding of culex tarsalis on nestling and adult birds*, Mosq News **18** (1958), 15–17
2. C. Bowman, A. B. Gumel, P. van den Driessche, J. Wu, and H. Zhu, *A mathematical model for assessing control strategies against west nile virus*, Bull. Math. Biol. **67** (2005), 1107–1133
3. N. F. Britton, *Spatial structures and periodic travelling waves in an integro-differential reaction-diffusion population model*, SIAM J. Appl. Math. **50** (1990), 1663–1688
4. S. A. Gourley and N. F. Britton, *A predator prey reaction diffusion system with nonlocal effects*, J. Math. Biol. **34** (1996), 297–333
5. S. A. Gourley, J. W. H. So, and J. Wu, *Non-locality of reaction–diffusion equations induced by delay: biological modeling and nonlinear dynamics*, J. Math. Sci. **124** (2004), 5119–5153
6. S. A. Gourley and J. Wu, *Delayed non-local diffusive systems in biological invasion and disease spread*, Fields Inst. Commun. **48** 137–200, Amer. Math. Soc., Providence, RI, 2006
7. S. A. Gourley, R. Liu, and J. Wu, *Some vector borne diseases with structured host populations: extinction and spatial spread*, SIAM J. Appl. Math. **67** (2006/07), 408–433
8. S. Gourley, R. Liu, and J. Wu, *Eradicating vector-borne diseases via age-structured culling*, J. Math. Biol. **54** (2007), 309–335
9. W. S. C. Gurney, S. P. Blythe, and R. M. Nisbet, *Nicholson's blowflies revisited*, Nature **287** (1980), 17–21
10. L. C. Harrington, T. W. Scott, K. Lerdthusnee, R. C. Coleman, A. Costero, G. G. Clark, J. J. Jones, S. Kitthawee, P. Kittayapong, R. Sithiprasasna, and J. D. Edman, *Dispersal of the dengue vector aedes aegypti within and between rural communities*, Am. J. Trop. Med. Hyg. **72**(2) (2005), 209–220
11. Y. Kuang, *Delay differential equations with applications in population dynamics*, vol. Mathematics in Science and Engineering 191, Academic Press, Boston, 1993
12. M. Lewis, J. Renclawowicz, and P. van den Driessche, *Traveling waves and spread rates for a west nile virus model*, Bull. Math. Biol. **68** (2006), 3–23

13. R. Liu, J. Shuai, H. Zhu, and J. Wu, *Modeling spatial spread of west nile virus and impact of directional dispersal of birds.*, Math. Biosci. Eng. **3** (2006), 145–160

14. C. C. Lord and J. F. Day, *Simulation studies of st. louis encephalitis and west nile virues: the impact of bird mortality*, Vector Borne Zoonotic Dis. **1** (4) (2001), 317–329

15. C. C. Lord and J. F. Day, *Simulation studies of st. louis encephalitis virus in south florida*, Vector Borne Zoonotic Dis. **1** (4) (2001), 299–315

16. P. J. McCall and D. W. Kelly, *Learning and memory in disease vectors*, Trends Parasitol. **18** (2002), 429–433

17. A. Okubo, *Diffusion-type models for avian range expansion*, University of Ottawa Press, Ottawa 1998

18. C. Ou and J. Wu, *Spatial spread of rabies revisited: influence of age-dependent diffusion on nonlinear dynamics*, SIAM J. Appl. Math. **67** (2006), 138–163

19. M. A. Pozio, *Behaviour of solutions of some abstract functional differential equations and application to predator-prey dynamics*, Nonlinear Anal. **4** (1980), 917–938

20. M. A. Pozio, *Some conditions for global asymptotic stability of equilibria of integro-differential equations*, J. Math. Anal. Appl. **95** (1983), 501–527

21. R. Redlinger, *Existence theorems for semilinear parabolic systems with functionals*, Nonlinear Anal. **8** (1984), 667–682

22. R. Redlinger, *On volterra's population equation with diffusion*, SIAM J. Math. Anal. **16** (1985), 135–142

23. M. R. Sardelis and M. J. Turell, *Ochlerotatus j. japonicus in frederick county, maryland: Discovery, distribution, and vector competence for west nile cirus*, J. Am. Mosq. Control Assoc. **17** (2001), 137–141

24. T. W. Scott, L. H. Lorenz, and J. D. Edman, *Effects of house sparrow age and arbovirus infection on attraction of mosquitoes*, J Med Entomol **27** (1990), 856–863

25. M. W. Service, *Effects of wind on the behaviour and distribution of mosquitoes and blackflies*, Int. J. Biometeorol. **24** (1980), 347–353

26. R. R. L. Simons and S. A. Gourley, *Extinction criteria in stage-structured population models with impulsive culling*, SIAM. J. Appl. Maths **66** (2006), 1853–1870

27. H. L. Smith and H. R. Thieme, *Strongly order preserving semiflows generated by functional-differential equations*, J. Diff. Eqns. **93** (1991), 332–363

28. J. W. H. So, J. Wu, and X. Zou, *A reaction diffusion model for a single species with age structure, i. travelling wave fronts on unbounded domains*, Proc. Roy. Soc. Lond. Ser. A. **457** (2001), 1841–1853

29. A. J. Tatem, S. I. Hay, and D. J. Rogers, *Global traffic and disease vector dispersal*, Proc. Nat. Acad. Sci. **103** (2006), 6242–6247

30. H. R. Thieme and X. -Q. Zhao, *Asymptotic speeds of spread and traveling waves for integral equations and delayed reaction-diffusion models*, J. Differ. Equ. **195** (2) (2003), 430–470

31. M. J. Turell, M. O'Guinn, and J. Oliver, *Potential for new york mosquitoes to transmit west nile virus*, Am. J. Trop. Med. Hyg. **62** (2002), 413–414

32. M. J. Turell, M. L. O'Guinn, D. J. Dohm, and J. W. Jones, *Vector competence north american mosquitoes (diptera: Cullocidae) for west nile virus*, J. Med. Entomol. **38** (2001), 130–134

33. P. van den Driessche and J. Watmough, *Reproduction numbers and sub-threshold endemic equilibria for compartmental models of disease transmission*, Math. Biosci. **180** (2002), 29–48
34. M. P. Ward, A. Raim, S. Yaremych-Hamer, R. Lampman, and R. J. Novak, *Does the roosting behavior of birds affect transmission dynamics of west nile virus?*, Am. J. Trop. Med. Hyg. **75** (2006), 350–355
35. M. J. Wonham, T. de Camino-Beck, and M. Lewis, *An epidemiological model for west nile virus: Invasion analysis and control applications*, Proc. R. Soc. Lond., Ser. B **271** (1538) (2004), 501–507
36. Y. Yamada, *Asymptotic stability for some systems of semilinear volterra diffusion equations*, J. Diff. Eqns. **52** (1984), 295–326

Aggregation of Variables and Applications to Population Dynamics

P. Auger[1,5], R. Bravo de la Parra[2,1], J.-C. Poggiale[3], E. Sánchez[4], and T. Nguyen-Huu[5,1,3]

[1] IRD UR Géodes, Centre IRD de l'Ile de France, 32, Av. Henri Varagnat, 93143 Bondy cedex, France
pierre.auger@bondy.ird.fr
[2] Departamento de Matemáticas, Universidad de Alcalá, 28871 Alcalá de Henares, Madrid, Spain
rafael.bravo@uah.es
[3] Laboratoire de Microbiologie, Géochimie et d'Ecologie Marines, UMR 6117, Centre d'Océanologie de Marseille (OSU), Université de la Méditerranée, Case 901, Campus de Luminy, 13288 Marseille Cedex 9, France
Jean-Christophe.Poggiale@univmed.fr
[4] Departamento de Matemáticas, E.T.S.I. Industriales, U.P.M., c/ José Gutiérrez Abascal, 2, 28006 Madrid, Spain
esanchez@etsii.upm.es
[5] IXXI, ENS Lyon, 46 allée d'Italie, 69364 Lyon cedex 07, France
tri.nguyen-huu@ens-lyon.fr

Summary. Ecological modelers produce models with more and more details, leading to dynamical systems involving lots of variables. This chapter presents a set of methods which aim to extract from these complex models some submodels containing the same information but which are more tractable from the mathematical point of view. This "aggregation" of variables is based on time scales separation methods. The first part of the chapter is devoted to the presentation of mathematical aggregation methods for ODE's, discrete models, PDE's and DDE's. The second part presents several applications in population and community dynamics.

5.1 Introduction

Ecology aims to understand the relations between living organisms and their environment. This environment constitutes a set of physical, chemical and biological constraints acting at the individual level. In order to deal with the complexity of an ecosystem, ecology has been developed on the basis of a wide range of knowledge starting from the molecular level (molecular ecology) to the ecosystem level. One of the current aims of ecological modelling is to use the mathematical formalism for integrating all this knowledge.

On the other hand, mathematical ecology provided a large amount of rather simple models involving a small number of state variables and parameters. The time continuous Lotka–Volterra models, published in the beginning of the twentieth century [59, 60, 94] as well as the discrete host–parasite Nicholson–Bailey models [69, 70] are classic examples and can be found in many bio-mathematical textbooks as the book by Edelstein-Keshet [48] and the book by Murray [66] in which many other examples and references are given. In such population dynamics models, the state variables are often chosen as the population densities and the model is a set of nonlinear coupled ordinary differential equations (ODE's) or discrete equations. The models describe the time variation of the interacting populations. Of course, mathematical ecologists proposed also more realistic models taking account of some populations structures (space, age, physiology, etc.). Mathematical methods have been developed to deal with these structured population models, but which may fail to get robust results for high dimensional systems.

During the last decades, supported by the fast development of computers, a new generation of ecological models has appeared. Nowadays, lots of ecological models consider more and more details. Lots of populations are involved in a community and in food webs. Furthermore, each population is not homogeneous in the sense that all individuals are identical but each individual has changing properties (physiology, metabolism, behaviour), according to its environment. For instance, functional ecology considers functional groups corresponding to the different functions of living organisms in the ecosystems. It follows that lots of models consider populations structured in subgroups.

Incorporating more details in models is necessary to advance toward a more realistic description of ecological systems and to understand how living organisms respond to the forcing imposed by their environment changes. The drawback of a detailed description of systems is the fact that models become more complex, involving an increasing number of variables and parameters. A mathematical study with general and robust results is then difficult to perform. For this reason, it is important to find which details are really relevant and must be incorporated in a model. An important goal of ecological modelling should thus be to describe tractable models.

In the context of terrestrial ecology, if we consider a forest dynamics for instance, we can just consider the total forest surface or how trees are distributed among species or globally. But the dynamics of these variables or indicators depend on the individuals properties of the trees (height, weight, basal area, metabolism, etc.). Should we take into account all the details? Is there a trade-off between the amount of details to be integrated and the relative simplicity required for understanding forest dynamics? In this case, the surface or the spatial structure indicators are global variables that we call macro or aggregated variables. These variables actually depend on the individuals descriptors, which we call the micro-variables.

The same approaches are considered in marine ecology. The simplest way should be to consider the concentrations of mineral matter, primary producers, zooplankton, top-predators and microbial loop with bacteria and detritus as the variables of the model. This point of view permits to summarize the biological components of a marine ecosystem with only a few macro-variables. However, each variable describes a set of lots of populations having different properties. This is the main reason to split them into different micro-variables, leading to a set of differential systems involving typically dozens of variables and parameters.

In ecology, the problem of aggregation of variables may be set in this way: when considering a detailed system with various interacting organization levels, is there a way to find, at each level, a reduced set of variables describing the dynamics of this level? How to find such variables? How to find the relations between these macro-variables and the micro-variables associated with the detailed description? Under which assumptions these questions could be dealt with? Do these assumptions have a realistic basis? This chapter aims to describe some mathematical methods of aggregation of variables which help to answer parts of these questions. Two main goals of variables aggregation are dealt with in this chapter. The first one is to reduce the dimension of the mathematical model to be handled analytically. The second one is to understand how different organization levels interact and which properties of a given level emerge at other levels.

Aggregation of variables is coming from economy and has been introduced in ecology by Iwasa, Andreasen and Levin, in [52]. In general, the aggregation of a system consists of defining a small number of global variables, functions of its state variables, and a system describing their dynamics. When the aggregated dynamics are consistent with the original dynamics in the sense that the global variables behave identically both in the initial system and in the aggregated one, it is called perfect aggregation [52]. Perfect aggregation is a very particular situation which is rarely possible since it requires very drastic conditions. Consequently, methods for approximate aggregation have been developed [53]. Approximate aggregation deals with methods of reduction where the consistency between the dynamics of the global variables in the complete system and the aggregated system is only approximate.

This chapter is devoted to approximate aggregation methods that are based on the existence of different time scales. It is common in ecology to consider different ecological levels of organization, the individual, population, community and ecosystem levels. In general, different characteristic time scales are associated with these levels of organization. For example, a fast time scale corresponds to individual processes while a slow time scale is associated with demographic ones. It is possible to take advantage of these two time scales in order to reduce the dimension of the initial complete model and to build a simplified system which describes the dynamics of a small number of global variables. Such methods originated in Auger [3] and were presented in a rigorous mathematical form for ODEs in Auger and Roussarie [16] and in Auger and Poggiale [12], extended to discrete models in Sanchez et al. [82] and in Bravo de la Parra et al. [29], to PDEs in Arino et al. [1] and to DDEs in Sanchez et al. in [81]. There are lots of examples in various applications fields where the intuitive ideas of the methods are used implicitly. It is for instance the case in epidemiology, when population dynamics is ignored at the epidemiological scale since the latter is much faster, see for example Chap. 3 Sect. 4 in this book. It is often correct but we shall give some examples where the intuitive ideas are not sufficient and the mathematical developments are useful.

The chapter is organized as follows. Three sections are devoted to mathematical aggregation methods associated to different mathematical formalisms while the last section illustrates these methods on particular ecological examples. The methods described in this chapter are not intended to be exhaustive and just address partially the problems suggested by the questions arisen above. Some open problems are discussed along the chapter. In the next section, we focus on aggregation methods for ODE's systems involving at least two different time scales. The third section proposes an approach for discrete time models. The fourth section is devoted to aggregation methods for Partial Differential Equation (PDE's) and Delayed Differential Equations (DDE's) systems involving different time scales. Finally, we illustrate the different methods presented in the previous sections by means of a set of examples from population dynamics and community dynamics.

5.2 Aggregation of Variables for ODE's Systems

5.2.1 Notation and Position of the Problem

Let us consider a population dynamics model describing the interactions between A populations and let us assume that each population is structured in subpopulations. We denote by n_i^α the abundance of subpopulation i in population α, $\alpha = 1, \ldots, A$ and $i = 1, \ldots, N_\alpha$ where N_α is the number of subpopulations in population α. We now assume that the dynamics of the subpopulation i in population α results from the interactions of a set of processes among which some are much faster than the other ones. The complete model reads:

$$\frac{dn_i^\alpha}{d\tau} = F_i^\alpha(\mathbf{n}) + \varepsilon f_i^\alpha(\mathbf{n}) \tag{5.1}$$

where $\mathbf{n}$ is the vector

$$\left(n_1^1, n_2^1, \ldots, n_{N_1}^1, n_1^2, n_2^2, \ldots, n_{N_2}^2, \ldots, n_1^\alpha, n_2^\alpha, \ldots, n_{N_\alpha}^\alpha, \ldots, n_1^A, n_2^A, \ldots, n_{N_A}^A\right)$$

F_i^α describes the fast processes affecting n_i^α and εf_i^α describes the slow processes affecting n_i^α. The parameter ε is small and means that the speed of the processes described in f_i^α are slow. This model is assumed to contain all the details that we want to include in the description. It governs the so called micro-variables n_i^α which are those associated to a detailed level. We denote by k the number of micro-variables, that is the dimension of $\mathbf{n}$. More precisely, we have:

$$k = \sum_{\alpha=1}^{A} N_\alpha$$

We want now to build a model which describes the system at the macro-level. We thus define a set of macro-variables. In this framework, a macro-variable is a variable varying slowly, that is a first integral of the fast dynamics. More precisely, let us define Y_j, $j = 1, \ldots, N$ the macro-variables. A such variable can be defined as a function of $\mathbf{n}$. The fact that Y_j is a slow variable means that its derivative with respect to τ is of order ε:

$$Y_j = \Phi_j(\mathbf{n}), \ j = 1, \ldots, N \tag{5.2a}$$

$$\frac{dY_j}{d\tau} = \sum_{\alpha=1}^{A} \sum_{i=1}^{N_\alpha} \frac{\partial \Phi_j(\mathbf{n})}{\partial n_i^\alpha} \frac{dn_i^\alpha}{d\tau} = O(\varepsilon) \tag{5.2b}$$

The second equation (5.2b), associated with the equation (5.1), implies the following equality:

$$\sum_{\alpha=1}^{A} \sum_{i=1}^{N_\alpha} \frac{\partial \Phi_j(\mathbf{n})}{\partial n_i^\alpha} F_i^\alpha(\mathbf{n}) = 0, \ j = 1, \ldots, N \tag{5.3}$$

Finally, the equations for the macro-variables read:

$$\frac{dY_j}{d\tau} = \varepsilon \sum_{\alpha=1}^{A} \sum_{i=1}^{N_\alpha} \frac{\partial \Phi_j(\mathbf{n})}{\partial n_i^\alpha} f_i^\alpha(\mathbf{n}) \tag{5.4}$$

Since the system is more detailed at the micro-level, we should have $N << k$. In order to use the macro-variables, we replace N micro-variables in the complete model (5.1) by some expressions depending of the macro-variables and this can be done under the following conditions. We suppose that the set of N equations (5.2a) permits to write N variables among the micro-variables n_i^α, $\alpha = 1, \ldots, A$, $i = 1, \ldots, N_\alpha$, as functions of the N macro-variables Y_j, $j = 1, \ldots, N$. We thus have to deal with k variables among which N are macro-variables and $k - N$ are micro-variables. This system is formed by $k - N$ equations of system (5.1) and the N equations of system (5.4). In other words, we perform a change of variables $(X, Y) \mapsto \mathbf{n}(X, Y)$ where X is a $k - N$ vector for which the coordinates are taken among the micro-variables n_i^α. With this change of variables, the complete system reads:

$$\frac{dX_i}{d\tau} = F_i(X, Y) + \varepsilon f_i(X, Y), \ i = 1, \ldots, k - N \tag{5.5a}$$

$$\frac{dY_j}{d\tau} = \varepsilon G_j(X, Y), \ j = 1, \ldots, N \tag{5.5b}$$

where

$$G_j(X, Y) = \sum_{\alpha=1}^{A} \sum_{i=1}^{N_\alpha} \frac{\partial \Phi_j(\mathbf{n}(X, Y))}{\partial n_i^\alpha} f_i^\alpha(\mathbf{n}(X, Y))$$

In this form, the model (5.5) is a so-called slow-fast system of differential equations, or slow-fast vector field. The Geometrical Singular Perturbation (GSP) theory provides some results to deal with such systems and the most important point is that, under some conditions, we can reduce the complete model to an aggregated model governing only the macro-variables. We now first recall some important points of this theory and then explain the conditions for the reduction and their consequences.

5.2.2 Normally Hyperbolic Manifolds and GSP Theory

There exists lots of results concerning the reduction of the dimension of a dynamical system in order to facilitate its study. For instance, we can find several statements of the centre manifold theorem in various contexts (ordinary differential equations, partial differential equations, delay differential equations, difference equations). Carr's book [32] gives a detailed description of the theorem with many applications. The centre manifold theorem states some conditions under which there exists a regular manifold containing the non trivial part of the dynamics. This kind of manifolds are associated to non hyperbolic singularities and are local ones. In 1971, Fenichel [49] stated a theorem which provides conditions under which an invariant manifold persists to small enough perturbations, in the case of vector fields. In the same time, Hirsch et al. in [50] gave some necessary conditions for the persistence and developed the normally hyperbolic manifolds theory. The perturbations of invariant manifolds theory originates from the works of Krylov and Bogoliubov [56]. Nowadays, this theory has lots of applications and some illustrations can be found in Pliss and Sell [72]. Furthermore, Wiggins [95] gives a complete description of the theory in finite dimension, this book is based on the Fenichel original paper. In these references, the conditions of normal hyperbolicity are based on geometrical considerations, which are not always useful in applications. Sakamoto [80] gave similar conditions by using eigenvalues of Jacobian matrices. His proof may also be obtained by Fenichel's methods. Our reduction method is based on this approach, see Auger et al. [9,11], and Auger and Poggiale [12–15]. Note that the Fenichel theorem has been extended to semi-groups on Banach spaces by Bates et al. [19,20].

5.2.3 Reduction Theorem

In order to perform the analysis, we add to system (5.5) the equation $\frac{d\varepsilon}{d\tau} = 0$:

$$\frac{dX_i}{d\tau} = F_i\left(X, Y\right) + \varepsilon f_i\left(X, Y\right), \, i = 1, \ldots, k - N \tag{5.6a}$$

$$\frac{dY_j}{d\tau} = \varepsilon G_j\left(X, Y\right), \, j = 1, \ldots, N \tag{5.6b}$$

$$\frac{d\varepsilon}{d\tau} = 0 \tag{5.6c}$$

The system (5.6) can be considered as an ε -perturbation of the system obtained with $\varepsilon = 0$. The situation where $\varepsilon = 0$ refers to the unperturbed problem. The conditions for the reduction are:

- (C1) When ε is null in system (5.6), then Y is a constant. We assume that, for each $Y \in I\!\!R^N$, there exists at least one equilibrium $(X = X^*(Y), Y, 0)$, defined by $F_i(X^*(Y), Y) = 0$, $i = 1, \ldots, k - N$. We define the set:

$$\mathcal{M}_0 = \{(X, Y, \varepsilon) \, ; X = X^*(Y) \, ; \varepsilon = 0\}$$

  This invariant set for the unperturbed system shall play the role of the invariant normally hyperbolic manifold mentioned in the GSP theory.

- (C2) Let us denote $J(Y)$ the linear part of system (5.6) around the equilibrium $(X^*(Y), Y, 0)$. We assume that the Jacobian matrix $J(Y)$ has $k - N$ eigenvalues with negative real parts and $N + 1$ null eigenvalues. With this condition, the set $\mathcal{M}_0$ is *normally hyperbolic* since, at each point in $\mathcal{M}_0$, the restriction of the linear part to the $\mathcal{M}_0$ normal space has negative eigenvalues. We now give the statement of the main theorem.

Theorem 1. *Under the conditions (C1) and (C2), for each compact subset Ω in $I\!\!R^N$ and for each integer $r > 1$, there exists a real number ε_0 and a C^r function Ψ,*

$$\begin{aligned} \Psi : \Omega \times [0; \varepsilon_0] &\longrightarrow & I\!\!R^{k-N} \\ (Y, \varepsilon) &\longmapsto & X = \Psi(Y, \varepsilon) \end{aligned}$$

such that:

  (1) $\Psi(Y, 0) = X^*(Y)$;

  (2) The graph $\mathcal{W}$ of Ψ is invariant under the flow defined by the vector field (5.6);

  (3) At each $(X^*(Y), Y, 0) \in \mathcal{M}_0$, $\mathcal{W}$ is tangent to the central eigenspace E^c associated with the eigenvalues of $J(Y)$ with null real parts.

This means that we can consider the restriction of the vector field to the invariant manifold which allows us to reduce the dimension of the model. The reduced system, called aggregated model, is:

$$\frac{dY_j}{dt} = G_j(\Psi(Y, \varepsilon), Y)$$

where $t = \varepsilon\tau$. Usually, since ε is small, we approximate the previous system by:

$$\frac{dY_j}{dt} = G_j(\Psi(Y, 0), Y)$$

Moreover, since Ψ is C^r, we can calculate a Taylor expansion of the invariant manifold with respect to the small parameter ε in order to increase the accuracy of the reduced system. The reduction and the Taylor expansion are illustrated in the following example. In this example, the zero order term in the expansion leads to a non structurally stable system. It means that the ε term is important to understand the real dynamics. This term is then calculated and the dynamics of the complete and reduced models are compared.

Nontrivial Example of Application

This example has been completely studied in Poggiale and Auger [77]. It illustrates the application of the previous theorem in a nontrivial case where a Taylor expansion of the application Ψ with respect to the small parameter ε is needed to understand how the reduced model is similar to the complete one. We consider a two patches predator–prey system. The prey can move on both patches while the predator remains on patch 1. The patch 2 is a refuge for the prey. We denote by n_i the prey density on patch i, $i = 1, 2$. We denote by p the predator density. On each patch, the prey population growth rate and the predator population death rate are linear, the predation rate is bilinear, that is proportional to prey and predator densities and the predator growth rate is proportional to the predation rate. The model is given by the following set of three ordinary differential equations:

$$\frac{dn_1}{d\tau} = m_2 n_2 - m_1 n_1 + \varepsilon n_1 (r_1 - ap) \tag{5.7a}$$

$$\frac{dn_2}{d\tau} = m_1 n_1 - m_2 n_2 + \varepsilon n_2 r_2 \tag{5.7b}$$

$$\frac{dp}{d\tau} = \varepsilon p(b n_1 - d) \tag{5.7c}$$

where m_i are respectively the proportions of prey populations leaving patch i by displacement per unit time, r_i is the prey population growth rate on patch i, d is the predator population death rate, a is the predation rate on patch 1 and bn_1 is the per capita predator growth rate. $\varepsilon << 1$ is a small parameter which means that movements have a larger speed than that associated to growth and death processes.

Let $n = n_1 + n_2$ be the total amount of prey. It follows that $u_1 = \frac{n_1}{n}$ and $u_2 = \frac{n_2}{n}$ are the proportions of prey on patch 1 and patch 2 respectively. With these variables, we can write the system (5.7) in the following equivalent way:

$$\frac{du_1}{d\tau} = m_2 - (m_1 + m_2)u_1 + \varepsilon u_1(1 - u_1)(r_1 - r_2 - ap) \tag{5.8a}$$

$$\frac{dn}{d\tau} = \varepsilon n \left(r_1 u_1 + r_2 u_2 - au_1 p \right) \tag{5.8b}$$

$$\frac{dp}{d\tau} = \varepsilon p \left(bu_1 n - d \right) \tag{5.8c}$$

We now build a two dimensional system governing the dynamics of the total populations densities n and p. Moreover, this system gives the same dynamics as that obtained for n and p in the system (5.8). This will facilitate the mathematical study of system (5.7).

Let us start to calculate the fast equilibrium, that is the equilibrium value of the fast variables u_i, $i = 1, 2$. In order to get this equilibrium value, we put $\varepsilon = 0$ in system (5.8). The result is:

$$u_1^* = \frac{m_2}{m_1 + m_2} \quad \text{and} \quad u_2^* = \frac{m_1}{m_1 + m_2} \tag{5.9}$$

By replacing u_i by u_i^* in (5.8b) and in (5.8c), we get the following two dimensional system:

$$\frac{dn}{dt} = n(r - a_1 p) \tag{5.10a}$$

$$\frac{dp}{dt} = p(b_1 n - d) \tag{5.10b}$$

where $t = \varepsilon\tau$, $r = r_1 u_1^* + r_2 u_2^*$, $a_1 = a u_1^*$ and $b_1 = b u_1^*$.

The system (5.10) is a classical Lotka–Volterra model. All the solutions of this system with initial conditions in the positive quadrant are periodic. There is a positive equilibrium which is a center (see Murray's book for instance, [66]). However, the dynamics of n and p in the system (5.7) do not match with the Lotka–Volterra dynamics, as illustrated on Figs. 5.1 and 5.2. Indeed, when we replace the fast variable by its equilibrium value, we make an error of order of ε. Since the Lotka–Volterra model is not structurally stable, the

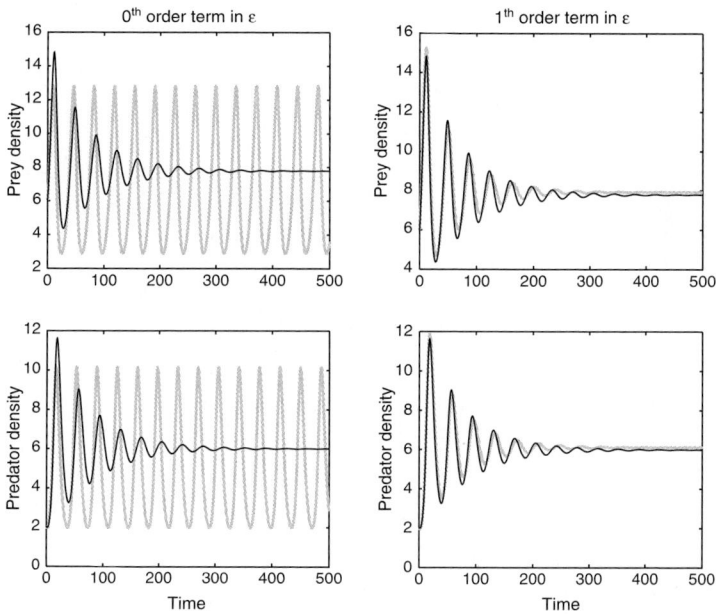

Fig. 5.1. Comparison between the dynamics of x and y given by the complete system (5.7) (*black line*) and that obtained with the two dimensional system (5.10) (*thick grey line*). Above are the prey densities and below are the predator densities. On the *left* column, the term of order of ε is neglected while on the *right* column, this term is taken into account, which improve the similarity between the reduced and complete systems simulations. The parameters values used in the simulation are: $m_1 = 2, m_2 = 1, r_1 = 1, r_2 = 2, a = 1, d = 2, b = 0.9$ and $\varepsilon = 0.1$

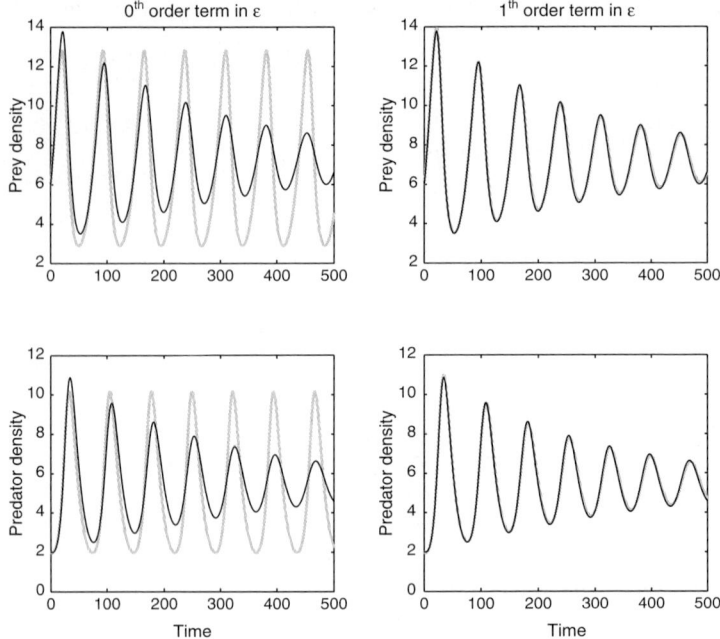

Fig. 5.2. Comparison between the dynamics of x and y given by the complete system (5.7) (*black line*) and that obtained with the two dimensional system (5.10) (*thick grey line*). Above are the prey densities and below are the predator densities. On the *left* column, the term of order of ε is neglected while on the *right* column, this term is taken into account, which improve the similarity between the reduced and complete systems simulations. The parameters values used in the simulation are: $m_1 = 2, m_2 = 1, r_1 = 1, r_2 = 2, a = 1, d = 2, b = 0.9$ and $\varepsilon = 0.05$

ε-error plays an important role in the dynamics. The Fenichel theorem claims that there is an invariant manifold $\mathcal{M}_\varepsilon = \{u_1 = u_1(n, p, \varepsilon)\}$ in the phase space (u_1, n, p, ε). Since the fast equilibrium is hyperbolically stable, the manifold $\mathcal{M}_0$ is normally hyperbolically stable. The previous approximation we made is thus a zero order approximation of the manifold $\mathcal{M}_\varepsilon$.

We now get a first order approximation of the manifold. Let us write:

$$u_1(n, p, \varepsilon) = u_1^* + \varepsilon w_1(n, p) + o(\varepsilon) \tag{5.11}$$

We have to determine w_1 and then to replace u_1 by its expression (5.11) in the system (5.7) in order to improve the approximate two dimensional model (5.10). We can note that the asymptotic expansion of the derivative $\frac{du_1}{d\tau}$ with respect to the small parameter ε, can be written in two different ways. The first one consists in replacing u_1 by the expression (5.11) in the equation (5.8a). The second way consists in writing:

$$\frac{du_1}{d\tau} = \frac{\partial u_1}{\partial n} \frac{dn}{d\tau} + \frac{\partial u_1}{\partial p} \frac{dp}{d\tau} = O\left(\varepsilon^2\right) \tag{5.12}$$

Then we identify the terms of order of ε in both formulas, we get:

$$-(m_2 + m_1)w_1(n, p) + u_1^*(1 - u_1^*)(r_1 - r_2 - ap) = 0 \qquad (5.13)$$

which allows us to conclude, in this case, that w_1 is a function depending only on p:

$$w_1(p) = \frac{u_1^*(1 - u_1^*)}{m_2 + m_1}(r_1 - r_2 - ap) \qquad (5.14)$$

It follows that the system on the invariant manifold is reduced to:

$$\frac{dn}{dt} = n(r - a_1 p) + \varepsilon n w_1(p)(r_1 - r_2 - a_1 p) \qquad (5.15a)$$

$$\frac{dp}{dt} = p(b_1 n - d) + \varepsilon n p b_1 w_1(p) \qquad (5.15b)$$

A numerical simulation has been performed and is shown on Figs. 5.1 and 5.2, in order to illustrate that this reduced model provides a good approximation of the dynamics of the total population densities governed by the four dimensional system (5.7). Those figures show that a decrease of ε increases the similarity between the complete and aggregated model.

5.2.4 Limits of the Method and Possible Extensions

How to Define the Slow and Fast Variables in a Given System

An ecosystem involves a large number of variables and processes. It is largely recognised that some processes are much faster than others. However, each variable may be affected by fast and slow processes. It follows that, when we write a model, the slow and fast processes are mixed in the differential systems and it is not clear that some variables are faster than other ones. According to the previous notations, the problem is, given the system (5.1), is there an algorithm allowing to define the slow variables Y? This is generally a crucial problem. From the mathematical point of view, a such algorithm permits to transform system (5.1) into system (5.5). Moreover, from the applied point of view, it would permit to define the variables of interest for the long time dynamics on the basis of the detailed description. In our framework, the slow variables are defined by the fact that they are first integrals of the fast processes. But it is not always easy to find such functions and this problem can limit practical applications.

Loss of Normal Hyperbolicity and Multiaggregated Models

There is an interesting phenomenon, which has a wide range of possible applications. It occurs when the invariant manifold loses its normal hyperbolicity. For instance, we can easily imagine that the normal attraction of the manifold is more or less important depending on the position of the manifold: some regions of the manifold are more attractive than other ones. We can even have the situation where there are some regions on the manifold in which it is normally repulsive. Let us now assume that a trajectory starting from an initial point in the phase space is going toward the invariant manifold in a region where it is normally hyperbolic and attracting. According to the previous theorems, the trajectory will stay along the invariant manifold but then it can reach a region where the manifold is normally repulsive. Before that region, the trajectory shall cross a line where the manifold is not normally hyperbolic, the normal hyperbolicity is lost. The precise description of the trajectory behavior is no longer trivial: shall the trajectory leave or stay for a while along the manifold? In the case of leaving, what is the global dynamics? Some works have been dedicated to this kind of analysis and the exchange lemma can be a useful tool for this [54]. In some situations, this loss of normal hyperbolicity leads to a strange process named delayed bifurcation. Indeed, a region of the invariant manifold where it is normally hyperbolic and attracting corresponds to the case where the Jacobian matrix associated to the linearised vector field at a point in the region has eigenvalues in the negative complex half plane in the normal directions. The loss of normal hyperbolicity corresponds to a situation where at least one of these eigenvalues is vanishing, leading to a bifurcation. A priori, if the trajectory close to the manifold enters into a region where it is normally repulsive (positive real parts of the previous eigenvalues), it should leave the vicinity of the manifold. However, in some cases, the trajectory stays along the manifold for a transient time and leaves it only after a while, leading to a delayed bifurcation. This phenomenon has also been named "canard" and has been described by Benoit in [21, 22] and Diener [43, 44] and [23]. More recently, a geometrical approach of this process has been provided by Dumortier and Roussarie [46, 47]. In these works, the method, based on blowing up of singularities, is presented through some examples but it is very general and can be extended. It provides an GSP theory approach of the "canard" phenomenon. A large number of possible applications of this method can be found in ecological works [42, 65, 91, 92], for instance).

This loss of normal hyperbolicity has another interesting consequence. Indeed, let us consider that the normally hyperbolic manifold is everywhere normally stable and contains an omega limit set. If a trajectory is entering in a small vicinity of the manifold, it can reach the above omega limit set and then stays close to the invariant manifold for an infinitely long time. In this case, the reduction applies without time limitation and the dynamics of the complete system can be analysed by the study of the dynamics reduced to the manifold. However, let us now suppose that the invariant manifold contains a region R_1 where it is normally hyperbolically attracting and another region R_2 where it is normally hyperbolically repulsive. If a trajectory approaches the manifold in the region R_1, it shall stay close to the manifold as long as it remains in R_1. We can apply the reduction method as long as the trajectory is close to the invariant manifold. But if, after a transient time, the trajectory leaves the vicinity of R_1 and enters in the vicinity of R_2, then it may leave the neighborhood of the manifold. The reduction loses its validity.

If there are several invariant normally hyperbolic manifolds, a trajectory of the complete system can visit the neighborhood of each of them successively. For each invariant manifold, we can define a reduced system. Then the complete system shall be approximated by different aggregated models when time is running. This means that even if the study of the complete model is simplified by considering reduced systems, the whole dynamics may remain complex. For example, the oscillation–relaxation phenomenon can be approached by this way, refer to [10] for an example.

Note that the method can easily be extended to the situation where the fast dynamics exhibits a limit cycle instead of an equilibrium [78].

5.2.5 Aggregation and Emergence

Relation Between Aggregation and Emergence

Aggregation not only provides a reduction of the dimension of the initial model and its simplification, but it also provides interesting information about the emergence of fast processes at a global level in the long run. Indeed, the invariant manifold on which the reduction is performed is a graph on the slow variables $X = \Psi(Y, \varepsilon)$. In other words, at a fast time scale, the fast variables reach an attractor, for instance an equilibrium, which depends on the slow variables. From the practical point of view, the reduced model is obtained by replacing the fast variables by $\Psi(Y, \varepsilon)$ in the slow variables equations. Consequently, if the fast part of the model is changed, then Ψ is modified and the aggregated model as well. This application Ψ contains the effects of the fast part on the long term dynamics. These effects can lead to emerging properties. The concept of emergence has been widely developed in ecology. We provide here two kinds of emergence properties by using the aggregation approach. These properties are then compared and we show that there are differences.

Functional and Dynamical Emergences

Let us consider a system in which the equations governing the slow variables have, for each of them, the same mathematical form. It means that if we do not consider the fast part, the models for the slow variables are identical, maybe with different parameters values. Now, by considering the fast processes, we shall find an aggregated model governing the slow variables. We shall say that there is *functional emergence* if the equations in the aggregated model do not have the same mathematical form as the slow part of the complete system. More precisely, let us consider the following complete system:

$$\frac{dX_i}{d\tau} = F_i\left(X, Y\right) + \varepsilon f_i\left(X, Y\right), \ i = 1, \ldots, k - N$$

$$\frac{dY_j}{d\tau} = \varepsilon G_j\left(X, Y\right), \ j = 1, \ldots, N$$

We suppose that, for each fixed X, the functions $Y \mapsto f_i\left(X, Y\right)$ are the same functions f, with potentially different parameters values. The aggregated model reads:

$$\frac{dY_j}{dt} = G_j\left(\Psi\left(Y, \varepsilon\right), Y\right)$$

Definition 1. *There is functional emergence if at least one of the functions $Y \mapsto G_j\left(\Psi\left(Y, \varepsilon\right), Y\right)$ do not have the same mathematical expressions as f.*

We now provide two examples, one without functional emergence, the other one with functional emergence.

Example 1. Let us consider a population on two patches. We denote by X_1 and X_2 the amount of individuals on patch 1 and 2 respectively. On each patch, we assume that the subpopulation has a logistic growth. The model reads:

$$\frac{dX_1}{d\tau} = m_2 X_2 - m_1 X_1 + \varepsilon r X_1 \left(1 - \frac{X_1}{K}\right)$$

$$\frac{dX_2}{d\tau} = -m_2 X_2 + m_1 X_1 + \varepsilon r X_2 \left(1 - \frac{X_2}{K}\right)$$

where m_1 and m_2 are the migration rates from patch 1 to patch 2 and from patch 2 to patch 1 respectively, r and K are the intrinsic growth rate and the carrying capacity respectively. It follows that the f function is a second degree polynomial of the form:

$$f\left(x\right) = r x \left(1 - \frac{x}{K}\right)$$

Let $Y = X_1 + X_2$, the previous system can be written as follows:

$$\frac{dX_1}{d\tau} = m_2 Y - \left(m_1 + m_2\right) X_1 + \varepsilon r X_1 \left(1 - \frac{X_1}{K}\right)$$

$$\frac{dY}{d\tau} = \varepsilon r \left(Y - \frac{2X_1^2 + Y^2 - 2X_1 Y}{K}\right)$$

The aggregated model is obtained by considering the equilibrium of the fast part:

$$X_1 = \frac{m_2}{m_1 + m_2} Y$$

and by replacing X_1 by this expression in the slow part, which gives the following aggregated model:

$$\frac{dY}{d\tau} = \varepsilon r \left(Y - \frac{2u_1^2 Y^2 + Y^2 - 2u_1 Y^2}{K} \right) = \varepsilon r Y \left(1 - \frac{Y}{\tilde{K}} \right)$$

where $u_1 = \frac{m_2}{m_1 + m_2}$ and $\tilde{K} = \frac{K}{2u_1^2 + 1 - 2u_1} = \frac{K}{1 - 2u_1(1 - u_1)}$. The aggregated model is a logistic equation thus it has the same mathematical formulation as those on each patch. In this case, there is no functional emergence.

Example 2. For the sake of simplicity, we shall consider a purely theoretical example similar to the previous one. The complete model reads:

$$\frac{dX_1}{d\tau} = m_2 X_2 - m_1 X_1 + \varepsilon r_1 X_1$$

$$\frac{dX_2}{d\tau} = -m_2 X_2 + m_1 X_1 + \varepsilon r_2 X_2$$

Let us suppose that the individuals have a repulsive behaviour on patch 1. This can be formulated by assuming that the migration rate from patch one to patch two is proportional to the amount of individuals on patch one, that is $m_1 = \alpha X_1$, making the individuals leave patch 1 faster when their number on this patch is higher. In this case, the equilibrium of the fast part is obtained by solving the equation:

$$m_1 X_1 = m_2 (Y - X_1)$$

where $Y = X_1 + X_2$. By writing $m_1 = \alpha X_1$, this equation reads:

$$\alpha X_1^2 + m_2 X_1 - m_2 Y = 0$$

This is a second order equation for which the discriminant is always positive ($\Delta = m_2^2 + 4\alpha m_2 Y$). Thus the equation has two distinct roots among which only one is positive and is the equilibrium:

$$X_1 = \frac{-m_2 + \sqrt{m_2^2 + 4\alpha m_2 Y}}{2\alpha}$$

We get the aggregated model by replacing the fast variable X_1 by its equilibrium value given above in the slow variable equation:

$$\frac{dY}{d\tau} = \varepsilon (r_1 X_1 + r_2 (Y - X_1)) = \varepsilon \left((r_1 - r_2) \frac{-m_2 + \sqrt{m_2^2 + 4\alpha m_2 Y}}{2\alpha} + r_2 Y \right)$$

The aggregated model is not linear while the mathematical models on each patch are linear. As a consequence, a new formulation occurs and it describes the impact of the repulsive behaviour of the individuals on the population dynamics. We call this functional emergence.

The previous definition describes the emergence of a new formulation for the long term processes induced by the fast processes.

We shall now give another definition which considers the situation where there is a new dynamics of the slow variables when the fast processes are taken into account. This is the dynamical emergence. More precisely, we consider the following system:

$$\frac{dY_j}{d\tau} = \varepsilon G_j\left(X, Y\right), \ j = 1, \ldots, N$$

where X is a fixed vector in R^{k-N}.

Definition 2. *There is dynamical emergence if the previous system is not topologically equivalent to the aggregated model.*

We now provide two examples, one without dynamical emergence, the other one with dynamical emergence.

Example 3. In the above Example 1, there is no dynamical emergence since on each patch there is a logistic growth and the slow variable also has a logistic growth and two logistic dynamics are topologically equivalent.

Example 4. In Example 2, if $r_1 < r_2$, there is a positive equilibrium while the dynamics on each patch are linear. Thus, if $r_1 < r_2$, there is dynamical emergence.

Despite the results presented on the previous examples, there is no direct link between functional and dynamical emergence. We can exhibit examples with functional emergence and no dynamical emergence and examples without functional emergence and with dynamical emergence (see [12] for instance).

5.3 Aggregation Methods of Discrete Models

Let us suppose in this section a population in which evolution is described in discrete time. Apart from that, the population is generally divided into groups, and each of these groups is divided into several subgroups. We will represent the state at time t of a population with q groups by a vector $\mathbf{n}(t) := (\mathbf{n}^1(t), \ldots, \mathbf{n}^q(t))^{\mathsf{T}} \in \mathbb{R}_+^N$ where T denotes transposition. Every vector $\mathbf{n}^i(t) := (n^{i1}(t), \ldots, n^{iN^i}(t)) \in \mathbb{R}_+^{N^i}$, $i = 1, \ldots, q$, represents the state of the ith group which is divided into N^i subgroups, with $N = N^1 + \cdots + N^q$. Following the terminology of the previous section n^{ij} are the micro-variables.

In the evolution of the population we will consider two processes which corresponding characteristic time scales, and consequently their projection intervals, that is their time units, are very different from each other. We will refer to them as the fast and the slow processes or, still, as the fast and the slow dynamics. We will start with the simplest case by considering both processes to be linear and go on with the presentation of a general nonlinear setting.

5.3.1 Linear Discrete Models

We present in details the results concerning the basic autonomous case as developed in Sánchez et al. [82] and Sanz and Bravo de la Parra [85].

We represent fast and slow processes by two different matrices F and S. The characteristic time scale of the fast process gives the projection interval associated to matrix F, that is, the state of the population due to the fast process, after one fast time unit, is $F\mathbf{n}(t)$. Analogously, the effect of the slow process after one slow time unit is calculated by multiplying by matrix S. In order to write a single discrete model combining both processes, and therefore their different time scales, we have to choose its time unit. Two possible and reasonable choices are the time units associated to each one of the two processes. We use here as time unit of the model the one corresponding to the slow dynamics, i.e., the time elapsed between times t and $t+1$ is the projection interval associated to matrix S. We then need to approximate the effect of the fast dynamics over a time interval much longer than its own. In order to do so we will suppose that during each projection interval corresponding to the slow process matrix F has operated a number k of times, where k is a big enough integer that can be interpreted as the ratio between the projection intervals corresponding to the slow and fast dynamics. Therefore, the fast dynamics will be modelled by F^k and the proposed model will consist in the following system of N linear difference equations that we will call *general system*:

$$\mathbf{n}_k(t+1) = SF^k\mathbf{n}_k(t). \tag{5.16}$$

In order to reduce the system we must make some assumptions on fast dynamics. We suppose that for each group $i = 1, \ldots, q$ the fast dynamics is internal, conservative of a certain global variable, macro-variable, for the group and with an asymptotically stable distribution among the subgroups. These assumptions are met if we represent the fast dynamics for each group i by an $N^i \times N^i$ projection matrix F_i which is primitive with 1 as strictly dominant eigenvalue, for example a regular stochastic matrix. The matrix F that represents the fast dynamics for the whole population is then $F := \text{diag}(F_1, \ldots, F_q)$. Every matrix F_i has, associated to eigenvalue 1, positive right and left eigenvectors, v_i and u_i, respectively column and row vectors, verifying $F_i v_i = v_i$, $u_i F_i = u_i$ and $u_i \cdot v_i = 1$. The Perron–Frobenius theorem applies to matrix F_i and we denote $\bar{F}_i := \lim_{k\to\infty} F_i^k = v_i u_i$, where F_i^k is the k-th power of matrix F_i. Denoting $\bar{F} := \text{diag}(\bar{F}_1, \ldots, \bar{F}_q)$, we also have that:

$$\bar{F} = \lim_{k\to\infty} F^k = VU. \tag{5.17}$$

where $V := \text{diag}(v_1, \ldots, v_q)_{N\times q}$ and $U := \text{diag}(u_1, \ldots, u_q)_{q\times N}$.

If we think that the ratio of slow to fast time scale tends to infinity, i.e. $k \to \infty$, or, in other words, that the fast process is instantaneous in relation to the slow process, we can approximate system (5.16) by the following so-called *auxiliary system*:

$$\mathbf{n}(t+1) = S\bar{F}\mathbf{n}(t), \tag{5.18}$$

which using (5.17) can be written as

$$\mathbf{n}(t+1) = SVU\mathbf{n}(t).$$

Here we see that the evolution of the system depends on $U\mathbf{n}(t) \in \mathbb{R}^q$, what suggests that dynamics of the system could be described in terms of a smaller number of variables, the global variables or macro-variables defined by

$$Y(t) := U\mathbf{n}(t). \tag{5.19}$$

The auxiliary system (5.18) can be easily transformed into a q-dimensional system premultiplying by matrix U, giving rise to the so-called aggregated system or macro-system $Y(t+1) = USVY(t)$, where we denote $\bar{S} = USV$ and obtain

$$Y(t+1) = \bar{S}Y(t). \tag{5.20}$$

The solutions to the auxiliary system can be obtained from the solutions to the aggregated system. It is straightforward that the solution $\{\mathbf{n}(t)\}_{t \in \mathbb{N}}$ of system (5.18) for the initial condition $\mathbf{n}_0$ is related to the solution $\{Y(t)\}_{t \in \mathbb{N}}$ of system (5.20) for the initial condition $Y_0 = U\mathbf{n}_0$ in the following way: $\mathbf{n}(t) = SVY(t-1)$ for every $n \geq 1$. The auxiliary system is an example of perfect aggregation in the sense of Iwasa et al. [52].

Once the task of building up a reduced system is carried out, the important issue is to see if the dynamics of the general system (5.16) can also be studied by means of the aggregated system (5.20). In Sánchez et al. [82] it is proved that the asymptotic elements defining the long term behaviour of system (5.16) can be approximated by those of the corresponding aggregated system when the matrix associated to the latter is primitive.

Hypothesis (H): $\bar{S}$ is a primitive matrix.

Assuming hypothesis (H), let $\lambda > 0$ be the strictly dominant eigenvalue of $\bar{S}$, and $\bar{w}_l$ and $\bar{w}_r$ its associated left and right eigenvectors, respectively. We then have that, given any non negative initial condition Y_0, system (5.20) verifies

$$\lim_{t \to \infty} \frac{Y(t)}{\lambda^n} = \frac{\bar{w}_l \cdot Y_0}{\bar{w}_l \cdot \bar{w}_r} \bar{w}_r$$

Concerning the asymptotic behaviour of the auxiliary system (5.18), it is proved that the same $\lambda > 0$ is the strictly dominant eigenvalue of $S\bar{F}$, $U^{\mathsf{T}}\bar{w}_l$ its associated left eigenvector and $SV\bar{w}_r$ its associated right eigenvector.

The asymptotic behaviour of the general system (5.16) could be expressed in terms of the asymptotic elements of the aggregated system (5.18) by considering SF^k as a perturbation of $S\bar{F}$. To be precise, let us order their eigenvalues of F according to decreasing modulus in the following way: $\lambda_1 = \ldots = \lambda_q = 1 > |\lambda_{q+1}| \geq \ldots \geq |\lambda_N|$. So, if $\| * \|$ is any consistent norm in the space $\mathcal{M}_{N \times N}$ of $N \times N$ matrices, then for every $\alpha > |\lambda_{q+1}|$ we have $\|SF^k - S\bar{F}\| = o(\alpha^k)$ $(k \to \infty)$. This last result implies, see [93], that matrix SF^k has a strictly dominant eigenvalue of the form $\lambda + O(\alpha^k)$, an associated left eigenvector $U^{\mathsf{T}}\bar{w}_l + O(\alpha^k)$ and an associated right eigenvector $SV\bar{w}_r + O(\alpha^k)$. Having in mind that α can be chosen to be less than 1, we see that the elements defining the asymptotic behaviour of the aggregated

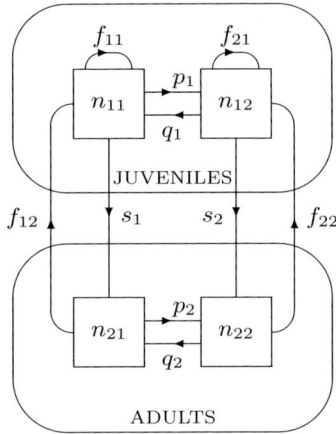

Fig. 5.3. Transition graph of a population structured in two age classes and two patches

and the general systems can be related in a precise way as a function of the separation between the two time scales.

Example 5. We consider a population with juveniles (age class 1) and adults (age class 2) in a two-patch environment. Let $n_{ij}(t)$ be the density of the subpopulation aged i on patch j at time t. On each patch, the population grows according to a Leslie model. Individuals belonging to a given age-class also move from patch to patch, see Fig. (5.3).

Let s_j be the survival rate of juveniles on patch j and f_{ij} the fertility rate of age class i on patch j. The matrix describing the demography of the population in both patches is the following:

$$L = \begin{pmatrix} f_{11} & 0 & f_{21} & 0 \\ 0 & f_{12} & 0 & f_{22} \\ s_1 & 0 & 0 & 0 \\ 0 & s_2 & 0 & 0 \end{pmatrix}.$$

The migration of individuals of age i is described by the following migration matrix

$$P_i = \begin{pmatrix} 1 - p_i & q_i \\ p_i & 1 - q_i \end{pmatrix},$$

where p_i (respectively q_i) is the migration rate from patch 1 to patch 2 (respectively from patch 2 to patch 1) for individuals of age i. So the matrix describing the migration process of the population is:

$$P = \begin{pmatrix} 1 - p_1 & q_1 & 0 & 0 \\ p_1 & 1 - q_1 & 0 & 0 \\ 0 & 0 & 1 - p_2 & q_2 \\ 0 & 0 & p_2 & 1 - q_2 \end{pmatrix}$$

Furthermore, it is assumed that the migration process is fast in comparison to the demographic process.

The dynamics of the four variables n_{11}, n_{12}, n_{21} and n_{22} is thus described by a discrete system of four equations which reads as follows:

$$\mathbf{n}(t+1)=LP^k\mathbf{n}(t)=\begin{pmatrix} f_{11} & 0 & f_{21} & 0 \\ 0 & f_{12} & 0 & f_{22} \\ s_1 & 0 & 0 & 0 \\ 0 & s_2 & 0 & 0 \end{pmatrix}\begin{pmatrix} 1-p_1 & q_1 & 0 & 0 \\ p_1 & 1-q_1 & 0 & 0 \\ 0 & 0 & 1-p_2 & q_2 \\ 0 & 0 & p_2 & 1-q_2 \end{pmatrix}^k\mathbf{n}(t).$$

$$(5.21)$$

where k represents the ratio between the projection intervals corresponding to the slow and fast processes.

We now proceed to reduce system (5.21). For that we need the matrices U and V used in expression (5.17) which are composed of left and right eigenvectors of matrices P_i associated to eigenvalue 1. So they can be expressed in the following way:

$$U=\begin{pmatrix} 1 & 1 & 0 & 0 \\ 0 & 0 & 1 & 1 \end{pmatrix} \quad\text{and}\quad V=\begin{pmatrix} \dfrac{q_1}{p_1+q_1} & 0 \\ \dfrac{p_1}{p_1+q_1} & 0 \\ 0 & \dfrac{q_2}{p_2+q_2} \\ 0 & \dfrac{p_2}{p_2+q_2} \end{pmatrix}$$

The aggregated system governing the total populations of juveniles and adults is the system of two equations $Y(t+1)=ULVY(t)$.

$$Y(t+1)=\begin{pmatrix} 1 & 1 & 0 & 0 \\ 0 & 0 & 1 & 1 \end{pmatrix}\begin{pmatrix} f_{11} & 0 & f_{21} & 0 \\ 0 & f_{12} & 0 & f_{22} \\ s_1 & 0 & 0 & 0 \\ 0 & s_2 & 0 & 0 \end{pmatrix}\begin{pmatrix} \dfrac{q_1}{p_1+q_1} & 0 \\ \dfrac{p_1}{p_1+q_1} & 0 \\ 0 & \dfrac{q_2}{p_2+q_2} \\ 0 & \dfrac{p_2}{p_2+q_2} \end{pmatrix}Y(t)$$

which is a classical two ages Leslie model.

$$Y(t+1)=\begin{pmatrix} F_1 & F_2 \\ S & 0 \end{pmatrix}Y(t)$$

where $Y(t)=(Y_1(t),Y_2(t))$, $Y_1(t)=n_{11}(t)+n_{12}(t)$ and $Y_2(t)=n_{21}(t)+n_{22}(t)$. In the following we use specific values for all the coefficients appearing in system (5.21):

$$\mathbf{n}(t+1) = \begin{pmatrix} 0 & 0 & 3 & 0 \\ 0 & 0.5 & 0 & 2 \\ 0.3 & 0 & 0 & 0 \\ 0 & 0.7 & 0 & 0 \end{pmatrix} \begin{pmatrix} 0.3 & 0.4 & 0 & 0 \\ 0.7 & 0.6 & 0 & 0 \\ 0 & 0 & 0.5 & 0.8 \\ 0 & 0 & 0.5 & 0.2 \end{pmatrix}^{k} \mathbf{n}(t)$$

the corresponding aggregated system is:

$$Y(t+1) = \begin{pmatrix} \dfrac{7}{22} & \dfrac{34}{13} \\[2ex] \dfrac{61}{110} & 0 \end{pmatrix} Y(t)$$

To illustrate the method we include below a table where we can see the dominant eigenvalue associated to the complete model for different values of k as well as the dominant eigenvalue associated to the aggregated model.

k	Dominant Eigenvalue
2	1.3641662953971971997
5	1.3740916698468009220
10	1.3738576343292195643
20	1.3738581986146791779
30	1.3738581986180111511
Aggregated	1.3738581986180111707

In Sanz and Bravo de la Parra [85] these results are extended to more general linear cases where the projection matrices F_i defining the fast dynamics in each group need not be primitive.

In Blasco et al. [25] the fast process is still considered linear but changing at the fast time scale. The fast dynamics is described by means of the first k terms of a converging sequence of different matrices. This case is called the fast changing environment case. Under certain assumptions the limit of the sequence of matrices plays the same role as the matrix $\bar{F}$ in (5.17) obtaining an aggregated system. Similar results to the already described relating the asymptotic properties of the complete and the aggregated systems are proved.

It is also possible to build the general system using as time unit the projection interval of the fast dynamics, see Sánchez et al. [82], Bravo de la Parra et al. [28, 29] and Bravo de la Parra and Sánchez [30]. As we are using the projection interval associated to matrix F we need, therefore, to approximate the effect of matrix S over a projection interval much shorter than its own. For that we use matrix $S_\varepsilon = \varepsilon S + (1 - \varepsilon)I$ where I is the identity matrix and ε a positive small number reflecting the ratio of slow to fast time scale. Matrix S_ε has the following property: if S has a dominant eigenvalue λ with an associated eigenvector $\bar{v}$, then S_ε has $\varepsilon\lambda + (1 - \varepsilon)$ as strictly dominant eigenvalue and $\bar{v}$ is also its associated eigenvector; what implies that the dynamics associated to S and S_ε have the same asymptotically stable stage distribution but S has a much greater growth rate than S_ε because $\varepsilon\lambda + (1 - \varepsilon)$ is closer to 1 than λ.

The complete model reads now as follows

$$\mathbf{n}_\varepsilon(t+1) = S_\varepsilon F \mathbf{n}_\varepsilon(t), \qquad (5.22)$$

the auxiliary system, supposing that fast dynamics has already attained its asymptotic state, is

$$\bar{\mathbf{n}}_\varepsilon(t+1) = S_\varepsilon \bar{F} \bar{\mathbf{n}}_\varepsilon(t). \qquad (5.23)$$

and the aggregated system, for the same global variables $Y(t) = U\mathbf{n}(t)$, becomes

$$Y_\varepsilon(t+1) = U S_\varepsilon V Y_\varepsilon(t) = \bar{S}_\varepsilon Y_\varepsilon(t) \qquad (5.24)$$

where $\bar{S}_\varepsilon = \varepsilon U S V + (1-\varepsilon)I = \varepsilon \bar{S} + (1-\varepsilon)I$ Assuming hypothesis (H), if $\lambda > 0$ is the strictly dominant eigenvalue of $\bar{S}$, and $\bar{w}_l$ and $\bar{w}_r$ its associated left and right eigenvectors, we have that $\varepsilon \lambda + (1-\varepsilon)$ is the strictly dominant eigenvalue of $\bar{S}_\varepsilon$, and $\bar{w}_l$ and $\bar{w}_r$ its associated left and right eigenvectors. For the auxiliary system (5.23), we conclude that the strictly dominant eigenvalue of $S_\varepsilon \bar{F}$ is also $\varepsilon \lambda + (1-\varepsilon)$, and $U^{\mathsf{T}} \bar{w}_l$ and $V \bar{w}_r$ its associated left and right eigenvectors. Finally, we obtain that the elements defining the asymptotic behaviour of the complete system (5.17) can be expressed in terms of those of the aggregated system in the following way: the strictly dominant eigenvalue of matrix $S_\varepsilon F$ is of the form $\varepsilon \lambda + (1-\varepsilon) + O(\varepsilon^2)$ and the corresponding left and right eigenvectors $U^{\mathsf{T}} \bar{w}_l + O(\varepsilon)$ and $V \bar{w}_r + O(\varepsilon)$.

The approximate aggregation methods for time discrete linear models have been extended to non-autonomous and stochastic cases. The complete model in all these extensions is written using the slow time unit.

The case of time varying environments, non-autonomous case, is treated in Sanz and Bravo de la Parra [84] where the variation in time is periodic or tending to a steady state. These two cases admit similar results to the autonomous case. In Sanz and Bravo de la Parra [87] it is studied the case of a general non-autonomous complete system. The property of weak ergodicity, which has to do with the capacity of a system to become asymptotically independent of initial conditions, is compared for the complete and aggregated systems. Related to that work, Sanz and Bravo de la Parra [88] obtained different bounds for the error we incur in when we describe the dynamics of the complete system in terms of the aggregated one. Finally the results in [87] are extended for fast changing environments in Blasco et al. [26].

Two papers of Sanz and Bravo de la Parra [86, 89] are devoted to extend previous results to simplify the study of discrete time models for populations that live in an environment that changes randomly with time. They present the reduction of a stochastic multiregional model in which the population, structured by age and spatial location, lives in a random environment and in which migration is fast with respect to demography. However, the technique could work in much more general settings. The state variables of the complete system and the global variables of the aggregated system are related in the case the pattern of temporal variation is Markovian. Strong stochastic ergodicity for the original and reduced systems are compared, as well as the different measures of asymptotic population growth for these systems.

5.3.2 Nonlinear Discrete Models

The previous framework can be extended to include general nonlinear fast and slow processes. We present the complete model which will be reduced. Both processes, fast and slow, are defined respectively by two mappings

$$S, F : \Omega_N \to \Omega_N; \quad S, F \in C^1(\Omega_N)$$

where $\Omega_N \subset \mathbb{R}^N$ is a nonempty open set.

We first choose a time step of the model that corresponds to the slow dynamics as we did in the linear case, see Sanz et al. [90]. We still assume that during this time step the fast process acts k times before the slow process acts. Therefore, denoting by $\mathbf{n}_k(t) \in \mathbb{R}^N$ the vector of state variables at time t, the complete system is defined by

$$\mathbf{n}_k(t+1) = S(F^k(\mathbf{n}_k(t))) \tag{5.25}$$

where F^k denotes the k-fold composition of F with itself.

In order to reduce the system (5.25), we have to impose some conditions to the fast process. In what follows we suppose that the following hypotheses are met. For each initial condition $X \in \Omega_N$, the fast dynamics tends to an equilibrium, that is, there exists a mapping $\bar{F} : \Omega_N \to \Omega_N$, $\bar{F} \in C^1(\Omega_N)$ such that for each $X \in \Omega_N$, $\lim_{k \to \infty} F^k(X) = \bar{F}(X)$. This equilibrium depends on a lesser number of variables in the following form: there exists a non-empty open set $\Omega_q \subset \mathbb{R}^q$ with $q < N$ and two mappings $G : \Omega_N \to \Omega_q$, $G \in C^1(\Omega_N)$, and $E : \Omega_q \to \Omega_N$, $E \in C^1(\Omega_q)$, such that $\bar{F} = E \circ G$.

Now, we proceed to define the so-called *auxiliary system* which approximates (5.25) when $k \to \infty$, i.e., when the fast process has reached an equilibrium. Keeping the same notation as in the linear case, this auxiliary system is

$$\mathbf{n}(t+1) = S(\bar{F}(\mathbf{n}(t))) \tag{5.26}$$

which can be also written as $\mathbf{n}(t+1) = S \circ E \circ G(\mathbf{n}(t))$.

The *global variables* in this case are defined through

$$Y(t) := G(\mathbf{n}(t)) \in \mathbb{R}^q. \tag{5.27}$$

Applying G to both sides of (5.26) we have $Y(t+1) = G(\mathbf{n}(t+1)) = G \circ S \circ E \circ G(\mathbf{n}(t)) = G \circ S \circ E(Y(t))$ which is an autonomous system in the global variables $Y(t)$. Summing up, we have approximated system (5.25) by the *reduced* or *aggregated* system defined by

$$Y(t+1) = \bar{S}(Y(t)) \tag{5.28}$$

where we denote $\bar{S} = G \circ S \circ E$.

Now we present some results relating the behavior of systems (5.25) and (5.28), for big enough values of parameter k. All the results in this section are presented in more general setting in [90]. First we compare the solutions of both systems for a fixed value of t. The next theorem states that the dynamics of the auxiliary system is completely determined by the dynamics of the reduced system and that the solution to the complete system, given mild extra assumptions, for each t fixed can be approximated by the solution to the aggregated model.

Theorem 2. *Let* $\mathbf{n}_0 \in \Omega_N$ *and let* $Y_0 = G(\mathbf{n}_0) \in \Omega_q$. *Then:*

(i) *The solution* $\{\mathbf{n}(t)\}_{t=1,2,...}$ *to (5.26) corresponding to the initial condition* $\mathbf{n}_0$ *and the solution* $\{Y(t)\}_{t=1,2,...}$ *to (5.28) corresponding to the initial condition* Y_0 *are related by the following expressions*

$$Y(t) = G(\mathbf{n}(t)) \text{ and } \mathbf{n}(t) = S \circ E(Y(t-1)), \ n = 1, 2, \ldots \quad (5.29)$$

(ii) *Let* t *be a fixed positive integer and let us assume that there exists a non-empty bounded open set* Ω *such that* $\bar{\Omega} \subset \Omega_N$, Ω *contains the points* $\mathbf{n}(0)$, $\mathbf{n}(i+1) = S \circ E(Y_i)$ $(i = 0, \ldots, n-1)$, *and such that* $\lim_{k \to \infty} F^k = \bar{F}$ *uniformly in* Ω. *Then the solution* $\mathbf{n}_k(t)$ *to (5.25) corresponding to the initial condition* $\mathbf{n}(0)$ *and the solution* $Y(t)$ *to (5.28) corresponding to the initial condition* Y_0 *are related by the following expressions*

$$Y(t) = \lim_{k \to \infty} G(\mathbf{n}_k(t)) \text{ and } \lim_{k \to \infty} \mathbf{n}_k(t) = S \circ E(Y(t-1)).$$

Now we turn our attention to the study of some relationships between the fixed points of the original and reduced systems. Concerning the auxiliary system, relations (5.29) in Theorem 2 yield straightforward relationships between the fixed points of the auxiliary and reduced systems: if $\mathbf{n}^* \in \Omega_N$ is a fixed point of (5.26) then $Y^* = G(\mathbf{n}^*) \in \Omega_q$ is a fixed point of (5.28); conversely, if Y^* is a fixed point of (5.28) then $\mathbf{n}^* = S \circ E(Y^*)$ is a fixed point of (5.26). The corresponding fixed points in the auxiliary and aggregated systems are together asymptotically stable or unstable.

In the following result, it is guaranteed that, under certain assumptions, the existence of a fixed point Y^* for the aggregated system implies, for large enough values of k, the existence of a fixed point $\mathbf{n}_k^*$ for the original system, which can be approximated in terms of Y^*. Moreover, in the hyperbolic case the stability of Y^* is equivalent to the stability of $\mathbf{n}_k^*$ and in the asymptotically stable case, the basin of attraction of $\mathbf{n}_k^*$ can be approximated in terms of the basin of attraction of Y^*.

Theorem 3. *Let us assume that* $\bar{F} \in C^1(\Omega_N)$ *and that* $\lim_{k \to \infty} F^k = \bar{F}$, $\lim_{k \to \infty} DF^k = D\bar{F}$ *uniformly on any compact set* $K \subset \Omega_N$.

Let $Y^* \in \mathbb{R}^q$ *be a hyperbolic fixed point of (5.28) which is asymptotically stable (respectively unstable). Then there exists* $k_0 \in \mathbb{N}$ *such that for each*

$k \geq k_0$, $k \in \mathbb{N}$, there exists a hyperbolic fixed point $\mathbf{n}_k^*$ of (5.25) which is asymptotically stable (respectively unstable) and that satisfies $\lim_{k \to \infty} \mathbf{n}_k^* = S \circ E(Y^*)$.

Moreover, let $\mathbf{n}_0 \in \Omega_N$, if the solution $Y(t)$ to (5.28) corresponding to the initial condition $Y_0 = G(\mathbf{n}_0)$ is such that $\lim_{t \to \infty} Y(t) = Y^*$ then for each $k \geq k_0$, $k \in \mathbb{N}$, the solution $\mathbf{n}_k(t)$ to (5.25) corresponding to $\mathbf{n}_0$ verifies $\lim_{t \to \infty} \mathbf{n}_k(t) = \mathbf{n}_k^*$.

Particular models where this last result applies are presented in Bravo de la Parra et al. [31] and the review paper Auger and Bravo de la Parra [8].

As in the linear case we can build the general system using as time unit the one of fast dynamics. In Bravo de la Parra et al. [28] and Bravo de la Parra and Sánchez [30] a system with linear fast dynamics and general nonlinear slow dynamics is reduced by means of a center manifold theorem.

The mapping representing fast dynamics is expressed in terms of a matrix F, $F(X) = FX$, where we are naming equally the map and the matrix. Matrix F is considered to have the same properties stated in the linear case, in particular the asymptotic behaviour associated to it is reflected in the following equality $\bar{F} = \lim_{k \to \infty} F^k = VU$. So, for each initial condition $X \in \mathbb{R}^N$, the fast dynamics tends to $\lim_{k \to \infty} F^k(X) = \lim_{k \to \infty} F^k X = \bar{F}X = \bar{F}(X)$ and we have $G : \mathbb{R}^N \to \mathbb{R}^q$, $G(X) = UX$ and $E : \mathbb{R}^q \to \mathbb{R}^N$, $E(Y) = VY$, such that $\bar{F} = E \circ G$.

Concerning the slow dynamics we represent it by a general mapping $S : \Omega_N \to \Omega_N$, $S \in C^1(\Omega_N)$. To approximate the effect of mapping S over the projection interval of fast dynamic we use mapping $S_\varepsilon(X) = \varepsilon S(X) + (1-\varepsilon)X$. The complete model reads now as follows

$$\mathbf{n}_\varepsilon(t+1) = S_\varepsilon(F\mathbf{n}_\varepsilon(t)) \tag{5.30}$$

In Bravo de la Parra et al. [28] it is developed a center manifold theorem which applies to system (5.30). It suffices to write it in the following form

$$\mathbf{n}_\varepsilon(t+1) = F\mathbf{n}_\varepsilon(t) + \varepsilon(S(F\mathbf{n}_\varepsilon(t)) - F\mathbf{n}_\varepsilon(t)) \tag{5.31}$$

For any small enough fixed ε there exists a locally attractive invariant manifold that allows us to study the dynamics of system (5.31) by means of its restriction to it. The system restricted to the center manifold is what we call the aggregated system. Though, in general, it is not possible to find out explicitly the map defining the aggregated system we can calculate its expansion in ε powers. Using the expansion to the first order we get the simplest form of the aggregated system,

$$Y_\varepsilon(t+1) = Y_\varepsilon(t) + \varepsilon(US(VY_\varepsilon(t)) - Y_\varepsilon(t)) + O(\varepsilon^2) \tag{5.32}$$

in common applications, for instance when studying hyperbolic fixed points, the term $O(\varepsilon^2)$ is negligible and the reduced system to be analysed is

$$Y_\varepsilon(t+1) = \varepsilon(US(VY_\varepsilon(t)) + (1-\varepsilon)Y_\varepsilon(t) = \bar{S}_\varepsilon(Y_\varepsilon(t)) \tag{5.33}$$

which has much the same aspect as its linear counterpart, system (5.24).

5.4 Aggregation of Partial Differential Equations (PDE) and Delayed Differential Equations (DDE)

In this section we will apply the aggregation of variables method in the linear case to structured population dynamics models formulated in terms of partial differential equations and to models formulated in terms of delayed differential equations. The different time scales introduce into the model a small parameter $\varepsilon > 0$ which gives rise to a singular perturbation problem. Although both contexts are mathematically very different, the underlying ideas in the construction of the aggregated model are similar in both cases, due to the structure of the fast dynamics: it is supposed that this dynamics is represented by a matrix K whose spectrum allows the decomposition of the space $\mathbf{R}^N$ in a direct sum of the eigenspace $\ker K$ associated to the eigenvalue 0 and generated by a vector ν, plus a complementary stable subspace S corresponding to the remaining part of the spectrum, which are eigenvalues with negative real part. The aggregated model is constructed by projecting the global dynamics onto $\ker K$. The theory of semi-groups allows a unified formulation of both situations aimed at obtaining approximation results for the solutions X_ε to the perturbed global model by the solutions s_0 to the aggregated model, using the same technique in both cases. Projecting the global system onto the subspaces $\ker K$ and S and using a variation-of-constants formula, this perturbed system can be transformed into a fixed point problem for the projection of X_ε onto $\ker K$. Roughly speaking, in both contexts the conclusion is reached that $X_\varepsilon = s_0 \nu + O(\varepsilon)$, $(\varepsilon \to 0_+)$.

5.4.1 Aggregation in Structured Population Models

In this section we apply aggregation of variables methods to a general linear structured population dynamics model with both a continuous age structure and a finite spatial structure. It is assumed that discrete migration processes take place between spatial patches at a frequency much higher than the demographic events, so high that one almost cannot see them. The impression is that of a spatially homogeneous age-dependent population governed by a Von Foerster equation with birth and death coefficients averaged from the original patch-dependent coefficients through a weighted average. The weights are computed in terms of a migration matrix and are in fact the mark of the hidden underlying spatial structure. See [1, 2, 27] for the details.

The Model

We consider an age-structured population, with age a and time t being continuous variables. The population is divided into N spatial patches. The evolution of the population is due to the migration process between the different patches at a fast time scale and to the demographic process at a slow time scale.

Let $n_i(a,t)$ be the population density in patch i $(i = 1, \ldots, N)$ so that $\int_{a_1}^{a_2} n_i(a,t)\,da$ represents the number of individuals in patch i whose age $a \in [a_1, a_2]$ at time t and

$$\mathbf{n}(a,t) := (n_1(a,t), \ldots, n_N(a,t))^T.$$

Let $\mu_i(a)$ and $\beta_i(a)$ be the patch and age-specific mortality and fertility rates respectively and

$$M(a) := \operatorname{diag}\{\mu_1(a), \ldots, \mu_N(a)\}; \quad B(a) := \operatorname{diag}\{\beta_1(a), \ldots, \beta_N(a)\}.$$

Let $k_{ij}(a)$ be the age-specific migration rate from patch j to patch i, $i \neq j$, and

$$K(a) := (k_{ij}(a))_{1 \leq i,j \leq N} \quad \text{with} \quad k_{ii}(a) := -\sum_{j=1, j \neq i}^{N} k_{ji}(a).$$

A crucial assumption is that the jump process is conservative with respect to the life dynamics of the population, that is, no death or birth is directly incurred by spatial migrations.

The model based upon the classical McKendrick-Von Foerster model for an age-structured population is as follows:

Balance law:

$$\frac{\partial \mathbf{n}}{\partial t}(a,t) + \frac{\partial \mathbf{n}}{\partial a}(a,t) = \left[-M(a) + \frac{1}{\varepsilon}K(a)\right]\mathbf{n}(a,t) \quad (a > 0,\ t > 0) \quad (5.34)$$

Birth law:

$$\mathbf{n}(0,t) = \int_0^{+\infty} B(a)\mathbf{n}(a,t)\,da \quad (t > 0) \quad (5.35)$$

where $\varepsilon > 0$ is a constant small enough and completed with an initial age distribution

$$\mathbf{n}(a,0) = \Phi(a) := (\Phi_1(a), \ldots, \Phi_N(a))^T, \quad (a > 0)$$

In what follows we assume that

Hypothesis 1 *The matrix $K(a)$ is irreducible for every $a > 0$.*

As a consequence 0 is a simple eigenvalue larger than the real part of any other eigenvalue. The left eigenspace of matrix $K(a)$ associated with the eigenvalue 0 is generated by vector $\mathbf{1} := (1, \ldots, 1)^T \in \mathbf{R}^N$. The right eigenspace is generated by vector $\nu(a) := (\nu_1(a), \ldots, \nu_N(a))^T$ and is unique if we choose it having positive entries and verifying $\mathbf{1}^T \nu(a) = 1$.

For each initial age distribution $\Phi \in X := L^1(\mathbf{R}_+; \mathbf{R}^N)$, the problem (5.34)–(5.35) has a unique solution $\mathbf{n}_\varepsilon$. Moreover we can associate with it a strongly continuous semi-group of linear bounded operators (C_0-semi-group) $\{T_\varepsilon(t)\}_{t \geq 0}$ on X, defined by $T_\varepsilon(t)\Phi := \mathbf{n}_\varepsilon(\cdot, t)$.

The Aggregated Model

We build up a model which describes the dynamics of the total population:

$$n(a,t) := \sum_{i=1}^{N} n_i(a,t) \quad \text{(global variable)}.$$

The exact model satisfied by the new variable $n(a,t)$ is obtained by adding up the variables $n_i(a,t)$ in system (5.34) and (5.35):

$$\frac{\partial n}{\partial t}(a,t) + \frac{\partial n}{\partial a}(a,t) = -\sum_{i=1}^{N} \mu_i(a)n_i(a,t), \quad (a,t>0)$$

$$n(0,t) = \int_0^{+\infty} \left(\sum_{i=1}^{N} \beta_i(a)n_i(a,t) \right) da, \quad (t>0)$$

together with the initial condition $n(a,0) = \varphi(a) := \sum_{i=1}^{N} \Phi_i(a) \ (a>0)$.

In order to obtain a system with the global variable as the unique state variable, we propose the following approximation:

$$\nu_i(a,t) := \frac{n_i(a,t)}{n(a,t)} \approx \nu_i(a) \quad i = 1, \ldots, N$$

which implies that

$$\sum_{i=1}^{N} \mu_i(a)n_i(a,t) \approx \left(\sum_{i=1}^{N} \mu_i(a)\nu_i(a) \right) n(a,t) := \mu^*(a)n(a,t)$$

$$\sum_{i=1}^{N} \beta_i(a)n_i(a,t) \approx \left(\sum_{i=1}^{N} \beta_i(a)\nu_i(a) \right) n(a,t) := \beta^*(a)n(a,t).$$

The *aggregated* model for the density of the total population is the following

$$\frac{\partial n}{\partial t}(a,t) + \frac{\partial n}{\partial a}(a,t) = -\mu^*(a)n(a,t), \quad (a,t>0) \tag{5.36}$$

$$n(0,t) = \int_0^{+\infty} \beta^*(a)n(a,t)\, da, \quad (t>0) \tag{5.37}$$

together with the initial condition $n(a,0) = \varphi(a)$, $(a>0)$.

It is a classical Sharpe–Lotka–McKendrick linear model to which the general theory applies. Under some technical conditions which are specified in [1] the solutions to this problem define a C_0-semi-group on $L^1(\mathbf{R}_+)$, which has the so-called *asynchronous exponential growth property*, namely

Proposition 1. *There exists a unique* $\lambda_0 \in \mathbf{R}$ *(malthusian parameter)*
such that

$$\lim_{t \to +\infty} e^{-\lambda_0 t} n(a, t) = c(\varphi)\theta_0(a)$$

where $c(\varphi) > 0$ *is a constant depending on the initial value* $\varphi \in L^1(\mathbf{R}_+)$ *and*

$$\theta_0(a) := e^{-\lambda_0 a} e^{-\int_0^a \mu^*(\sigma)\,d\sigma}$$

is the asymptotic distribution.

Approximation Result

Let us say a few words about the nature of the convergence of the solutions
to the perturbed problem (5.34) and (5.35) when $\varepsilon \to 0_+$ to the solutions to
the aggregated model (5.36) and (5.37).

To this end, let us consider the following direct sum decomposition, whose
existence is ensured by Hypothesis 1:

$$\mathbf{R}^N = [\nu(a)] \oplus S \tag{5.38}$$

where $[\nu(a)]$ is the one-dimensional subspace generated by the vector $\nu(a)$ and
$S := \{\mathbf{v} \in \mathbf{R}^N \ ; \ \mathbf{1}^T \cdot \mathbf{v} = 0\}$. Notice that S is the same for all a, and moreover
$K_S(a)$, the restriction of $K(a)$ to S is an isomorphism on S with spectrum
$\sigma(K_S(a)) \subset \{\lambda \in \mathbf{C} \ ; \ \mathrm{Re}\lambda < 0\}$.

We decompose the solutions to the perturbed problem according to (5.38),
that is

$$\mathbf{n}_\varepsilon(a, t) := p_\varepsilon(a, t)\nu(a) + \mathbf{q}_\varepsilon(a, t); \quad \mathbf{q}_\varepsilon(a, t) \in S$$

giving

$$\frac{\partial p_\varepsilon}{\partial t}(a, t) + \frac{\partial p_\varepsilon}{\partial a}(a, t) = -\mu^*(a)p_\varepsilon(a, t) - \mathbf{1}^T M(a)\mathbf{q}_\varepsilon(a, t) \tag{5.39}$$

$$\frac{\partial \mathbf{q}_\varepsilon}{\partial t}(a, t) + \frac{\partial \mathbf{q}_\varepsilon}{\partial a}(a, t) = -[M_S(a)\nu(a) + \nu'(a)]p_\varepsilon(a, t)$$
$$+ \left[\frac{1}{\varepsilon}K_S(a) - M_S(a)\right]\mathbf{q}_\varepsilon(a, t) \tag{5.40}$$

$$p_\varepsilon(a, t) = \int_0^{+\infty} \beta^*(a)p_\varepsilon(a, t)\,da + \int_0^{+\infty} \mathbf{1}^T B(a)\mathbf{q}_\varepsilon(a, t)\,da \tag{5.41}$$

$$\mathbf{q}_\varepsilon(0, t) = \int_0^{+\infty} B_S(a)\nu(a)p_\varepsilon(a, t)\,da + \int_0^{+\infty} B_S(a)\mathbf{q}_\varepsilon(a, t)\,da \tag{5.42}$$

where $M_S(a)$, $B_S(a)$ are the projections of $M(a)$ and $B(a)$ respectively onto S.

Notice that the solutions to the homogeneous problem

$$\frac{\partial \mathbf{q}}{\partial t}(a, t) + \frac{\partial \mathbf{q}}{\partial a}(a, t) = \left[\frac{1}{\varepsilon} K_S(a) - M_S(a)\right] \mathbf{q}(a, t)$$

$$\mathbf{q}(0, t) = \int_0^{+\infty} B_S(a)\mathbf{q}(a, t)\, da$$

define a C_0-semi-group $\{\mathcal{U}_\varepsilon(t)\}_{t \geq 0}$ which satisfies the estimation

$$\|\mathcal{U}_\varepsilon(t)\| \leq C_1 e^{(C_2 - C_3/\varepsilon)t}, \quad t \geq 0, \quad (C_1, C_2, C_3 > 0).$$

Under some technical conditions and using a variation-of-constants formula, we can express the solution to the nonhomogeneous system (5.40)–(5.42) in terms of $\mathcal{U}_\varepsilon(t)$ and p_ε. Then substituting this expression in (5.39)–(5.41), we can transform these equations into a fixed point problem for p_ε which can be solved giving the following result of approximation:

Theorem 4. *For each $\varepsilon > 0$ small enough, we have*

$$\mathbf{n}_\varepsilon(a, t) = n(a, t)\nu(a) + (\mathcal{U}_\varepsilon(t)\mathbf{q}_0)(a) + O(\varepsilon)$$

where $n(a, t)$ is the solution to the aggregated model corresponding to the initial age distribution p_0 and $\Phi := p_0\nu + \mathbf{q}_0$ ($\mathbf{q}_0 \in S$) is the initial age distribution for the perturbed system.

We point out that the above formula is of interest mainly in the case when $\lambda_0 \geq 0$. In this case, it can be concluded from the above formula that $\mathbf{n}_\varepsilon(\cdot, t) \approx n(\cdot, t)\nu(\cdot)$ as $t \to +\infty$ uniformly with respect to $\varepsilon > 0$ small enough. Also, if $\lambda_0 < 0$ then $\mathbf{n}_\varepsilon(\cdot, t) \to 0$ as $t \to +\infty$ and this is again uniform with respect to $\varepsilon > 0$ small enough. In this case, however, $n(\cdot, t)\nu(\cdot)$ does not dominate in general the terms $O(\varepsilon)$.

See [1] for the details. In this chapter it is also shown that the semi-group $\{T_\varepsilon(t)\}_{t \geq 0}$ has the asynchronous exponential growth property. Roughly speaking, it is shown that for $\varepsilon > 0$ small enough, each solution $\mathbf{n}_\varepsilon(a, t)$ of the perturbed system is such that

$$\mathbf{n}_\varepsilon(a, t) \approx C(\Phi)e^{\lambda_\varepsilon t}\Psi_\varepsilon(a) \quad (t \to +\infty)$$

where $C(\Phi) > 0$ is a constant depending on the initial age distribution and

$$\lim_{\varepsilon \to 0_+} \lambda_\varepsilon = \lambda_0; \quad \lim_{\varepsilon \to 0_+} \Psi_\varepsilon = \nu\theta_0$$

where λ_0 and θ_0 are, respectively, the Malthus parameter and the associated asymptotic distribution of the aggregated system mentioned in Proposition 1.

5.4.2 Aggregation of Variables in Linear Delayed Differential Equations

Let us describe in some detail the aggregation of variables method in a simple linear model with a discrete delay. On one side, this case is interesting in itself and clarifies other abstract formulations while on the other, it has its own methods for the step-by-step construction of the solution.

The Model

The model consists of the following system of linear delayed differential equations, depending on a small parameter $\varepsilon > 0$, that we call the *perturbed system*:

$$\begin{cases} X'(t) = (1/\varepsilon)KX(t) + AX(t) + BX(t-r), & t > 0 \\ X(t) = \Phi(t) , \ t \in [-r,0] \ ; \ \Phi \in C([-r,0]; \mathbf{R}^N) \end{cases} \tag{5.43}$$

where $X(t) := (\mathbf{x}_1(t), \ldots, \mathbf{x}_q(t))^T$, $\mathbf{x}_j(t) := \left(x_j^1(t), \ldots, x_j^{N_j}(t)\right)^T$, $j = 1, \ldots, q$.

K, A and B are $N \times N$ real constant matrices with $N = N_1 + \cdots + N_q$. As usual, $C([-r,0]; \mathbf{R}^N)$ represents the Banach space of $\mathbf{R}^N$-valued continuous functions on $[-r,0]$, $(r > 0)$, endowed with the norm $\|\varphi\|_C := \sup_{\theta \in [-r,0]} \|\varphi(\theta)\|$.

System (5.43) can be solved by the classical step-by-step procedure.

Throughout this section, we suppose that matrix K is a block-diagonal matrix

$$K := \operatorname{diag}\{K_1, \ldots, K_q\}$$

in which each diagonal block K_j has dimensions $N_j \times N_j$, $j = 1, \ldots, q$ and satisfies the following hypothesis

Hypothesis 2 *For each $j = 1, \ldots, q$, the following holds:*

(i) $\sigma(K_j) = \{0\} \cup \Lambda_j$, *with* $\Lambda_j \subset \{z \in \mathbf{C}; \ Re\, z < 0\}$, *where* $\sigma(K_j)$ *is the spectrum of matrix K_j.*
(ii) 0 *is a simple eigenvalue of K_j.*

As a consequence, $\ker K_j$ is generated by an eigenvector associated to eigenvalue 0, which will be denoted $\mathbf{v}_j$. The corresponding left eigenspace is generated by a vector $\mathbf{v}_j^*$ and we choose both vectors verifying the *normalization condition*: $(\mathbf{v}_j^*)^T \mathbf{v}_j = 1$.

The Aggregated Model

In order to build the *aggregated system* of system (5.43), we define the following matrices:

$$\mathcal{V}^* := \operatorname{diag}\{(\mathbf{v}_1^*)^T, \ldots, (\mathbf{v}_q^*)^T\}; \quad \mathcal{V} := \operatorname{diag}\{\mathbf{v}_1, \ldots, \mathbf{v}_q\}.$$

As a consequence of Hypothesis 2, we can consider the following direct sum decomposition of space $\mathbf{R}^N$:

$$\mathbf{R}^N = \ker K \oplus S \qquad (5.44)$$

where $\ker K$ is a q-dimensional subspace generated by the columns of matrix $\mathcal{V}$ and $S := \operatorname{Im} K = \{\mathbf{v} \in \mathbf{R}^N;\ \mathcal{V}^*\mathbf{v} = \mathbf{0}\}$.

We now define the q *aggregated variables*:

$$\mathbf{s}(t) := (s_1(t), \ldots, s_q(t))^T = \mathcal{V}^* X(t)$$

which satisfy a linear differential system obtained by premultiplying both sides of (5.43) by $\mathcal{V}^*$. We get the aggregated variables on the left-hand side but we fail on the right-hand side. To avoid this difficulty, we write $X(t)$ according to the decomposition (5.44) so that $X(t) = \mathcal{V}\mathbf{s}(t) + X_S(t)$ and then

$$\mathbf{s}'(t) = \mathcal{V}^* A \mathcal{V}\mathbf{s}(t) + \mathcal{V}^* B \mathcal{V}\mathbf{s}(t-r) + \mathcal{V}^* A X_S(t) + \mathcal{V}^* B X_S(t-r).$$

Let us observe that for $t \in [0, r]$ we have

$$\mathbf{s}'(t) = \mathcal{V}^* A \mathcal{V}\mathbf{s}(t) + \mathcal{V}^* A X_S(t) + \mathcal{V}^* B \Phi(t-r).$$

Therefore, we propose as *aggregated model* the following approximated system

$$\mathbf{s}'(t) = \overline{A}\mathbf{s}(t) + \overline{B}\mathbf{s}(t-r), \quad t > r \qquad (5.45)$$

where $\overline{A} := \mathcal{V}^* A \mathcal{V}$, $\overline{B} := \mathcal{V}^* B \mathcal{V}$, and

$$\mathbf{s}'(t) = \overline{A}\mathbf{s}(t) + \mathcal{V}^* B \Phi(t-r), \quad t \in [0, r]. \qquad (5.46)$$

Equation (5.45) is a delayed linear differential system of equations which can be solved by a standard step-by-step procedure from an initial data in $[0, r]$ which is the solution to (5.46), that is:

$$\mathbf{s}(t) = e^{t\overline{A}} \left[\mathcal{V}^* \Phi(0) + \int_0^t e^{-\sigma \overline{A}} \mathcal{V}^* B \Phi(\sigma - r) \, d\sigma \right]. \qquad (5.47)$$

Comparison Between the Solutions to Systems (5.43) and (5.45)

Decomposing the system (5.43) according to the direct sum decomposition (5.44) and solving it with the help of a variation-of-constants formula in a similar way of the previous section, we can obtain a comparison between the solutions of both systems (5.43) and (5.45). See [81] for the details.

Theorem 5. *Under Hypothesis 2, for each initial data $\Phi \in C([-r, 0]; \mathbf{R}^N)$, $\Phi = \mathcal{V}\psi + \varphi$, the corresponding solution X_ε to system (5.43) can be written as:*

$$\forall t \geq r, \quad X_\varepsilon(t) = \mathcal{V}\mathbf{s}_0(t) + \mathbf{r}_\varepsilon(t)$$

where $\mathbf{s}_0$ *is the solution to the aggregated system (5.45) for* $t \geq r$, *with the initial data defined* $\forall t \in [0, r]$ *by (5.47).*

Moreover, there exist three constants $C > 0$, $C^* > 0$, $\gamma > 0$, *such that*

$$\forall t \geq r, \quad \|\mathbf{r}_\varepsilon(t)\| \leq \varepsilon(C + C^* e^{\gamma t})\|\Phi\|_C.$$

Therefore, for each $T > r$, $\lim_{\varepsilon \to 0_+} X_\varepsilon = \mathcal{V} \mathbf{s}_0$ *uniformly in the interval* $[r, T]$.

This approximation result is similar to that obtained in the previous section for continuous time structured models formulated in terms of partial differential equations, but we have to point out that the delay introduces significant differences due to the influence of the initial data on the solution in the interval $[0, r]$. In fact, the approximation when $\varepsilon \to 0$ is valid only for $t \geq r$ and hence the initial data in $[0, r]$ for the aggregated system is $\mathcal{V}^* X_\varepsilon(t)$, which is the projection on $\ker K$ of the exact solution to system (5.43), constructed in $[0, r]$ from an initial data $\Phi \in C([-r, 0]; \mathbf{R}^N)$.

The above procedure can be generalized to the following *perturbed system* of linear delayed differential equations:

$$\begin{cases} X'(t) = L(X_t) + (1/\varepsilon)KX(t), & t > 0 \\ X_0 = \Phi \in C([-r, 0]; \mathbf{R}^N) \end{cases} \tag{5.48}$$

where $L : C([-r, 0]; \mathbf{R}^N) \longrightarrow \mathbf{R}^N$ is a bounded linear operator and X_t, $(t \geq 0)$, is the *section of function* X *at time* t, namely, $X_t(\theta) := X(t + \theta)$, $\theta \in [-r, 0]$.

The *aggregated model* is

$$\mathbf{s}'(t) = \overline{L}(\mathbf{s}_t), \quad t \geq r$$

where $\overline{L}$ is the linear bounded operator defined by:

$$\overline{L} : C([-r, 0]; \mathbf{R}^q) \longrightarrow \mathbf{R}^q, \quad \overline{L}(\psi) := \mathcal{V}^* L(\mathcal{V}\psi).$$

As in the previous case, the initial data in $[0, r]$ should be constructed, but in this abstract setting it presents higher mathematical difficulties. In particular, we should use the Riesz representation theorem of bounded linear operators on $C([-r, 0]; \mathbf{R}^N)$. Operator L can be written as a Riemann–Stieltjes integral:

$$\forall \Phi \in C([-r, 0]; \mathbf{R}^N); \quad L(\Phi) = \int_{[-r, 0]} [d\eta(\theta)] \Phi(\theta)$$

where $\eta(\theta)$ is a bounded variation $N \times N$ matrix-valued function. It can be shown that the contributions of sections of the initial data $\Phi = \mathcal{V}\psi + \varphi$ to the aggregated model in $[0, r]$ is given by the Riemann–Stieltjes integral:

$$I(t, \varphi) := \int_{[-r, -t]} [d\eta(\theta)] \varphi(t + \theta).$$

5.5 Applications to Population Dynamics

In this section, we present several examples which illustrate how to use the methods described above and what kind of results can be expected with these methods. We have chosen examples where the fast and the slow parts are usual models which are combined. For instance, if we are interested in the effect of the behaviour of predator individuals on the population dynamics, we choose in this section a classical model for the behaviour of the predators and a classical model for the population dynamics and we analyse their interaction. The aggregation methods permit to analyse the models and to understand how the behavioral interaction emerges at the population level. All our examples are simple enough to make the calculations rather easy. However, some of them are not trivial and give an idea of the problems which occur when playing with these time scales arguments. In order to illustrate the whole chapter, we gave examples corresponding to the different sections. Other applications can be found in the literature. We would like to say that the method described in the ODE's section may for instance be applied to get some general results on the model proposed in Sect. 6 of Chap. 4, even if some of the dispersal rates are null or small. Indeed, only some of the dispersal rates must be high enough to satisfy the reduction conditions.

5.5.1 Aggregation for Ordinary Differential Equations

In this subsection, we give two examples which show the effect of prey or predator individuals behaviours on the populations dynamics. The first case deals with the assumption under which the prey individuals attract the predators. This is done by considering that the predator movements are prey density dependent. By using the aggregated model ,we show that a supercritical Hopf bifurcation can occur at the population level. The second example illustrates the situation where the prey avoids the predator, by considering that the prey movements are predator dependent. In this case, there is also a Hopf bifurcation. However, the bifurcation is degenerate for the aggregated model and it is not obvious *a priori* that the bifurcation has the same properties for the complete model. We show that this is the case. Thus these examples illustrates how to deal with bifurcation analysis for the complete model by using the bifurcation analysis of the aggregated model.

Preys Attracting Predators

In Auger et al. [10] and in Auger and Lett [6], a single population dynamics in a two-patch environment connected by fast migrations was studied. We extended this approach to the case of a predator–prey community in a two-patch environment. In a series of papers, Mchich et al. [62–64] we investigated the effects of density dependent dispersal of prey (respectively predator) with respect to predator (respectively prey) density on the global stability of the system. Let us consider a predator–prey system in a two-patch environment. This example is based on the paper Mchich et al. [64]. The model reads as follows:

$$\begin{cases} \frac{dn_1}{d\tau} = (k_{12}n_2 - k_{21}n_1) + \varepsilon \left(r_1 n_1 (1 - \frac{n_1}{K_1}) - a_1 n_1 p_1 \right) \\ \frac{dn_2}{d\tau} = (k_{21}n_1 - k_{12}n_2) + \varepsilon \left(r_2 n_2 (1 - \frac{n_2}{K_2}) - a_2 n_2 p_2 \right) \\ \frac{dp_1}{d\tau} = (m_{12}(n_2)p_2 - m_{21}(n_1)p_1) + \varepsilon \left(-\mu_1 p_1 + b_1 n_1 p_1 \right) \\ \frac{dp_2}{d\tau} = (m_{21}(n_1)p_1 - m_{12}(n_2)p_2) + \varepsilon \left(-\mu_2 p_2 + b_2 n_2 p_2 \right) \end{cases} \tag{5.49}$$

where the prey migration rates k_{12} and k_{21} are constant and predator dispersal rates are assumed to be prey density dependent:

$$m_{21}(n_1) = \frac{1}{k_0 + k n_1} \tag{5.50}$$

and:

$$m_{12}(n_2) = \frac{1}{k_0 + k n_2} \tag{5.51}$$

where k_0 and k are positive parameters. These prey density dependent dispersal rates for predators assume that:

–   If few preys are present in a patch, predators leave this patch
–   If many preys are available in a patch, predators remain on that patch

The slow part of the model assumes a Lotka–Volterra model on each patch with logistic growth of the prey. r_i and K_i are respectively the growth rate and the carrying capacity of the prey on patch i. μ_i is the death rate for predator on patch i. a_i and b_i are predation parameters on patch i. These assumptions can be justified as follows: we consider the heterogeneous environment as a set of homogeneous patches and then the interaction on each homogeneous patch are based on the Mass Action Law, which claims that the reaction rates are proportional to the meeting rates. Using aggregation methods described in the first section, this model can be aggregated as follows:

$$\begin{cases} \frac{dn}{dt} = rn(1 - \frac{n}{K}) - A(n)np + O(\varepsilon) \\ \frac{dp}{dt} = -\mu(n)p + B(n)np + O(\varepsilon) \end{cases} \tag{5.52}$$

where r and K are global growth rate and carrying capacity for the prey. $A(n)$, $\mu(n)$ and $B(n)$ are not constant but depend on total prey density according to functions which are not given here. For details we refer to [64]. This case shows an example of functional emergence and of qualitative emergence. Indeed, it can be shown, [64], that the dynamics of the aggregated model is qualitatively different from the local dynamics on each patch. The local model predicts either predator extinction or predator–prey coexistence at a positive gas equilibrium. The aggregated model predicts the same situations but also periodic solutions for the total prey and predator densities. A stable limit cycle can occur via a supercritical Hopf bifurcation, see Fig. 5.4.

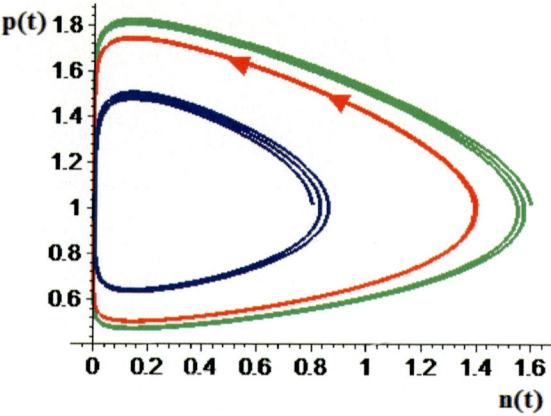

Fig. 5.4. Phase portrait of the aggregated model (5.52) exhibiting a limit cycle

Repulsive Effects of Predators on Preys

Let us consider a predator–prey system in a two-patch environment. The complete model, proposed in [62], reads as follows:

$$
\begin{cases}
\frac{dn_1}{d\tau} = (k_{12}(p_2)n_2 - k_{21}(p_1)n_1) + \varepsilon\,(r_1n_1 - a_1n_1p_1) \\
\frac{dn_2}{d\tau} = (k_{21}(p_1)n_1 - k_{12}(p_2)n_2) + \varepsilon\,(r_2n_2 - a_2n_2p_2) \\
\frac{dp_1}{d\tau} = (m_{12}p_2 - m_{21}p_1) + \varepsilon\,(-\mu_1p_1 + b_1n_1p_1) \\
\frac{dp_2}{d\tau} = (m_{21}p_1 - m_{12}p_2) + \varepsilon\,(-\mu_2p_2 + b_2n_2p_2)
\end{cases}
\tag{5.53}
$$

where the prey dispersal rate is assumed to be predator density dependent:

$$
k_{21}(p_1) = \alpha_0 + \alpha p_1
\tag{5.54}
$$

and:

$$
k_{12}(p_2) = \alpha_0 + \beta p_2
\tag{5.55}
$$

where m_0 and m are positive parameters. This predator density dependent dispersal rate for preys assumes that:

– If few predators are present in a patch, preys remain on this patch.
– If many predators are located in a patch, preys leave this patch.

The slow model is a classical Lotka–Volterra model with linear growth rate on each patch for the prey. All parameters have the same meaning as in previous section.

Using aggregation methods, this model can be aggregated as follows:

$$
\begin{cases}
\frac{dn}{dt} = \frac{1}{d_0+dp}\,(rn + anp - bnp^2) + O\,(\varepsilon) \\
\frac{dp}{dt} = -\mu p + \frac{1}{d_0+dp}\,(bnp + cnp^2) + O\,(\varepsilon)
\end{cases}
\tag{5.56}
$$

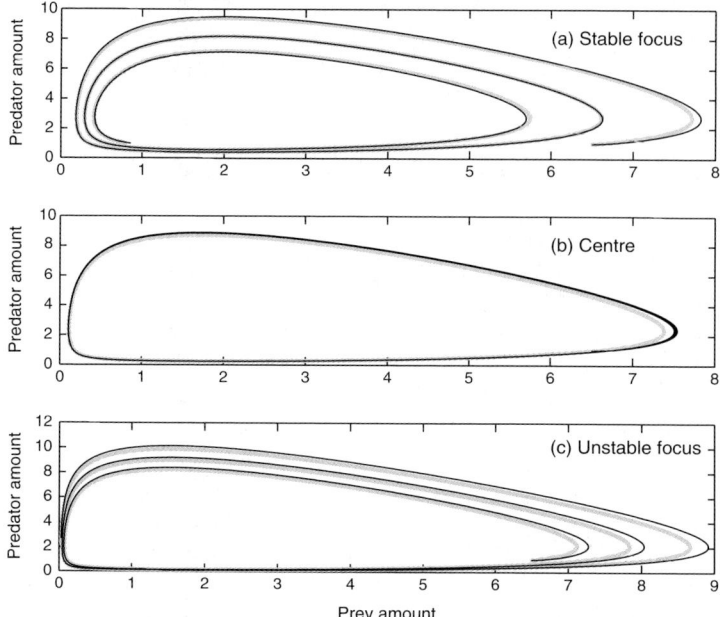

Fig. 5.5. This figure illustrates the degenerate Hopf bifurcation of the complete model. (**a**) corresponds to a stable focus, (**b**) is a centre and (**c**) is an unstable focus. One curve is obtained with the complete model while the other curve is obtained with the aggregated model. The set of parameters used for these simulations is: $\alpha_0 = 1$, $\alpha = 1$, $\beta = 0.5$, $m_{12} = 2$, $m_{21} = 1$, $r_1 = 1.1$, $r_2 = 1.2$, $a_1 = 1$, $a_2 = 0.9$, $\mu_1 = 0.9$, $\mu_2 = 0.8$, $b_1 = 1$, $b_2 = 1$ and $\varepsilon = 0.1$

where global parameters r, a, b, c, d_0 and d are expressed in terms of local parameters. For details we refer to [62]. This case also shows an example of functional emergence and of qualitative emergence. It can be shown that the dynamics of the aggregated model is qualitatively different from the local dynamics on each patch. The local model is a classical Lotka–Volterra models and therefore predicts periodic solutions according to center trajectories. For the aggregated model, [62], one can show that there is a degenerate Hopf bifurcation. Therefore, as it is presented on Fig. 5.5 and according to the parameters values:

– Prey and predator can coexist at constant densities
– The predator–prey system is not persistent
– At bifurcation, there exists periodic solutions (centers)

To conclude this section devoted to spatial predator–prey dynamics, let us mention that the most general case combining the two previous effects (attraction of predators by preys and repulsion of preys by predators) is under study.

We mention here a series of articles where the effects of different individual decisions on the global dynamics of a prey–predator system, in an heterogeneous environment, have been studied [4, 24, 40, 61, 79].

Effects of Competitive Behaviour of Predators on a Predator–Prey System Dynamics

In a previous contribution [7], we investigated the effects of contests between predators disputing preys on the stability of a predator–prey Lotka–Volterra model. Roughly, it was assumed that when a predator captures a prey, another predator is coming and the two predators come into contest. Predators can be aggressive (hawk) or non aggressive (dove).

In another article [5], we considered a detailed version of a predator–prey model with hawk-dove contests between predators at the fast time scale.

On the individual level, predator individuals [51] have three possible states of behaviour: they can be searching for prey, finding a prey or defending it. Individuals in each of these subpopulations can play the hawk or dove tactics. We denote by p_{SD}, p_{FD}, p_{DD}, p_{SH}, p_{FH} and p_{DH} the biomass of searching and dove predators, finding and dove predators, defending and dove predators, searching and hawk predators, finding and hawk predators and defending and hawk predators respectively. The individuals can change their tactics only in the defending subpopulation. Let:

$$p_S = p_{SD} + p_{SH}, \qquad (5.57a)$$

$$p_F = p_{FD} + p_{FH}, \qquad (5.57b)$$

$$p_D = p_{DD} + p_{DH}, \qquad (5.57c)$$

be the biomass of searching predators, finding predators and defending predators, respectively. We denote n the prey density.

In this model, searching predators can capture preys according to the mass action law. When a predator has captured a prey, it comes into the finding state. It takes some time for a finding predator to manipulate a prey before returning to the searching state $(1/\beta)$. However, another searching predator can find a predator when it manipulates its prey. We assume the mass action law for searching and finding predators encountering rates. When a manipulating predator is found by a searching one, both predators come into the defending state, come into contest and dispute for the prey. Defending predators play against each other using hawk and dove strategies. After some time $(1/\gamma)$, defending predators return to the searching state. The complete model is thus a set of seven ODE's governing the prey and six predator densities, obtained by coupling the predator behavioural model at a fast time scale, see Fig. 5.6, to a predator–prey model at a slow time scale as follows:

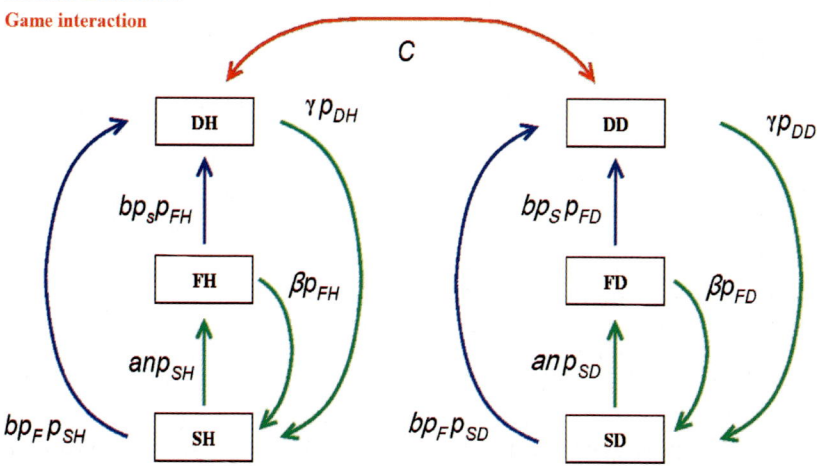

Fig. 5.6. Scheme of the possible states for the predator individuals and flux between these states

$$\frac{dp_{SD}}{d\tau} = -bp_F p_{SD} - anp_{SD} + \beta p_{FD} + \gamma p_{DD}$$

$$+ \varepsilon\big(\alpha\left(\beta p_{FD} + (\mathbf{Au})_D p_{DD}\right) - \mu p_{SD}\big), \tag{5.58a}$$

$$\frac{dp_{FD}}{d\tau} = -bp_S p_{FD} + anp_{SD} - \beta p_{FD} - \varepsilon\mu p_{FD}, \tag{5.58b}$$

$$\frac{dp_{DD}}{d\tau} = bp_F p_{SD} - \gamma p_{DD} + bp_S p_{FD} + cp_{DD}\big((\mathbf{Au})_D - \mathbf{u}^T\mathbf{Au}\big) - \varepsilon\mu p_{DD}, \tag{5.58c}$$

$$\frac{dp_{SH}}{d\tau} = -bp_F p_{SH} - anp_{SH} + \beta p_{FH} + \gamma p_{DH}$$

$$+ \varepsilon\big(\alpha\left(\beta p_{FH} + (\mathbf{Au})_H p_{DH}\right) - \mu p_{SH}\big), \tag{5.58d}$$

$$\frac{dp_{FH}}{d\tau} = -bp_S p_{FH} + anp_{SH} - \beta p_{FH} - \varepsilon\mu p_{FH}, \tag{5.58e}$$

$$\frac{dp_{DH}}{d\tau} = bp_F p_{SH} - \gamma p_{DH} + bp_S p_{FH} + cp_{DH}\big((\mathbf{Au})_H - \mathbf{u}^T\mathbf{Au}\big) - \varepsilon\mu p_{DH}, \tag{5.58f}$$

$$\frac{dn}{d\tau} = \varepsilon\big(rn(1 - \frac{n}{K}) - anp_S\big). \tag{5.58g}$$

where $(\mathbf{Au})_D$ and $(\mathbf{Au})_H$ respectively represent the gain of dove and hawk individuals. The meaning of the different parameters of the fast part of the model can be understood from the flows shown in Fig. 5.6. The slow part of the model, of order ε, contains a logistic growth for preys, a type I functional response, a constant predator natural mortality and a predator growth

depending on the consumption of preys either in the finding state or in the
defending state. For more details see [5].

Using aggregation methods, this complete model can be aggregated into
two different systems of 2 ODE's governing the prey and the total predator,
denoted p, densities. At fast equilibrium, gain is γ and cost is δ. These two
systems correspond either to the mixed hawk-dove fast equilibrium or to the
pure hawk fast equilibrium:

– If $\gamma < \delta$ then we have the dimorphic case (mixed hawk and dove predators)
which we call model I:

$$\frac{dn}{dt} = rn\left(1 - \frac{n}{K}\right) - anp_S^* ,$$ (5.59a)

$$\frac{dp}{dt} = -\mu p + \alpha\left(\beta p_F^* + \frac{\gamma}{2}(1 - \frac{\gamma}{\delta})p_D^*\right).$$ (5.59b)

– If $\gamma > \delta$ then we have the monomorphic case (only hawk predators) which
we call model II:

$$\frac{dn}{dt} = rn\left(1 - \frac{n}{K}\right) - anp_S^* ,$$ (5.60a)

$$\frac{dp}{dt} = -\mu p + \alpha\left(\beta p_F^* + \frac{\gamma}{2}(1 - \frac{\delta}{\gamma})p_D^*\right),$$ (5.60b)

where the values of p_S^*, p_F^* and p_D^* are fast equilibrium values which can
be expressed in terms of n and p.

The aggregated model has been studied by bifurcation analysis. Two important parameters have been chosen, the cost of an escalated contest C and
the carrying capacity of the prey K. Using these two parameters, all other
parameters being fixed, one can capture the essential of the dynamics, see
Fig. 5.7.

As expected in a prey–predator model, at constant cost, when the carrying
capacity of the prey increases, predator invades (TC), and then, for small and
large cost values where there is no coexistence of two limit cycles, there is a
supercritical Hopf bifurcation with the appearance of a stable limit cycle [5].
This is the so called "paradox of enrichment". However, if one assumes that
the prey carrying capacity K is bounded above, there always exists a cost-
window in which there is no "paradox of enrichment". Indeed, as shown on
Fig. 5.7, there is a cost-domain where for any value of K, predator and prey
can coexist at constant densities. This stability domain occurs for pure hawk
as well as for a mixed predator population. This last result shows that contests
between predators can make the predator–prey system more stable.

We also mention earlier articles on the dynamics of a population of two
competing populations using fast game dynamics (Auger and Pontier [17],
Sánchez et al. [83], Auger et al. [18]).

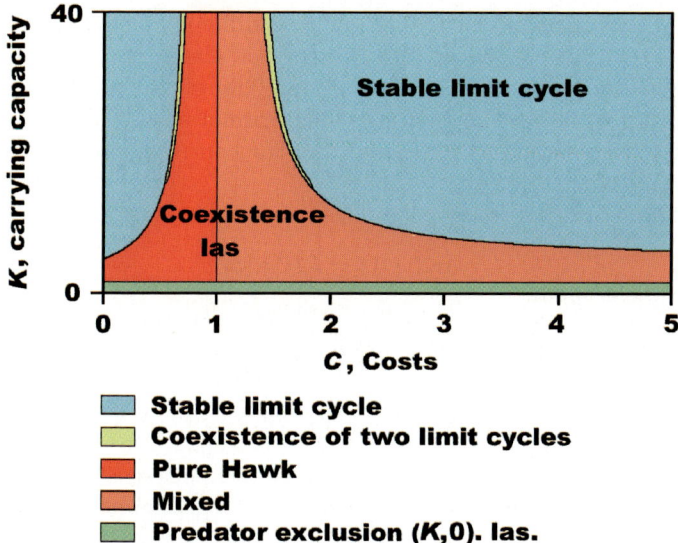

Fig. 5.7. Bifurcation diagram in the parameters space (C, K). Note that there is a domain around $C = 1$ where, for every K, the prey and its predator coexist at fixed densities

5.5.2 Discrete Models

We provide two discrete models examples. We have chosen examples for which aggregation methods presented in the previous section of discrete models can be applied. The first example is linear (linear fast model and linear slow model) and, the second one is nonlinear (linear fast model but, nonlinear slow model). The first one concerns the effects of habitat fragmentation on an insect population dynamics. It is based on the article by Pichancourt et al. [71]. The second example deals with the problem of spatial synchrony of a host–parasitoid systems. It is based on two articles by Nguyen Huu et al. [67, 68]. In this second example, we analyze the effect of the time scale factor, that is the ratio between fast time scale unit and slow time scale unit, on spatial synchrony which is needed to proceed to spatial aggregation. Among other results, we show that this ratio does not need to be very high and that aggregation methods can thus be useful in very realistic and concrete ecological situations.

Effects of Habitat Fragmentation on Insect Population Dynamics

In Pichancourt et al. [71], we studied the effect of habitat fragmentation on insect population dynamics, *Abax parallelepipedus* (*coleoptera, carabidae*). This insect population is considered to have a metapopulation structure in the agricultural landscape in Brittany. Roughly speaking, the landscape can be represented by a network of patches of four different kinds, the agricultural matrix, pieces of woods, lanes and hedgerows. The agricultural matrix usually represents a large proportion of about 0.80 of the total landscape and is an unfavorable habitat for insects which cannot survive for a long time in this habitat (as well as in hedgerows). On the contrary, woods and lanes are favorable habitats.

The complete model reads as follows:

$$\mathbf{n}(t+1) = LP^k \mathbf{n}(t) \qquad (5.61)$$

where $\mathbf{n}(t)$ is the population vector structured by age and habitat at time step t. k is a parameter which represents the number of dispersal events that are performed during one time step. This number can be assumed to be large and thus migration between patches is assumed to be fast with respect to demography. The time step of the model corresponds to one year which is also the duration of each stage. We considered three stages, larvae (L), adults aged 1 (A1) which do not reproduce and adults aged 2 and more (A2) which can reproduce. There are four different types of habitat, agricultural matrix (M), wood (W), lane (L) and, hedgerow (H). As we consider 3 stages and 4 habitats, the complete model deals with 12 variables.

$P = \mathbf{diag}\,[I, P_A, P_A]$ is a migration matrix between different types of habitat. It is a block diagonal matrix. Each block matrix represents the migration process between the different types of habitats for individuals belonging to a given stage, with constant proportions of migrants from patch to patch: I is the identity matrix corresponding to the larvae L which do not move, while P_A corresponds to both A_1 and A_2 movements. Therefore, the fast model is linear.

L is a Leslie multiregional matrix (Caswell [33]) which describes the insect life cycle in each type of habitat. The matrix L incorporates survival as well as fecundities for each stage and for each habitat. All parameters of the Leslie matrix have been obtained from experimental data and for details we refer to [71].

To summarize the life cycle of *Abax parallelepipedus* in the different habitats, insects cannot reproduce in agricultural matrix and hedgerows. Fecundity is high in woods and lanes. Larvae cannot survive in agricultural matrix and hedgerows. Larvae can survive only in woods and lanes in proportion up to 0.50 from year to year. Adult insects cannot survive in agricultural matrix. Adult insects survive in a proportion up to 0.45 in woods and lanes. Survival as well as fecundity are constant. Therefore, the slow model is also linear. Figure 5.8 shows life cycles in (a) favorable and (b) unfavorable habitats.

In [71], we investigated the effects of habitat fragmentation on viability of the overall insect population. In this model, two parameters allow to take into account an increase of habitat fragmentation. First, the landscape is more fragmented when the overall proportion of favorable habitat decreases. Secondly, for a constant proportion of each habitat, fragmentation increases when the average size of a favorable habitat decreases. In other words, at

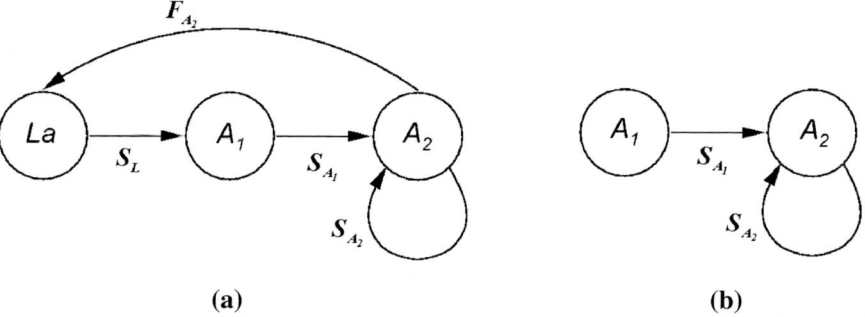

Fig. 5.8. Life cycles in (**a**) favourable and (**b**) unfavourable habitats for *Abax paral-lelepipedus*. S_L, S_{A_1} and S_{A_2}, represent respectively the proportions of larvae, adults aged 1 and adults aged 2 of a generation which survive to the next generation. F_{A_2} is the fecundity of adults aged 2

constant total wood proportion, a landscape with many small pieces of woods is more fragmented than a landscape with a single large piece of wood.

The first effect can be taken into account in the model because migration rates, for example from wood to agricultural matrix, depend on the propor-tion of the different habitats, for details see [71]. The second effect can also be taken into account by increasing parameter k of the complete model pre-sented below. Indeed, if the average size of a patch of favorable habitat is small, insects come more frequently to a boundary with another habitat and are more likely to change habitat. Thus, parameter k represents at constant proportion of habitats, the degree of fragmentation of the landscape, from small fragmentation (small k) to high fragmentation (large k).

In the limit case, $k \gg 1$, which corresponds to a highly fragmented land-scape, one can perform a "spatial aggregation". Indeed, let us first consider the fast system which reads as follows, $\mathbf{n}_a$ being either adults 1 or adults 2 of *Abax parallelepipedus*:

$$\mathbf{n}_a(t+1) = P_A \mathbf{n}_a(t) \tag{5.62}$$

This fast model is conservative because the total insect population does not vary at the fast time scale. Therefore, it has a dominant eigenvalue equal to 1. The corresponding eigenvector has non negative components. When nor-malized to 1, these components represent constant proportions of insects of the different stages in the different habitats. Following the method presented in the previous section of aggregation of discrete models, we can build an ag-gregated model governing two global variables representing the overall adult 1 and adult 2 densities, obtained by summation over all habitats of the land-scape, and two more variables for larvae densities in the two habitats where they can survive, i.e (W) and (L).

The aggregated model reads as follows:

$$\mathbf{n}(t+1) = \bar{L}\mathbf{n}(t) \tag{5.63}$$

where $\mathbf{n}(t)$ is a dimension 4 population vector structured by stage at time step t. $\bar{L}$ is an aggregated matrix with dimension 4. Thus, using aggregation methods, we reduced the dimension from 12 (complete model) to 4 (aggregated model). As shown in the above section about aggregation of discrete models, when k tends to infinity, dominant eigenvalues of complete and aggregated models tend to the same value. Therefore, the aggregated model can be used to find the asymptotic behavior of the complete model. Remember that the dominant eigenvalue has a significant ecological interpretation, as it represents the asymptotic growth rate of the overall insect population. When the dominant eigenvalue is bigger than 1, the total population grows, when it is smaller than 1, the total population goes extinct. In [71], we have shown that the dominant eigenvalues of complete and aggregated models are close at less than 0.05 for a k-value larger than 12, which corresponds to one dispersal event per month and is rather realistic for insects in Brittany landscape.

In [71], we studied the particular case of only two types of habitats, a favorable habitat (W) and an unfavorable one (L), see Fig. 5.9. Figure 5.10

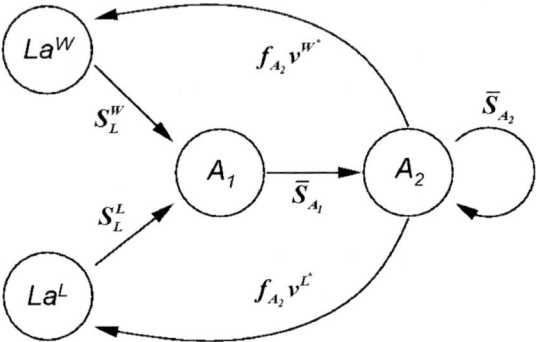

Fig. 5.9. Life cycles of *Abax parallelepipedus* for the aggregated model with two habitats (W) and (L). ν^{W^*} and ν^{L^*} are the proportion of adults on habitat (W) and (L) respectively. $\bar{S}_{A_1}$ and $\bar{S}_{A_2}$ represent the proportion of adults 1 and 2 surviving to the next generation, and S_L^W and S_L^L the survival of larvae in both habitats

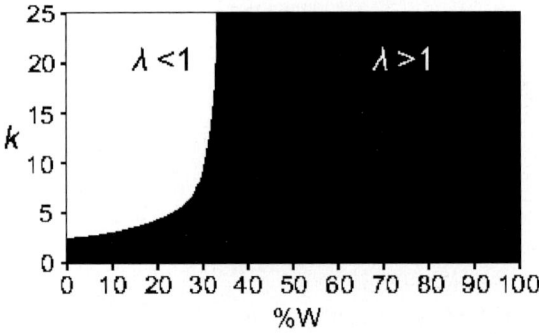

Fig. 5.10. Effect of W-fragmentation on asymptotic population growth rate λ for landscapes with only W and M

shows domains where the overall asymptotic growth rate of the insect population is bigger than 1 (global survival) and smaller than 1 (global extinction) as a function of global proportion of favorable habitat (W) and of parameter k. Results showed that fragmentation has a negative effect on overall population viability. For large k-values, a proportion of $1/3$ of woods is needed for insect persistence. Below this critical wood proportion, insect population goes extinct. Moreover, for very small k-values, the insect population can survive even for small proportion of woods. This is due to the fact that a very large patch of wood (even with small overall wood proportion) can promote viability because insects can always remain in the favorable patch from year to year and never go to the matrix where they die. The article [71] had also investigated several other cases with four types of habitats in different proportions.

Aggregation of a Spatial Model for a Host–Parasitoid Community

We investigated the effects of fast migration on the global stability of a host–parasitoid community in a two patch model [57] and in the case of a linear chain of patches [58]. We have also been studying the case of a host–parasitoid community in a 2-Dimensional network of patches ([67] and [68]).

Let us consider the Nicholson–Bailey model with logistic growth of the hosts:

$$\begin{cases} n(t+1) = n(t) \exp\left(r\left(1 - \frac{n(t)}{K}\right)\right) \exp\left(-ap(t)\right) \\ p(t+1) = cn(t)\left(1 - \exp\left(-ap(t)\right)\right) \end{cases} \tag{5.64}$$

where r is the host growth rate, K its carrying capacity. a is a positive parameter, the searching efficiency of the parasitoid. c is a positive parameter, the average number of hosts merging from a single infected host. This model has a unique positive equilibrium which can be stable or unstable according to parameters values. The model can exhibit periodic solutions and chaotic dynamics. Trajectories can also tend asymptotically to an invariant curve, see [48].

Now, we consider a two-dimensional network of patches on a square lattice. The size of the network is A^2. We further assume that individuals can move to the eight neighbouring patches according to the following dispersal model at any patch (i, j) of the network:

$$\begin{cases} n(t+1) = (1 - \mu_n)\, n(t) + \frac{\mu_n}{8} \sum n(t) \\ p(t+1) = (1 - \mu_p)\, p(t) + \frac{\mu_p}{8} \sum p(t) \end{cases} \tag{5.65}$$

where the sum holds for the eight nearest patches around a given patch. For simplicity, we omit the patch index position (i, j). μ_n (respectively μ_p) is the proportion of host (respectively parasitoid) which disperses during a time step of dispersal. Numerical simulations are made by considering that during a generation, individuals first disperse according to the dispersal submodel and then locally interact according to the logistic Nicholson–Bailey submodel. The dynamics of this spatial host–parasitoid community has been widely studied and we refer to an early paper [41]. According to mobility values, after some transient dynamics, the system can exhibit crystal structures, spirals or else chaotic dynamics.

In order to introduce time scales, fast dispersal with respect to local dynamics, we are going to consider a new complete model written in the same form as in equation (5.25):

$$\mathbf{X}(\mathbf{t} + \mathbf{1}) = S\left(F^k \mathbf{X}(t)\right) \tag{5.66}$$

where $\mathbf{X}(t) = (\mathbf{n}(t), \mathbf{p}(t))$ is the spatial host and parasitoid population density vector. $\mathbf{n}(t)$ (respectively $\mathbf{p}(t)$) is the host (respectively parasitoid) density vector with A^2 components. In the previous system, S represents the local Nicholson–Bailey dynamics. F is the dispersal matrix corresponding to the previous dispersal submodel. k is an integer which is assumed to be large, $k \gg 1$. During one generation, individuals disperse k times and interact locally one time. Therefore, when k is large, the dispersal process becomes fast in comparison to local interactions.

Figure 5.11 shows the effect of an increase of k on the spatial distribution which is "Gaussian" centered on the initial position with a variance increasing

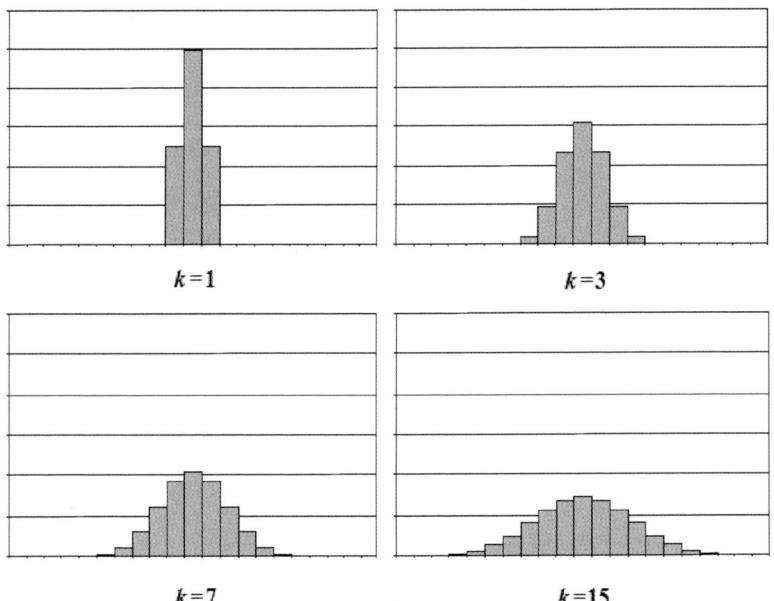

$k=1$ $k=3$

$k=7$ $k=15$

Fig. 5.11. Distribution of the distance from an initial position after one generation with respect to the parameter k when a proportion of $1/2$ of the individuals leave the patch

with k. Hosts and parasitoids are flying insects and it is realistic to assume that during one generation an insect can fly not only to the nearest patches but farther.

Using aggregation methods, one can obtain an aggregated model for the total density of host $(N(t) = \sum n(t))$ and parasitoids $(P(t) = \sum P(t))$, where the sum holds for any patch of the network. This aggregated model assumes that a fast dispersal equilibrium is reached. This fast dispersal equilibrium corresponds to a situation of spatial synchrony with constant and equal proportions of insects on any patch which is simply $\frac{1}{A^2}$, i.e. the inverse of the total number of patches. The use of the aggregated model implies that the local insect density is proportional to the total density at each patch. In that case, the aggregated model reads:

$$\begin{cases} N(t+1) = \exp\left(r\left(1 - \frac{N(t)}{KA^2}\right)\right)\exp\left(-\frac{aP(t)}{A^2}\right) \\ P(t+1) = cN(t)\left(1 - \exp\left(-\frac{ap(t)}{A^2}\right)\right) \end{cases} \quad (5.67)$$

The aggregated model depends on the size of the two-dimensional network. We compare the aggregated model and the complete model with parameters $r = 0.5$, $a = 0.2$, $K = 14.47$, $c = 1$, $A = 50$, $\mu_n = 0.2$ and $\mu_p = 0.89$ leading to a stable equilibrium. We use an initial condition with a few hosts and parasitoids located on the same patch at $t = 0$.

This example illustrates that even with a low value of k, the dynamics of the aggregated model and the complete model are very close. Both dynamics tend toward the same equilibrium, as shown in Fig. 5.12.

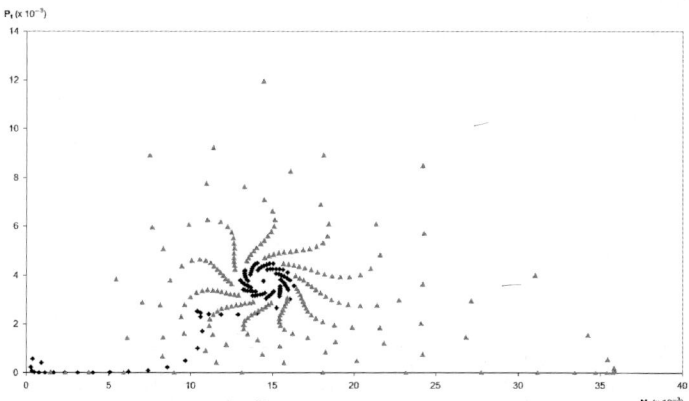

Fig. 5.12. Dynamics of the complete model (*in black*) and the aggregated model (*in grey*) for $k = 10$ and parameters $r = 0.5$, $a = 0.2$, $K = 14.47$, $c = 1$, $A = 50$, $\mu_n = 0.2$ and $\mu_p = 0.89$. To make the figure more readable, points near the equilibrium have not been represented

5.5.3 Aggregation of Variables in Linear Delayed Differential Equations

Example: A Structured Model of Population Dynamics with Two Time Scales

Let us consider a continuous-time two-stage structured model of a population living in an environment divided into two different sites. Let us refer to the individuals in the two stages as juveniles and adults, so that $j_i(t)$ and $n_i(t)$ denote the juvenile and adult population respectively at site i, $i = 1, 2$. Changes in the juvenile population at site i occur through birth, maturation to the adult stage and death. Therefore, in absence of migrations, the growth rate is expressed as $\beta_i n_i(t) - e^{-\mu_i^* r_i}\beta_i n_i(t - r_i) - \mu_i^* j_i(t)$ where $\beta_i, \mu_i^*, \mu_i \geq 0$ are the fecundities and per capita death rates of juveniles and adults respectively and $r_i > 0$ is the juvenile-stage duration in site i. Without loss of generality, we suppose $0 < r_1 < r_2$.

In a similar way, the adult population growth rate in site i must contain recruitment and mortality terms so that in absence of migrations reads $e^{-\mu_i^* r_i}\beta_i n_i(t - r_i) - \mu_i n_i(t)$.

We consider a model which includes the demographic processes described below, together with a fast migration process between sites for the adult population defined by two parameters: $m_1 > 0$ represents the migration rate from site 1 to site 2 and $m_2 > 0$ is the migration rate from site 2 to site 1.

The difference between the two time scales: slow (demography) and fast (migration) is represented by a small parameter $\varepsilon > 0$:

$$\begin{cases} j_1'(t) = \beta_1 n_1(t) - e^{-\mu_1^* r_1}\beta_1 n_1(t - r_1) - \mu_1^* j_1(t) \\ j_2'(t) = \beta_2 n_2(t) - e^{-\mu_2^* r_2}\beta_2 n_2(t - r_2) - \mu_2^* j_2(t) \\ n_1'(t) = (1/\varepsilon)[m_2 n_2(t) - m_1 n_1(t)] + e^{-\mu_1^* r_1}\beta_1 n_1(t - r_1) - \mu_1 n_1(t) \\ n_2'(t) = (1/\varepsilon)[m_1 n_1(t) - m_2 n_2(t)] + e^{-\mu_2^* r_2}\beta_2 n_2(t - r_2) - \mu_2 n_2(t) \end{cases}$$

As we notice, the last two equations of the above system are autonomous, so we can reduce the system into them:

$$\mathbf{n}'(t) = \frac{1}{\varepsilon}K\mathbf{n}(t) + A\mathbf{n}(t) + B_1\mathbf{n}(t - r_1) + B_2\mathbf{n}(t - r_2) \qquad (5.68)$$

where

$$\mathbf{n}(t) := \begin{pmatrix} n_1(t) \\ n_2(t) \end{pmatrix}; \quad K := \begin{pmatrix} -m_1 & m_2 \\ m_1 & -m_2 \end{pmatrix}; \quad A = \begin{pmatrix} -\mu_1 & 0 \\ 0 & -\mu_2 \end{pmatrix}$$

$$B_1 := \begin{pmatrix} e^{-\mu_1^* r_1}\beta_1 & 0 \\ 0 & 0 \end{pmatrix}; \quad B_2 := \begin{pmatrix} 0 & 0 \\ 0 & e^{-\mu_2^* r_2}\beta_2 \end{pmatrix}$$

together with an initial condition $\Phi(t) := (\Phi_1(t), \Phi_2(t))^T$, $t \in [-r_2, 0]$.

Matrix K satisfies Hypothesis 2 and in order to build the *aggregated model* of (5.68) we choose the right and left eigenvectors associated to eigenvalue

$\lambda = 0$ of K as $\mathbf{v} := 1/(m_1 + m_2) \, (m_2, m_1)^T$, $\mathbf{v}^* := (1, 1)^T$, so that we construct an aggregated model for the total adult population:

$$n(t) := (\mathbf{v}^*)^T \mathbf{n}(t) = n_1(t) + n_2(t).$$

Due to the two different delays this model does not fit in the formulation given by (5.43). Therefore we apply (5.48) so that we have

$$\forall \Phi \in C([-r_2, 0]; \mathbf{R}^2), \quad L(\Phi) := A\Phi(0) + B_1\Phi(-r_1) + B_2\Phi(-r_2)$$

and then, $\forall \psi \in C([-r_2, 0]; \mathbf{R})$:

$$\overline{L}(\psi) := -\mu^*\psi(0) + \nu_1^*\psi(-r_1) + \nu_2^*\psi(-r_2)$$

with

$$\mu^* := \frac{\mu_1 m_2 + \mu_2 m_1}{m_1 + m_2}; \quad \nu_1^* := \frac{e^{-\mu_1^* r_1}\beta_1 m_2}{m_1 + m_2}; \quad \nu_2^* := \frac{e^{-\mu_2^* r_2}\beta_2 m_1}{m_1 + m_2}.$$

The aggregated model is, for $t \geq r_2$:

$$n'(t) = -\mu^* n(t) + \nu_1^* n(t - r_1) + \nu_2^* n(t - r_2) \tag{5.69}$$

together with the initial condition defined by:

$$n(t) = \Phi_1(t) + \Phi_2(t), \ t \in [-r_2, 0]$$
$$n'(t) = \mu^* n(t) + e^{-\mu_1^* r_1}\beta_1\Phi_1(t - r_1) + e^{-\mu_2^* r_2}\beta_2\Phi_2(t - r_2), \ t \in [0, r_1]$$
$$n'(t) = \mu^* n(t) + \nu_1^* n(t - r_1) + e^{-\mu_2^* r_2}\beta_2\Phi_2(t - r_2), \ t \in [r_1, r_2]$$

We have reduced the initial complete system of four equations to a single equation governing the total adult population. If the solution to this equation is given, then the juvenile population densities can be derived from it.

It can be shown that, for each $T > r_2$, the solution to system (5.68) satisfies

$$\lim_{\varepsilon \to 0_+} \begin{pmatrix} n_{1\varepsilon}(t) \\ n_{2\varepsilon}(t) \end{pmatrix} = \frac{1}{m_1 + m_2} \begin{pmatrix} m_2 \\ m_1 \end{pmatrix} n(t)$$

uniformly in $[r_2, T]$, $n(t)$ being the solution to the aggregated model (5.69).

5.6 Perspectives and Conclusions

Regarding applications to population dynamics, aggregation methods have also been used in the following cases:

– Modelling the effect of migrations processes on the population dynamics (Poggiale et al. [73–77]).

- Modelling a trout fish population in an arborescent river network composed of patches connected by fast migrations (Charles et al. [34–36], Chaumot et al. [37–39]).
- Modelling a sole larvae population with a continuous age with fast migration between different spatial patches (Bravo et al. [27]).
- Modelling food chain structures (Kooi et al. [55]).

In this chapter, we have presented aggregation methods in several contexts, ODE's, Discrete models, PDE's and DDE's. However, there are still works to be done to show that a complete detailed mathematical model can be replaced by a reduced aggregated model. For example, there is no doubt that more work should be devoted to the case of stochastic models.

In our opinion, a major problem relates to the understanding of mechanisms which are responsible for the emergence of individual behaviour at the population and community level. In many cases, biologists prefer to use an Individual Based Model (IBM) because they want to take into account many individuals of different kinds and to model how they interact at the individual level. The IBM is then simulated with a computer and is used to look for global emerging properties at the level of the population and of the community. However, it can be difficult to obtain robust and general results from a complete and detailed IBM.

In this chapter, we have shown that in some cases, interactions between individuals can also be taken into account by classical models with differential equations. When two time scales are involved in the model, aggregation methods allow to proceed to a significant reduction of the dimension of the model and sometimes to a complete analysis of the aggregated model. This reduced model is useful to understand how the individual behaviour can influence the dynamics of the total population and its community (Dubreuil et al. [45]). Therefore, aggregation methods can be considered as a new and promising tool for the study of emergence of global properties in complex systems with many potential applications in ecological dynamics.

Acknowledgements

P. Auger, R. Bravo de la Parra and E. Sánchez were supported by Ministerio de Educación y Ciencia (Spain), proyecto MTM2005-00423 and FEDER. E. Sánchez was partially supported by CCG06-UPM/MTM-242. J.-C. Poggiale was supported by the French Program of Coastal Environment (IFREMER - CNRS, project AMPLI). T. Nguyen Huu was supported by the French Ministery for Research and Education. Figure 5.4 adapted from Mchich et al. [64], Figures 5.6 and 5.7 adapted from Auger et al. [5] and Figures 5.8, 5.9 and 5.10 adapted from Pichancourt et al. [71] with permission from Elsevier.

References

1. Arino, O., Sánchez, E., Bravo de la Parra, R., Auger, P.: A singular perturbation in an age-structured population model. SIAM J. Appl. Math., **60**, 408–436 (1999)
2. Arino, O., Sánchez, E., Bravo de la Parra, R.: A model of an age-structured population in a multipatch environment. Math. Comput. Model., **27**, 137–150 (1998)
3. Auger, P.: Dynamics and thermodynamics in hierarchically organized systems. Pergamon Press, Oxford (1989)
4. Auger, P., Benoit, E.: A prey–predator model in a multi-patch environment with different time scales. J. Biol. Syst., **1**(2), 187–197 (1993)
5. Auger, P., Kooi, B., Bravo de la Parra, R., Poggiale, J.C.: Bifurcation analysis of a predator–prey model with predators using hawk and dove tactics. J. Theor. Biol., **238**, 597–607 (2006)
6. Auger, P., Lett. C.: Integrative biology: linking levels of organization. C. R. Acad. Sci. Paris, Biol., **326**, 517–522 (2003)
7. Auger, P., Bravo de la Parra, R., Morand, S., Sánchez, E.: A predator–prey model with predators using hawk and dove tactics. Math. Biosci., **177/178**, 185–200 (2002)
8. Auger, P., Bravo de la Parra, R.: Methods of aggregation of variables in population dynamics. C. R. Acad. Sci. Paris, Sciences de la vie, **323**, 665–674 (2000)
9. Auger, P., Charles, S., Viala, M., Poggiale, J.C.: Aggregation and emergence in ecological modelling: integration of the ecological levels. Ecol. Model., **127**, 11–20 (2000)
10. Auger, P., Poggiale, J.C., Charles, S.: Emergence of individual behaviour at the population level: effects of density dependent migration on population dynamics. C. R. Acad. Sci. Paris, Sciences de la Vie, **323**, 119–127 (2000)
11. Auger, P., Chiorino, G., Poggiale, J.C.: Aggregation, emergence and immergence in hierarchically organized systems. Int. J. Gen. Syst., **27**(4–5), 349–371 (1999)
12. Auger, P., Poggiale, J.C.: Aggregation and Emergence in Systems of Ordinary Differential Equations. Math. Comput. Model., **27**(4), 1–22 (1998)
13. Auger, P., Poggiale, J.C.: Aggregation and emergence in hierarchically organized systems: population dynamics. Acta Biotheor., **44**, 301–316 (1996)
14. Auger, P., Poggiale, J.C.: Emergence of population growth models: fast migration and slow growth. J. Theor. Biol., **182**, 99–108 (1996)
15. Auger, P., Poggiale, J.C.: Emerging properties in population dynamics with different time scales. J. Biol. Syst., **3**(2), 591–602 (1995)
16. Auger, P., Roussarie, R.: Complex ecological models with simple dynamics: from individuals to populations. Acta Biotheor., **42**, 111–136 (1994)
17. Auger, P., Pontier, D.: Fast game theory coupled to slow population dynamics: the case of domestic cat populations. Math. Biosci., **148**, 65–82 (1998)
18. Auger, P., Bravo de la Parra, R., Sánchez, E.: Hawk-dove game and competition dynamics. Math. Comput. Model., Special issue Aggregation and emergence in population dynamics. Antonelli, P., Auger, P., guest-Editors, **27**(4), 89–98 (1998)
19. Bates, P.W., Lu, K., Zeng, C.: Invariant foliations near normally hyperbolic invariant manifolds for semiflows. Trans. Am. Math. Soc., **352**, 4641–4676 (2000)
20. Bates, P.W., Lu, K., Zeng, C.: Existence and persistence of invariant manifolds for semiflow in Banach space. Memoir. Am. Math. Soc., **135**, 129 (1998)

21. Benoît, E.: Canards et enlacements. Extraits des Publications Mathématiques de l'IHES, **72**, 63–91 (1990)
22. Benoît, E.: Systèmes lents-rapides dans R3 et leurs canards. Astérisque, **109/110**, 159–191 (1983)
23. Benoît, E., Callot, J.L., Diener, F., Diener, M.: Chasse au canard. Collection Mathématique, **31/32**(1–3), 37–119 (1981)
24. Bernstein, C., Auger, P.M., Poggiale, J.C.: Predator migration decisions, the ideal free distribution and predator–prey dynamics. Am. Nat., (1999), **153**(3), 267–281 (1999)
25. Blasco, A., Sanz, L., Auger, P., Bravo de la Parra, R.: Linear discrete population models with two time scales in fast changing environments II: non autonomous case. Acta Biotheor., **50**(1), 15–38 (2002)
26. Blasco, A., Sanz, L., Auger, P., Bravo de la Parra, R.: Linear discrete population models with two time scales in fast changing environments I: autonomous case. Acta Biotheor., **49**, 261–276 (2001)
27. Bravo de la parra, R., Arino, O., Sánchez, E., Auger, P.: A model of an age-structured population with two time scales. Math. Comput. Model., **31**, 17–26 (2000)
28. Bravo de la Parra, R., Sánchez, E., Auger, P.: Time scales in density dependent discrete models. J. Biol. Syst., **5**, 111–129 (1997)
29. Bravo de la Parra, R., Auger, P., Sánchez, E.: Aggregation methods in discrete models. J. Biol. Syst., **3**, 603–612 (1995)
30. Bravo de la Parra, R., Sánchez, E.: Aggregation methods in population dynamics discrete models. Math. Comput. Model., **27**(4), 23–39 (1998)
31. Bravo de la Parra, R., Sánchez, E., Arino, O., Auger, P.: A Discrete Model with Density Dependent Fast Migration. Math. Biosci., **157**, 91–110 (1999)
32. Carr, J.: Applications of centre manifold theory. Springer, Berlin Heidelberg New York (1981)
33. Caswell, H.: Matrix population models. Sinauer Associates, Sunderland, MA, USA (2001)
34. Charles, S., Bravo de la Parra, R., Mallet, J.P., Persat, H., Auger, P.: Population dynamics modelling in an hierarchical arborescent river network: an attempt with *Salmo trutta*. Acta Biotheor., **46**, 223–234 (1998)
35. Charles, S., Bravo de la Parra, R., Mallet, J.P., Persat, H., Auger, P.: A density dependent model describing *Salmo trutta* population dynamics in an arborescent river network: effects of dams and channelling. C. R. Acad. Sci. Paris, Sciences de la vie, **321**, 979–990 (1998)
36. Charles, S., Bravo de la Parra, R., Mallet, J.P., Persat, H., Auger, P.: Annual spawning migrations in modeling brown trout population dynamics inside an arborescent river network. Ecol. Model., **133**, 15–31 (2000)
37. Chaumot, A., Charles, S., Flammarion, P., Garric, J., Auger, P.: Using aggregation methods to assess toxicant effects on population dynamics in spatial systems. Ecol. Appl., **12**(6), 1771–1784 (2002)
38. Chaumot, A., Charles, S., Flammarion, P., Auger, P.: Ecotoxicology and spatial modeling in population dynamics: an attempt with brown trout. Environ. Toxicol. Chem., **22**(5), 958–969 (2003)
39. Chaumot, A., Charles, S., Flammarion, P., Auger, P.: Do migratory or demographic disruptions rule the population impact of pollution in spatial networks? Theor. Pop. Biol., **64**, 473–480 (2003)

40. Chiorino, O., Auger, P., Chasse, J.L., Charles, S.: Behavioral choices based on patch selection: a model using aggregation methods. Math. Biosci., **157**, 189–216 (1999)
41. Comins, H.N., Hassell, M.P., May, R.M.: The spatial dynamics of host–parasitoid systems. J. Anim. Ecol., **61**, 735–748 (1992)
42. De Feo, O., Rinaldi, S.: Singular homoclinic bifurcations in tritrophic food chains. Math. biosci., **148**, 7–20 (1998)
43. Diener, M.: Canards et bifurcations. In: Outils et modèles mathématiques pour l'automatique, l'analyse des systèmes et le traitement du signal, vol. 3, Publication du CNRS, 289–313 (1983)
44. Diener, M.: Etude générique des canards. Thesis, Université de Strasbourg (1981)
45. Dubreuil, E., Auger, P., Gaillard, J.M., Khaladi, M.: Effects of aggressive behaviour on age structured population dynamics. Ecol. Model., **193**, 777–786 (2006)
46. Dumortier, F., Roussarie, R.: Geometric singular perturbation theory beyond normal hyperbolicity. In: Jones, C.K.R.T., Khibnik, A.I. (eds) Multiple time scale dynamical systems. Springer, Berlin Heidelberg New York (2000)
47. Dumortier, F., Roussarie, R.: Canard cycles and center manifolds. Memoir. Am. Math. Soc., **121**(577), 1–100 (1996)
48. Edelstein-Keshet, L.: Mathematical models in biology. Random House, New York (1989)
49. Fenichel, N.: Persistence and Smoothness of Invariant Manifolds for Flows. Indiana Univ. Math. J., **21**(3), 193–226 (1971)
50. Hirsch, M.W., Pugh, C.C., Shub, M.: Invariant manifolds. Lecture notes in mathematics vol. 583. Springer, Berlin Heidelberg New York (1977)
51. Hofbauer, J., Sigmund, K.: Evolutionary games and population dynamics. Cambridge University Press, Cambridge (1998)
52. Iwasa, Y., Andreasen, V., Levin, S.: Aggregation in model ecosystems. I. Perfect aggregation. Ecol. Model., **37**, 287–302 (1987)
53. Iwasa, Y., Levin, S., Andreasen, V.: Aggregation in model ecosystems. II. Approximate Aggregation. IMA. J. Math. Appl. Med. Biol., **6**, 1–23 (1989)
54. Kaper, T.J., Jones, C.K.R.T.: A primer on the exchange lemma for fast-slow systems. In: Jones, C.K.R.T., Khibnik, A.I. (eds) Multiple time scale dynamical systems. Springer, Berlin Heidelberg New York (2000)
55. Kooi, B.W., Poggiale, J.C., Auger, P.M.: Aggregation methods in food chains. Math. Comput. Model., **27**(4), 109–120 (1998)
56. Krylov, N., Bogoliubov, N.: The application of methods of nonlinear mechanics to the theory of stationary oscillations. Publication 8 of the Ukrainian Academy of Science, Kiev (1934)
57. Lett, C., Auger, P., Bravo de la Parra, R.: Migration frequency and the persistence of host–parasitoid interactions. J. Theor. Biol., **221**, 639–654 (2003)
58. Lett, C., Auger, P., Fleury, F.: Effects of asymmetric dispersal and environmental gradients on the stability of host–parasitoid systems. Oikos, **109**, 603–613 (2005)
59. Lotka, A.J.: Undamped oscillations derived from the mass action law. J. Am. Chem. Soc., **42**, 1595–1599 (1920)
60. Lotka, A.J.: Elements of physical biology. William and Wilkins, Baltimore (1925)

61. Michalski, J., Poggiale, J.C., Arditi, R., Auger, P.: Effects of migrations modes on patchy predator–prey systems. J. Theor. Biol., **185**, 459–474 (1997)
62. Mchich, R., Auger, P., Poggiale, J.C.: Effect of predator density dependent dispersal of prey on stability of a predator–prey system. Math. Biosci., **206**, 343–356 (2007)
63. Mchich, R., Auger, P., Raïssi, N.: The stabilizability of a controlled system describing the dynamics of a fishery. C. R. Acad. Sci. Paris, Biol., **329**, 337–350 (2005)
64. Mchich, R., Auger, P., Bravo de la Parra, R., Raïssi, N.: Dynamics of a fishery on two fishing zones with fish stock dependent migrations: aggregation and control. Ecol. Model., **158**, 51–62 (2002)
65. Muratori, S., Rinaldi, S.: Low and high frequency oscillations in three dimensional food chain systems. SIAM J. Appl. Math., **52**(6), 1688–1706 (1992)
66. Murray, J.D.: Mathematical biology. Springer, Berlin Heidelberg New York (1989)
67. Nguyen Huu, T., Lett, C., Poggiale J.C., Auger, P.: Effects of migration frequency on global host–parasitoid spatial dynamics with unstable local dynamics. Ecol. Model., **177**, 290–295 (2006)
68. Nguyen-Huu, T., Lett, C., Auger, P., Poggiale, J.C.: Spatial synchrony in host–parasitoid models using aggregation of variables. Math. Biosci., **203**, 204–221 (2006)
69. Nicholson, A.J.: The balance of animal populations. J. Anim. Ecol., **2**, 132–178 (1933)
70. Nicholson, A.J., Bailey, V.A.: The balance of animal populations, part I. Proc. Zool. Soc. Lond., **3**, 551–598 (1935)
71. Pichancourt, J.B., Burel, F., Auger, P.: Assessing the effect of habitat fragmentation on population dynamics: an implicit modelling approach. Ecol. Model., **192**, 543–556 (2006)
72. Pliss, V.A., Sell, G.R.: Perturbations of normally hyperbolic manifolds with applications to the Navier–Stokes equations. J. Differ. Equat., **169**, 396–492 (2001)
73. Poggiale, J.C.: Lotka–Volterra's model and migrations: breaking of the well-known center. Math. Comput. Model., **27**(4), 51–62 (1998)
74. Poggiale, J.C.: From behavioural to population level: growth and competition. Math. Comput. Model., **27**(4), 41–50 (1998)
75. Poggiale, J.C.: Predator–prey models in heterogeneous environment: emergence of functional response. Math. Comp. Model., **27**(4), 63–71 (1998)
76. Poggiale, J.C., Michalski, J., Arditi, R.: Emergence of donor control in patchy predator–prey systems. Bull. Math. Biol., **60**(6), 1149–1166 (1998)
77. Poggiale, J.C., Auger, P.: Impact of spatial heterogeneity on a predator–prey system dynamics. C. R. Biol., **327**, 1058–1063 (2004)
78. Poggiale, J.C., Auger, P.: Fast oscillating migrations in a predator–prey model. Methods Model. Meth. Appl. Sci., **6**(2), 217–226 (1996)
79. Poggiale, J.C., Auger, P., Roussarie, R.: Perturbations of the classical Lotka–Volterra system by behavioural sequences. Acta Biotheor., **43**, 27–39 (1995)
80. Sakamoto, K.: Invariant manifolds in singular perturbations problems for ordinary differential equations. Proc. Roy. Soc. Ed., **116A**, 45–78 (1990)
81. Sánchez, E., Bravo de la Parra, R., Auger, P., Gómez-Mourelo, P.: Time scales in linear delayed differential equations. J. math. Anal. Appl., **323**, 680–699 (2006)

82. Sánchez, E., Bravo de la Parra, R., Auger, P.: Discrete models with different time-scales. Acta Biotheor., **43**, 465–479 (1995)
83. Sánchez, E., Auger, P., Bravo de la Parra, R.: Influence of individual aggressiveness on the dynamics of competitive populations. Acta Biotheor., **45**, 321–333 (1997)
84. Sanz, L., Bravo de la Parra, R.: Variables aggregation in time varying discrete systems. Acta Biotheor., **46**, 273–297 (1998)
85. Sanz, L., Bravo de la Parra, R.: Variables aggregation in a time discrete linear model. Math. Biosci., **157**, 111–146 (1999)
86. Sanz, L., Bravo de la Parra, R.: Time scales in stochastic multiregional models. Nonlinear Anal. R. World Appl., **1**, 89–122 (2000)
87. Sanz, L., Bravo de la Parra, R.: Time scales in a non autonomous linear discrete model. Math. Model. Meth. Appl. Sci., **11**(7), 1203–1235 (2001)
88. Sanz, L., Bravo de la Parra, R.: Approximate reduction techniques in population models with two time scales: study of the approximation. Acta Biotheor., **50**(4), 297–322 (2002)
89. Sanz, L., Bravo de la Parra, R.: Approximate reduction of multiregional models with environmental stochasticity. Math. Biosci., **206**, 134–154 (2007)
90. Sanz, L., Bravo de la Parra, R., Sánchez, E.: Approximate reduction of nonlinear discrete models with two time scales. J. Differ. Equ. Appl., DOI: 10.1080/10236190701709036 (2008)
91. Scheffer, M., Rinaldi, S., Kuztnetsov, Y.A., Van Nes, E.H.: Seasonal dynamics of Daphnia and algae explained as a periodically forced predator–prey system. Oikos, **80**(3), 519–532 (1997)
92. Scheffer, M., De Boer, R.J.: Implications of spatial heterogeneity for the paradox of enrichment. Ecol., **76**(7), 2270–2277 (1995)
93. Stewart, G.W, Guang Sun, J.I.: Matrix perturbation theory. Academic Press, Boston (1990)
94. Volterra, V.: Variazioni e fluttuazioni del numero d'individui in specie animali conviventi. Mem. R. Accad. Naz. dei Lincei. Ser. VI, **2**, 31–113 (1926)
95. Wiggins, S.: Normally Hyperbolic invariant manifolds in dynamical systems. Springer, Berlin Heidelberg New York (1994)

6

The Biofilm Model of Freter: A Review

M. Ballyk[1], D. Jones[2], and H.L. Smith[3]

[1] Department of Mathematical Sciences, New Mexico State University,
Las Cruces, NM 88003-8001, USA
`mballyk@nmsu.edu`
[2] Department of Mathematics and Statistics, Arizona State University, Tempe,
AZ 85287-1804, USA
`dajones@math.asu.edu`
[3] Supported by NSF Grant DMS 0414270, Department of Mathematics and
Statistics, Arizona State University, Tempe, AZ 85287-1804, USA
`halsmith@asu.edu`

Summary. R. Freter et al. (1983) developed a simple chemostat-based model of
competition between two bacterial strains, one of which is capable of wall-growth,
in order to illuminate the role of bacterial wall attachment on the phenomenon of
colonization resistance in the mammalian gut. Together with various collaborators,
we have re-formulated the model in the setting of a tubular flow reactor, extended
the interpretation of the model as a biofilm model, and provided both mathematical
analysis and numerical simulations of solution behavior. The present paper provides
a review of the work in [4–6, 31–35, 45, 46].

6.1 Introduction

The ability of bacteria to colonize surfaces forming biofilms and thereby to
create a refuge from the vagaries of fluid advection has stimulated a great deal
of recent research in a variety of disciplines. The importance of wall growth
was first made apparent to us from the work of microbiologist Rolf Freter
and his colleagues [23–26]. Their mathematical models of the phenomenon
of colonization resistance in the mammalian gut showed that bacterial wall
attachment could play a crucial role in the observed stability of the natural
microflora of the gut to invasion by non-indigenous microorganisms. The authors had the pleasure of learning of this work first hand from a lecture by
Freter at the Microbial Ecology Workshop organized by Frank Hoppensteadt
and Smith at Arizona State University in 1997. Our own work can be traced
to a collaboration that began at this workshop.

While Freter's model was formulated in a CSTR (chemostat) setting which
is natural since this reactor mimics the mouse cecum, the animal model for
gut research, we felt that another natural setting was the plug flow reactor

(PFR) where bacterial motility, fluid advection, and other spatial effects could play a larger role. The flow reactor more accurately reflects the environment of the large intestine of humans. Thus, Ballyk and Smith formulated a family of models for microbial growth and competition for limiting substrate and wall-attachment sites in our first work [5]. This paper focused on describing the model equations which consist of a system of parabolic partial differential equations for a diffusing substrate and for randomly motile bacteria in the fluid environment of a thin (one-dimensional) tube coupled to a system of ordinary differential equations for the immobile, wall-attached bacteria growing on the tubular surface. It was natural to begin a mathematical analysis of the model by considering only a single bacterial strain in the reactor in [6].

The phenomena of colonization resistance in the gut remained the primary motivation for our early work in [4–6, 45, 46]. However, we quickly became aware of the rapidly developing literature on biofilms and began to view our models in this more general context. Our subsequent work [29, 31–35] was largely motivated by our view that the Freter model is a crude model of a biofilm. A biofilm is simply a layer of material coating a surface, usually immersed in a fluid environment, made up of bacteria and an extra-cellular matrix exuded by the bacteria which provides an environment for growth. From the bacterial point of view, a biofilm is a comfortable refuge. Examples of a biofilm include the scum that grows on a rock in a stream, dental plaque on teeth, the surface slime that forms on the inside surface of water pipes, and a similar coating of the surface of the large intestine of mammals. These bacterial layers can have serious negative consequences in many man made environments. They contaminate medical devices such as contact lenses, implants, catheters, and stints; they can contaminate food and medicinal production facilities and air-conditioning and water circulation systems. Biofilms are notoriously difficult to eradicate once established.

The study of biofilms has exploded in the last two decades, largely driven by recent advances in noninvasive microscopy, staining techniques, and genetic probes. Contributing to this explosion of interest was the realization that many human disease processes are essentially biofilm infections of organs, e.g., Periodontitus in teeth and gums, Otitus media in the ear, and Cystic fibrosis pneumonia in the lung, to name only a few [27]. In ways not yet completely understood, these biofilm infections are difficult to treat using antibiotics [47]. It is hoped that increased understanding of the biofilm environment will lead to improved modes of therapy for these diseases.

From this recent attention focused on biofilms has come the realization that what scientists had learned about bacteria, based on more than a century of studying them in suspension as isolated cells, e.g. in a chemostat, or in simplified batch culture environments, did not prepare them for the observed level of complexity these organisms display in the biofilm setting. It is now recognized that almost all bacteria live in biofilm communities of remarkable structure which are attached to surfaces and in which individual bacteria are capable of expressing whole suites of genes not previously known that allow them to communicate and coordinate their activities. Quorum sensing, a bacterial communication mechanism based on the exchange of an internally produced small molecule and the detection of its extracellular concentration, has been shown to control aspects of biofilm formation and development and, in cases of biofilm infections of tissue, to control the expression of genes controlling virulence factors. For example, it is advantageous for bacteria not to express virulence until a suitable defence, a mature biofilm environment, has been established. The reader interested in more background information concerning biofilms should consult the many review articles of Costerton et al. [15–18] and the monograph of Bryers [11]. For an update on current directions in the study of biofilms consult the special issue (January 2005) of the journal Trends in Microbiology devoted to biofilm research.

Along with this increase in the interest in biofilms there has been increasing interest in mathematical models of biofilms. Below we give a necessarily limited review of some of this literature. A classical but now somewhat dated reference is the volume "Biofilms" edited by Characklis and Marshall [12] where partial differential equations are used to model steady state (time independent) bacterial densities. An updated edited volume "Biofilms II" [11] indicates some recent directions of research. Most recent modeling in the field has been directed towards understanding the mechanisms underlying the remarkable variety of spatial structure observed in different biofilms. These structures can be as simple as flat layers with little or no structure to fields of mushroom-like structures protruding from the surface. Modeling the evolution of a biofilm in detail requires dealing with the physical forces giving rise to fluid motion, advection and diffusion of nutrients and wastes, the forces acting on the moving interface between the biofilm and bulk fluid on the one hand and also including the biological aspects including growth, production of extra-cellular matrix, and attachment and detachment of cells and matrix from the biofilm. Classically, these considerations lead naturally to large systems of partial differential equations. Recent papers of Cogan and Keener [14], Eberl et al. [22], Dockery and Klapper [21], examine various aspects of biofilm morphology using continuum models that lead to systems of partial differential equations with a moving biofilm-fluid interface for the growing biofilm. Cellular automaton models and related Individual-based models have recently been used by Wimpenny and Colasanti [51], Kreft et al. [36], and Laspidou and Rittmann [38] to model biofilm development. Hybrid models employing a mixture of continuum modeling and cellular automata have been employed by Picioreanu et al. [41] and Noguera et al. [40]. Dillon et al. [20] combine partial differential equations for fluid flow, chemical densities with individual-based modeling of cells moving in response to forces to model biofilm formation. An up to date review of biofilm modeling by Noguera et al. [39] is useful. With the exception of [14, 21], most of these models are so complex that they can be investigated only using sophisticated numerical simulations.

There are few simple, conceptual biofilm models which are amenable to mathematical analysis yet which yield significant and useful results. Simple models do not attempt to provide much detail on the spatial structure of biofilms but aim to provide information on conditions suitable for biofilm establishment and maintenance and which model the formation of biofilms directly, starting from an inoculum of planktonic bacteria. Among the more widely known of these, we mention Topiwala and Hamer [49], Baltzis and Fredrickson [7], Bakke et al. [8], and Pilyugin and Waltman [42]. Less known is the work of Freter and his group (Freter [23,24] and Freter et al. [25,26]), who formulated a mathematical model to understand the phenomena of colonization resistance in the mammalian gut (stability of resident microflora to colonization). Essentially, their model can be viewed as a crude biofilm model. In contrast to state of the art biofilm models cited above, the Freter model completely ignores the three-dimensional spatial structure of the biofilm.

Our view is that the Freter model provides a useful yet mathematically tractable model of a biofilm which focuses primarily on the bacterial interactions. The present chapter provides an overview of our previous work in [4–6, 31–35, 45, 46]. More recent work on gene transfer in biofilms [29, 30], based on the biofilm models considered here is reviewed in [30]. The authors wish to acknowledge collaborators Dung, Kojouharov, Stemmons, Zhao who have contributed substantially towards portions of the work presented here.

As our models use nonlinear partial differential equations to capture random motility of cell populations and diffusion of nutrients, we will employ some of the same techniques used in Chaps. 3 and 4 where spatial movement also plays an important role.

6.2 The Freter Model

The Freter model describes a microbial population in a fluid environment with a bounding surface, a portion of which may be colonized by the bacteria forming a biofilm. At any given moment of time, the bacterial population can be viewed as consisting of cells suspended in the fluid, usually called planktonic cells, and those adhering to the surface, called adherent cells. Planktonic cells can adhere to the biofilm becoming adherent cells and adherent cells may detach from the biofilm becoming planktonic cells. The Freter model is based on the following assumptions:

1. There are a finite number of available colonization sites on the wall and thus a maximum attainable areal density of adherent bacteria.
2. Planktonic bacteria are attracted to the wall at a rate proportional to planktonic cell density and the fraction of unoccupied colonization sites on the wall.
3. Adherent cells are sloughed off into the fluid at a rate proportional to their density.
4. Daughter cells of adherent bacteria compete for space on the wall: A fraction G of the daughter cells find attachment sites and the fraction $1 - G$ do not and are forced into the fluid. G is a decreasing function of wall occupancy.

Table 6.1. Model parameters and functions

Symbol	Description	Dimension
u	Biomass concentration of planktonic bacteria	ml^{-3}
w	Areal biomass density of adherent bacteria	ml^{-2}
w_{max}	Maximum areal biomass density of adherent bacteria	ml^{-2}
$W = w/w_{max}$	Wall occupancy fraction	$-$
$G(W)$	Fraction of daughter cells that find wall sites	$-$
β	Sloughing rate of adherent bacteria	t^{-1}
α	Rate constant of adhesion	t^{-1}
S	Concentration of limiting substrate	ml^{-3}
S_0	Concentration of the substrate in the feed	ml^{-3}
γ	Yield constant	$-$
$f_u(S)$	Specific growth rate of planktonic bacteria	t^{-1}
$f_w(S)$	Specific growth rate of adherent bacteria	t^{-1}
k	Planktonic cell death rate	t^{-1}
k_w	Adherent cell death rate	t^{-1}

Model parameters are described in the Table 6.1.

If w denotes the areal density of wall-adherent bacteria and w_{max} the maximum attainable density, then the wall occupancy is given by

$$W = w/w_{max} \, .$$

Planktonic cells, with density u, are attracted to the wall at rate $\alpha u(1 - W)$, proportional to their density and to the unoccupied fraction $1 - W$ of wall sites (see also Baltzis and Fredrickson [7]). It is reasonable to assume that $G = G(W)$ is a decreasing function of the occupation fraction W because a more fully saturated wall provides less chance for a daughter cell to find space on it. Freter [24] employs the rational function

$$G(W) = \frac{1 - W}{1.1 - W} \, .$$

Bacteria consume substrate S at (per unit biomass) rate

$$f_u(S) = \frac{m_u S}{a_u + S}, \text{ planktonic cells,} \tag{6.1}$$

$$f_w(S) = \frac{m_w S}{a_w + S}, \text{ adherent cells,} \tag{6.2}$$

with conversion to biomass with yield constant γ. Cell death rates are introduced for later use when anti-microbials are considered.

6.3 The Chemostat-Based Model

The Freter model was originally proposed in the setting of a CSTR of volume V, colonizable surface area A and flow rate F (Freter [24], Freter et al. [25]) where it takes the form ($D = F/V$, $\delta = A/V$):

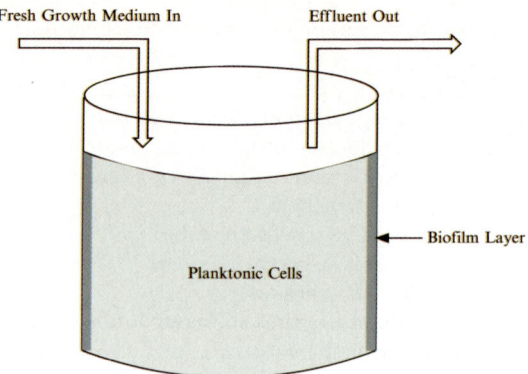

Fig. 6.1. Wall growth in the chemostat. Planktonic cells reside in well-mixed fluid; wall-adherent cells reside in the biofilm layer

$$S' = D(S^0 - S) - \gamma^{-1}[uf_u(S) + \delta w f_w(S)]$$
$$u' = u[f_u(S) - D - k] + \beta\delta w + \delta w f_w(S)[1 - G(W)] - \alpha u[1 - W] \quad (6.3)$$
$$w' = w[f_w(S)G(W) - \beta - k_w] + \alpha u[1 - W]\delta^{-1} \ .$$

The meaning of the terms in (6.3) are now described. Nutrient S enters the chemostat in the feed stream at rate DS^0 and unused nutrient leaves at rate DS. Inside the chemostat (Fig. 6.1), nutrient is consumed by both planktonic cells and wall attached cells. Planktonic cells u grow in response to the uptake of nutrient, are washed out of the chemostat at rate Du and suffer cell death rate k. Wall attached cells slough into the fluid at rate $\beta\delta w$ and a fraction $1 - G(W)$ of the output $f_w(S)w$ of daughter cells of wall attached bacteria fail to find wall sites and thus enter the fluid. Finally, planktonic cells attach to the wall at rate $\alpha u[1 - W]$. Wall attached cells grow (finding wall sites) at rate $f_w(S)G(W)$, suffer cell death k_w, are lost due to sloughing at rate β, and gain via the attachment of planktonic cells at rate $\alpha u[1 - W]$.

The classical Monod (no wall attachment) model for planktonic cell growth (see e.g. Herbert et al. [28]; Smith and Waltman [44]; Bailey and Ollis [2]) is recovered on setting $\alpha = w = 0 = k = k_w$.

If we set $w_{max} = \infty$ making $W = 0$, then we remove the hypothesis that the wall environment is finite and that daughter cells of wall adherent bacteria compete for wall sites. If also $G(0) = 1$ then the model reduces to that of Pilyugin and Waltman [42].

$$S' = D(S^0 - S) - \gamma^{-1}[uf_u(S) + \delta w f_w(S)]$$
$$u' = u[f_u(S) - D - k] + \beta\delta w - \alpha u \quad (6.4)$$
$$w' = w[f_w(S) - \beta - k_w] + \alpha u\delta^{-1} \ .$$

A similar reduction can be made in the setting of a flow reactor in the sections to follow.

A mathematical analysis of the system (6.3) is given in Stemmons and Smith [46] in the case that $f_u = f_w$, $k = k_w = 0$. The more general model considered here allows for different substrate uptake rates for planktonic and adherent cells and for nontrivial cell death rates.

The washout steady state

$$(S, u, w) = (S^0, 0, 0)$$

always exists. Its stability, is determined by linearizing (6.3) about this equilibrium. The three-by-three Jacobian has one-one entry $-D$ below which are zeros. Therefore, stability depends on the eigenvalues of the 2×2 lower-right sub-matrix of the Jacobian given by

$$A = \begin{pmatrix} f_u(S^0) - D - k - \alpha & f_w(S^0)(1 - G(0)) + \beta \\ \alpha & f_w(S^0)G(0) - k_w - \beta \end{pmatrix}. \tag{6.5}$$

This matrix is quasipositive (nonnegative off-diagonal entries), as are other matrices encountered throughout this work, and hence we will find it convenient to employ Perron–Frobenius theory [10]. The washout state is locally asymptotically stable if its eigenvalues have negative real part (they are real) and unstable if at least one is positive. We use the notation $s(A)$ for the largest eigenvalue of A. With this notation, the washout steady state is stable when $s(A) < 0$ and unstable when $s(A) > 0$. Although an exact (and ugly) expression may be written for $s(A)$ in terms of the entries, the following inequalities, the result of simple estimates, shed more light on the biology:

$$\max\{f_u(S^0) - D - k - \alpha, f_w(S^0)G(0) - k_w - \beta\}$$
$$< s(A) \leq \max\{f_u(S^0) - D - k, f_w(S^0) - k_w\} . \tag{6.6}$$

Therefore, $s(A) < 0$ and the washout state is (globally) stable if both

$$f_w(S^0) - k_w < 0 \text{ and } f_u(S^0) - k - D < 0 ; \tag{6.7}$$

$s(A) > 0$ and the washout state is unstable if either

$$f_w(S^0)G(0) - k_w - \beta \geq 0 \tag{6.8}$$

or if

$$f_u(S^0) - k - D - \alpha \geq 0 . \tag{6.9}$$

Also, $s(A) > 0$ if both

$$f_u(S^0) - k - D > 0 \quad \text{and } f_w(S^0) - k_w \geq 0 \tag{6.10}$$

because if both (6.8) and (6.9) fail yet (6.10) holds, then the determinant of A is negative.

The following result is proved in [34], relying on earlier work of [46].

Theorem 1. *If $s(A) < 0$, then the washout state is stable. It is globally attracting if, in addition to $s(A) < 0$, either (6.7) or $f_w(S^0) - k_w > f_u(S^0) - k - D$ or $f_u = f_w$ holds.*

If $s(A) > 0$ then at least one nontrivial steady state (S^*, u^*, w^*) exists and any nontrivial steady state satisfies

$$u^* > 0, \ w_{max} \geq w^* > 0, \ 0 < S^* < S^0 \ .$$

If $f_u(S) - k - D > 0$ implies $f_w(S) - k_w \geq 0$ or if $f_u = f_w$, then at most one nontrivial steady state may exist. Finally, $s(A) > 0$ implies that u and w persist: there exists $\epsilon > 0$, independent of initial data, such that if $w(0) + u(0) > 0$, there is a $T > 0$ (depending on the initial data) such that

$$u(t) + w(t) > \epsilon, \quad t > T \ .$$

In the case considered by Stemmons and Smith [46], sufficient conditions are given for the unique nontrivial steady state to attract all nontrivial initial data. The Pilyugin and Waltman system (6.4) is considerably simpler and global stability of the unique nontrivial equilibrium can be deduced in the special case that $f_u = f_w$ [42]. Unfortunately, this condition is not realistic.

Persistence of u, w means that there is an initial-condition-independent lower bound for the ultimate bacterial density. In view of our inability to show that a nontrivial steady state is globally attracting, this says that at least the microbial population survives when $s(A) > 0$. The proof of this assertion follows that given in Proposition 7.9 of [46].

Diekmann and Heesterbeek [19] (see Theorem 6.13) show that the stability modulus of A can be related to the spectral radius of a related matrix and that this relation can be used to define a basic reproductive ratio which is more useful for interpreting our results. Specifically, they show that

$$\text{sign } s(A) = \text{sign } [\rho(GT^*) - 1] \tag{6.11}$$

where ρ denotes spectral radius and T^* denotes the transpose of T. Here,

$$G = \begin{pmatrix} f_u(S^0) & f_w(S^0)(1 - G(0)) \\ 0 & f_w(S^0)G(0) \end{pmatrix} \tag{6.12}$$

and

$$T = \frac{1}{\Delta} \begin{pmatrix} k_w + \beta & \alpha \\ \beta & D + k + \alpha \end{pmatrix} \tag{6.13}$$

and

$$\Delta = (\beta + k_w)(D + k + \alpha) - \alpha\beta = (D + k)(\beta + k_w) + k_w\alpha \ .$$

The factors G and T have natural biological interpretations. If we identify indices $\{1, 2\} = \{u, w\}$ then

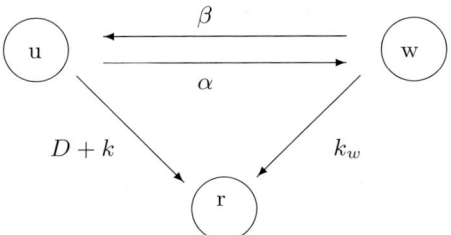

Fig. 6.2. Continuous-time Markov chain with absorbing state r = removal by washout or death

$$G_{ij} = \begin{array}{l} \text{rate of production of daughter cells in compartment i by cells in} \\ \text{compartment j in washout environment} \end{array}$$

and

$$T_{ij} = \begin{array}{l} \text{mean time spent in state j before washout or death by cell born in} \\ \text{state i.} \end{array}$$

The matrix T is calculated using the algorithm described in [29] (based on suggestions by Schrieber) identifying an appropriate continuous-time Markov chain model with rate matrix

$$B = \begin{pmatrix} -D - k - \alpha & \beta \\ \alpha & -k_w - \beta \end{pmatrix} \tag{6.14}$$

as described in Fig. 6.2. In [29] it is also argued that the mean residence time of a cell in the chemostat is given by

$$\text{Mean Residence Time} = -\frac{1}{s(B)} = \rho(T) .$$

Hence

$$GT^* = \begin{pmatrix} f_u(S^0)T_{uu} + f_w(S^0)(1 - G(0))T_{uw} & f_u(S^0)T_{wu} + f_w(S^0)(1 - G(0))T_{ww} \\ f_w(S^0)G(0)T_{uw} & f_w(S^0)G(0)T_{ww} \end{pmatrix} \tag{6.15}$$

gives the production of daughter cells into state i by mother cell in state j before removal by washout or cell death. Thus we are lead to define the *Basic Reproductive Number* R_0 as

$$R_0 = \rho(GT^*) . \tag{6.16}$$

If $v = (\tau, 1 - \tau)^*$, $0 < \tau < 1$, is the normalized positive eigenvector corresponding to the simple eigenvalue R_0 for GT^*, then $GT^*v = R_0 v$ and adding these two equations gives

$$R_0 = f_u(S^0)[\tau T_{uu} + (1 - \tau)T_{wu}] + f_w(S^0)[\tau T_{uw} + (1 - \tau)T_{ww}] .$$

The first convex combination in brackets can be interpreted as a mean time spent by a cell (regardless of where born) in the fluid environment while the second is a mean time spent in the biofilm. Thus, R_0 can be interpreted as the replacement ratio in the washout state.

Our main result implies the existence of a threshold for survival of the organism in CSTR. Indeed, it is easy to see that $s(A)$ and $\rho(GT^*)$ are strictly increasing with supply substrate concentration S^0 and, obviously, $s(A) < 0$ for very small S^0. Thus, there exists a critical substrate supply $0 < S_c^0 \leq \infty$ such that washout is stable if $S^0 < S_c^0$ and unstable if $S^0 > S_c^0$. However, the threshold may be so low as to be unobservable as it is likely that under most operating conditions, $k_w, \beta << f_w(S^0)$ and therefore (6.8) holds. This explains the ubiquity of biofilms; whereas in a fluid environment organisms must grow fast enough to overcome washout, in biofilm they simply have to grow fast enough to exceed sloughing.

The lack of effectiveness of antimicrobial agents in controlling bacteria in biofilms has been noted in the literature. See for example Costerton et al. [16] and Stewart et al. [47, 48]. As noted in [16], resistance to antimicrobials is likely to have multiple causes, one being "the failure of an agent to penetrate the full depth of the biofilm. Polymeric substances like those that make up the matrix of a biofilm are known to retard the diffusion of antibiotics...". An extensive discussion may be found in [47]. In [34] we examined the extreme case of a biofilm layer which is impenetrable to an antibiotic A introduced into the CSTR at concentration A^0. Following Stewart et al. [47], (see (11.19)), we assume that the antibiotic increases planktonic death rates as its concentration increases: the death rate of planktonic bacteria $k = k(A)$ is an increasing function of A. Adherent bacteria are assumed to be unaffected because the antibiotic cannot penetrate the biofilm layer. Assuming that A is not significantly depleted by its action, it satisfies

$$A' = D(A^0 - A)$$

and, obviously,

$$A(t) \to A^0, \quad t \to \infty .$$

Therefore the long-term effect of the antibiotic is essentially to adjust the death rate of planktonic bacteria to $k = k(A^0)$. However, it is easy to see from (6.6) that $s(A)$ is relatively insensitive to parameter k. If (6.8) holds, then $s(A) > 0$ and the washout steady state is unstable regardless of the value of k. Therefore, we expect the bacterial population to survive in the CSTR regardless of the planktonic cell death rate. The density of planktonic bacteria may be driven quite low by the antibiotic but one expects that the wall-attached cell density is largely unaffected. Thus we conclude that if (6.8) holds then introduction of antibiotic into the flow reactor will not be effective in eradicating the bacteria.

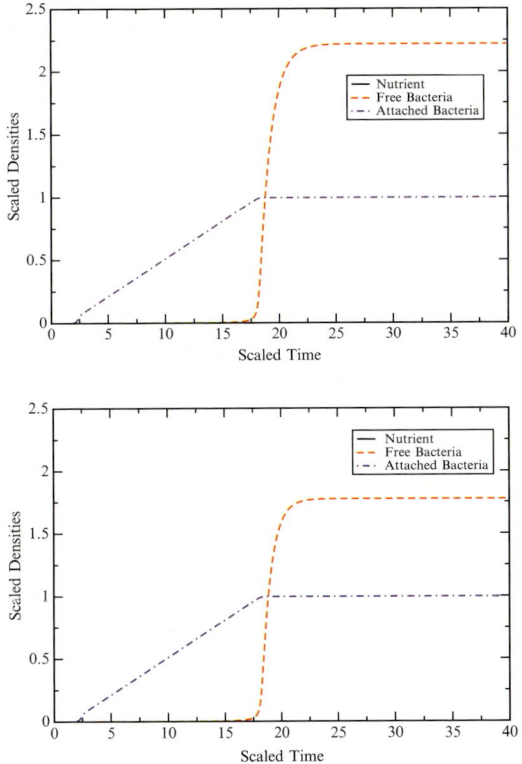

Fig. 6.3. *Top*: Biofilm formation in CSTR. *Bottom*: Biocide raises planktonic cell death rate

The effect of various biocides on biofilms is a topic of great interest and has spawned a great deal of mathematical modeling. See for example Hunt et al. [48] and the references therein and Cogan [13].

The top of Fig. 6.3, taken from [34], shows the formation of biofilm in CSTR. Both cell death rates are zero: $k = k_w = 0$. Dependent and independent variables are scaled as follows: $\bar{t} = Dt$, $\bar{S} = S/a$, $\bar{u} = u/(\gamma a)$, $\bar{w} = W = w/w_{max}$. Initial data are chosen to simulate an inoculum of planktonic cells introduced into a sterile CSTR with substrate at equilibrium with inflow. They are: $\bar{S} = S^0/a$, $\bar{u} = 10^{-5}$, $W = 0$.

The time series can be decomposed into three stages. During the first, substrate remains near its initial value while an adherent population slowly accumulates. The second stage is characterized by a linear growth of the adherent population which rapidly depletes the substrate. Finally, a mature adherent population, fully occupying the wall ($w \approx 1$), casts off significant numbers of planktonic cells. Note that one scaled time-unit equals $4hr$.

The bottom of Fig. 6.3, with the same initial data, shows the effect of a (very large) planktonic cell death rate of 25% of the dilution rate ($k = .25D$,

$k_w = 0$) resulting from the introduction of antibiotic that does not penetrate the biofilm. The only noticeable effect is a smaller planktonic cell density; the antibiotic cannot inhibit biofilm formation.

6.4 One-Dimensional Thin Tube Flow Reactor

The flow model treated in this section was first formulated by Ballyk and Smith in [5] and later analyzed in more detail in [6]. It builds on earlier work in which wall growth was not treated in Ballyk, Jones, Le and Smith [3].

Consider a thin tube with inner circumference C and cross-sectional area A extending along the x-axis. See Fig. 6.4. The reactor occupies the portion of the tube from $x = 0$ to $x = L$. It is fed with growth medium at a constant rate at $x = 0$ by a laminar flow of fluid in the tube in the direction of increasing x and at velocity v (a constant). The external feed contains all nutrients in near optimal amounts except one, denoted by S, which is supplied in a constant, growth limiting concentration S^0. The flow carries medium, depleted nutrients, cells, and their byproducts out of the reactor at $x = L$. Nutrient S is assumed to diffuse with diffusivity d_0 while free microbial cells are assumed to be capable of random run and tumble motion which can be modeled by diffusion with diffusivity (sometimes called random motility coefficient) d. See the classic monograph of Berg [9] for more on bacterial motility and our earlier work in [3]. Wall attached bacteria are assumed to be immobile. We assume negligible variation of free bacteria and nutrient concentration transverse to the axial direction of the tube.

The model accounts for the density of free bacteria (bacteria suspended in the fluid) $u(x,t)$, the density of wall-attached bacteria $w(x,t)$ and the density of nutrient $S(x,t)$. The total free bacteria at time t is given by

$$A \int_0^L u(x,t)dx$$

and the total bacteria on the wall at time t is given by

$$C \int_0^L w(x,t)dx .$$

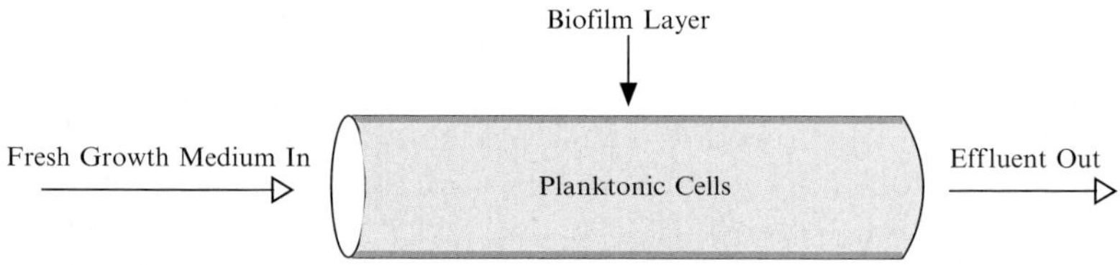

Fig. 6.4. Flow Reactor with biofilm

Let $\delta = C/A$, not to be confused with δ of the chemostat model. Then S, u, w satisfy the following system of equations.

$$\begin{aligned}
S_t &= d_0 S_{xx} - v S_x - \gamma^{-1} u f_u(S) - \gamma^{-1} \delta w f_w(S) \\
u_t &= d u_{xx} - v u_x + u(f_u(S) - k) + \delta w f_w(S)(1 - G(W)) \qquad (6.17) \\
&\quad - \alpha u(1 - W) + \delta \beta w \\
w_t &= w(f_w(S) G(W) - k_w - \beta) + \alpha \delta^{-1} u(1 - W),
\end{aligned}$$

with boundary conditions

$$\begin{aligned}
v S^0 &= -d_0 S_x(0, t) + v S(0, t), \quad S_x(L, t) = 0 \\
0 &= -d u_x(0, t) + v u(0, t), \quad u_x(L, t) = 0, \qquad (6.18)
\end{aligned}$$

and initial conditions

$$S(x, 0) = S_0(x), \quad u(x, 0) = u_0(x), \quad w(x, 0) = w_0(x), \quad 0 \le x \le L. \quad (6.19)$$

The boundary conditions (6.18), referred to as Danckwerts' boundary conditions by Aris [1], are often misunderstood. To understand their implications, it is useful to integrate the equations over the domain $[0, L]$ to obtain the mass balance

$$\begin{aligned}
\frac{d}{dt} A \int_0^L S \, dx &= A \int_0^L d_0 S_{xx} - v S_x - \gamma^{-1} [u f_u(S) + \delta w f_w(S)] dx \\
&= -A[-d_0 S_x(L, t) + v S(L, t)] + A[-d_0 S_x(0, t) + v S(0, t)] \\
&\quad - A \int_0^L \gamma^{-1} [u f_u(S) + \delta w f_w(S)] dx \\
&= -v A S(L, t) + v A S^0 - \int_0^L \gamma^{-1} [A u f_u(S) + C w f_w(S)] dx \ .
\end{aligned}$$

Thus, the change in the total amount of nutrient in the tube is due to the flow bringing fresh nutrient in at $x = 0$ at rate $v A S^0$, taking out unused nutrient at $x = L$ at rate $v A S(L, t)$ and due to the consumption of nutrient by planktonic and wall attached cells in the reactor. By contrast, a similar calculation for the planktonic bacteria u yields no counterpart to the influx of fresh nutrient into the reactor at $x = 0$ since we assume the inflow is sterile. In summary, the Danckwerts' boundary conditions say that advection alone mediates the interaction of the reactor with the external environment. See [2, 37, 50] for other uses of these boundary conditions and particularly the latter for alternative conditions.

System (6.17)–(6.19) has a trivial steady state

$$S \equiv S^0, \quad u = w \equiv 0$$

which we refer to as the 'washout steady state' since no organisms are present. The linearization of (6.17)–(6.19) about the washout steady state is given by:

$$S_t = d_0 S_{xx} - v S_x - \gamma^{-1} u f_u(S^0) - \delta\gamma^{-1} w f_w(S^0)$$
$$u_t = d u_{xx} - v u_x + u(f_u(S^0) - k) + \delta w f_w(S^0)(1 - G(0)) \qquad (6.20)$$
$$ -\alpha u + \delta\beta w$$
$$w_t = w(f_w(S^0)G(0) - k_w - \beta) + \delta^{-1}\alpha u ,$$

with the homogeneous boundary conditions:

$$0 = -d_0 S_x(0, t) + v S(0, t), \quad S_x(L, t) = 0$$
$$0 = -d u_x(0, t) + v u(0, t), \quad u_x(L, t) = 0 .$$

Introducing $(S, u, w) = \exp(\lambda t)(\bar{S}(x), \bar{u}(x), \bar{w}(x))$ into (6.20), we arrive at the eigenvalue problem relevant for the stability of the washout steady state

$$\lambda\bar{S} = d_0\bar{S}'' - v\bar{S}' - \gamma^{-1}\bar{u} f_u(S^0) - \delta\gamma^{-1}\bar{w} f_w(S^0)$$
$$\lambda\bar{u} = d\bar{u}'' - v\bar{u}' + \bar{u}(f_u(S^0) - k) + \delta\bar{w} f_w(S^0)(1 - G(0)) \qquad (6.21)$$
$$\phantom{\lambda\bar{u} =} -\alpha\bar{u} + \delta\beta\bar{w}$$
$$\lambda\bar{w} = \bar{w}(f_w(S^0)G(0) - k_w - \beta) + \delta^{-1}\alpha\bar{u} ,$$

with

$$0 = -d_0\bar{S}'(0) + v\bar{S}(0), \quad \bar{S}'(L) = 0$$
$$0 = -d\bar{u}'(0) + v\bar{u}(0), \quad \bar{u}'(L) = 0. \qquad (6.22)$$

It turns out that the eigenvalues of (6.21)–(6.22) determine the stability of the washout steady state despite the fact that the spectrum of the differential-algebraic operator appearing on the righthand side of (6.21), with the boundary conditions determining its domain, may not consist solely of eigenvalues.

The following result is proved in [6].

Theorem 2. *Let*

$$\hat{A} = \begin{pmatrix} f_u(S^0) - k - \alpha - \frac{v}{L}\lambda_{\bar{d}} & f_w(S^0)(1 - G(0)) + \beta \\ \alpha & f_w(S^0)G(0) - k_w - \beta \end{pmatrix}. \qquad (6.23)$$

where $\bar{d} = d/Lv$ and $-\lambda_{\bar{d}} < 0$ is the largest eigenvalue of the scaled eigenvalue problem

$$\lambda\phi = \bar{d}\phi'' - \phi'$$
$$0 = -\bar{d}\phi'(0) + \phi(0), \quad \phi'(1) = 0. \qquad (6.24)$$

Let $s(\hat{A})$ be the stability modulus, i.e., the largest of the distinct real eigenvalues of matrix $\hat{A}$. If $s(\hat{A}) < 0$ then all eigenvalues of (6.21) are negative and the washout steady state is asymptotically stable; the washout steady state is unstable whenever $s(\hat{A}) > 0$.

Note the similarity of the two stability determining matrices (6.5) and (6.23). The only difference is that D in (6.5) is replaced by $\frac{v}{L}\lambda_{\bar{d}}$ in (6.23). The term $\frac{v}{L}\lambda_{\bar{d}}$ should be viewed as an effective washout or removal rate from the bio-reactor. For the CSTR, $1/D$ is the mean residence time in the chemostat so there is a close correspondence between the two matrices.

It may seem striking that stability boils down to the sign of the leading eigenvalue of a 2×2 matrix exactly as in the case of the continuous culture model. We would argue that it is quite natural on biological grounds. There are two habitats for the bacteria, the wall and the bulk fluid. To survive, the organism must be able to establish itself in at least one of the habitats sufficiently well to overcome the constant leakage to the other, possibly less suitable, habitat.

It is evident that the basic reproductive number defined in (6.16) can be adopted to the tubular reactor. All formulas obtained for CSTR carry over to the tubular reactor by merely replacing the dilution rate D by its counterpart $\frac{v}{L}\lambda_{\bar{d}}$.

The following global results are special cases of Theorems 3.2 and 4.1 in [6].

Theorem 3. *If $s(\hat{A}) < 0$ and $f_w(S^0) - k_w > f_u(S^0) - k - \lambda_d$ then*

$$\int_0^1 [u(x,t) + w(x,t)]dx \to 0, \quad t \to \infty .$$

If $s(\hat{A}) > 0$ and

$$f_w(S^0)G(0) - k_w - \beta \neq 0 ,$$

then there exists a steady state solution (S, u, w) satisfying

$$0 < S(x) < S^0, \quad S'(x) < 0, \quad u(x) > 0, and$$
$$0 < w(x) < w_{max}, \quad 0 \leq x \leq L.$$

Figure 6.5 from [6] depicts the steady state solution of (6.17)–(6.19) where cell death rates are zero. Observe that wall attached cells fully occupy the front lip near $x = 0$ of the tube ($w_{max} = 1$), dropping rapidly to zero when nutrient is depleted downstream in the 150 cm tube. The shape of the suspended (planktonic) cell density profile implies that these cells are being sloughed off the wall at a rate just balancing their washout.

6.4.1 Advection Dominated Flow Reactor

If diffusion and cell motility are small compared to advection in (6.17)–(6.19) then we obtain the hyperbolic system studied by Jones and Smith in [35].

$$
\begin{aligned}
S_t + vS_x &= -\gamma^{-1}uf_u(S) - \gamma^{-1}\delta w f_w(S) \\
u_t + vu_x &= u(f_u(S) - k) + \delta w f_w(S)(1 - G(W)) \\
&\quad -\alpha u(1 - W) + \delta\beta w \\
w_t &= w(f_w(S)G(W) - k_w - \beta) + \alpha\delta^{-1}u(1 - W)
\end{aligned}
\qquad (6.25)
$$

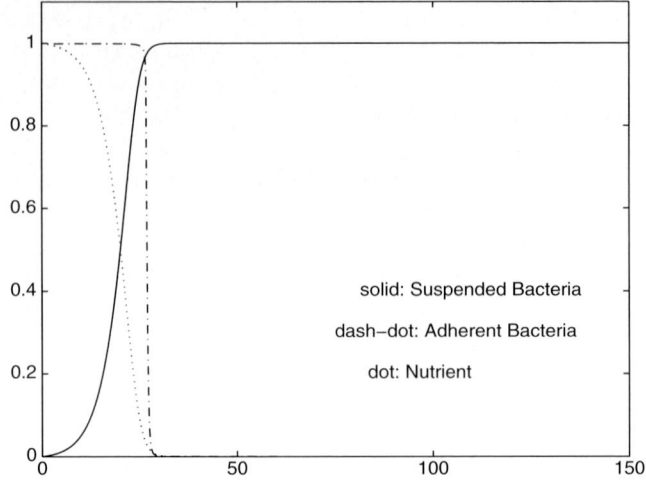

Fig. 6.5. Steady state solution of (6.17)–(6.19) with $k = k_w = 0.0$

Boundary conditions, representing the influx of nutrient in a sterile feed are given by

$$vS^0 = vS(0, t),$$
$$0 = vu(0, t), \tag{6.26}$$

and initial conditions are

$$S(x, 0) = S_0(x), \quad u(x, 0) = u_0(x), \quad w(x, 0) = w_0(x), \quad 0 \le x \le L. \tag{6.27}$$

We assume that S_0, u_0, w_0 are nonnegative, continuous and $w_0 \le w_{max}$.

It is shown in [35] that the system (6.25)–(6.27) has a unique global continuous solution provided that $S_0, u_0 \ge 0$ and $0 \le w_0 \le w_{max}$ and the compatibility conditions $S_0(0) = S^0$ and $u_0(0) = 0$ hold. A more thorough treatment of well-posedness for a general class of hyperbolic systems including (6.25)–(6.27) is carried out in [43].

As above, the focus is on the stability of the washout equilibrium solution

$$S = S^0, \quad u = w = 0.$$

Remarkably, the wall-attached population density $W(t) \equiv w(0, t)/w_{max}$ at $x = 0$ where nutrient concentration is highest can be computed from the scalar ordinary differential equation (since $u(0, t) = 0$):

$$\frac{dW}{dt} = W[f_w(S^0)G(W) - k_w - \beta]. \tag{6.28}$$

As G is strictly decreasing on $[0, 1]$, it is easily seen that

$$f_w(S^0)G(0) - k_w - \beta < 0 \implies W(t) \to 0$$

and if $W(0) > 0$ then

$$f_w(S^0)G(0) - k_w - \beta > 0 \implies W(t) \to W^*$$

where W^* is the unique solution of

$$f_w(S^0)G(W) - k_w - \beta = 0 \ .$$

This motivates the following result, proved in [35], which establishes that the sign of $f_w(S^0)G(0) - k_w - \beta$ determines the asymptotic behavior of (6.25)–(6.27).

Theorem 4. *If*

$$f_w(S^0)G(0) - k_w - \beta < 0$$

holds then every solution of (6.25)–(6.27) converges to the washout steady state, uniformly in $x \in [0, L]$.

If

$$f_w(S^0)G(0) - k_w - \beta > 0$$

then there is a unique steady state solution $(\bar{S}, \bar{u}, \bar{w})$ satisfying

$$0 < \bar{S}(x) < S^0, \ 0 < \bar{u}(x), \ 0 < \bar{w}(x) \le w_{max}, \ 0 < x \le L,$$

where $W^* = \bar{w}(0)/w_{max}$. Moreover, it is asymptotically stable in the linear approximation.

Again we see that biofilm bacteria simply must grow fast enough to out pace any losses due to cell death and sloughing in order to survive. The steady state plots in Fig. 6.6 show the strong decline in nutrient concentration as one moves downstream and the wall adherent cells occupying the front lip near $x = 0$ of the tubular reactor. Planktonic cells are sloughed off this front lip at a rate which balances washout. Parameter values for the simulation in Fig. 6.6 are as follows: $k_w = .05$, $\alpha = \beta = .2$, $\delta = 10.0$, $k = .2$, $f(S) = f_w(S) = S/(1 + S)$. Initial data are $S_0(x) = 1$, $u_0(x) = x$, $w_0(x) = 0$.

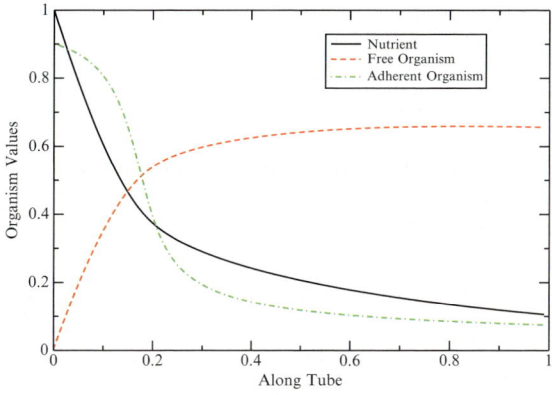

Fig. 6.6. Steady-state solution $(\bar{S}, \bar{u}, \bar{w})$ of (6.25)–(6.27)

6.4.2 Mobile Wall-Adherent Cells

Smith and Zhao [45] consider the case that wall-adherent bacteria are also mobile. The equations now become more mathematically tractable since the added diffusion term has a smoothing effect. The system is given by:

$$
\begin{aligned}
S_t &= d_0 S_{xx} - v S_x - \gamma^{-1} u f_u(S) - \gamma^{-1} \delta w f_w(S) \\
u_t &= d u_{xx} - v u_x + u(f(S) - k) + \delta w f_w(S)(1 - G(W)) \\
&\quad - \alpha u (1 - W) + \delta \beta w \\
w_t &= d_1 w_{xx} + w(f_w(S)G(W) - k_w - \beta) + \alpha \delta^{-1} u(1 - W),
\end{aligned}
\tag{6.29}
$$

where d_1 denotes the motility of wall-attached bacteria, with boundary conditions

$$
\begin{aligned}
v S^0 &= -d_0 S_x(0, t) + v S(0, t), \quad S_x(L, t) = 0 \\
0 &= -d u_x(0, t) + v u(0, t), \quad u_x(L, t) = 0 \\
0 &= w_x(0, t) = w_x(L, t)
\end{aligned}
\tag{6.30}
$$

and initial conditions

$$
S(x, 0) = S_0(x), \quad u(x, 0) = u_0(x), \quad w(x, 0) = w_0(x), \quad 0 \le x \le L.
\tag{6.31}
$$

This system is well-posed for continuous nonnegative initial data and induces a semiflow with a compact global attractor on the nonnegative cone of the space of triples of continuous functions on $[0, L]$. See Theorem 3.2 in [45]. The focus is again on the stability of the washout equilibrium

$$
S = S^0, \quad u = w = 0 .
$$

The associated eigenvalue problem is given by

$$
\begin{aligned}
\lambda \bar{S} &= d_0 \bar{S}_{xx} - v \bar{S}_x - \gamma^{-1} \bar{u} f_u(S^0) - \delta \gamma^{-1} \bar{w} f_w(S^0) \\
\lambda \bar{u} &= d \bar{u}_{xx} - v \bar{u}_x + \bar{u}(f(S^0) - k) + \delta \bar{w} f_w(S^0)(1 - G(0)) \\
&\quad - \alpha \bar{u} + \delta \beta \bar{w} \\
\lambda \bar{w} &= d_1 \bar{w}_{xx} + \bar{w}(f_w(S^0)G(0) - k_w - \beta) + \delta^{-1} \alpha \bar{u} ,
\end{aligned}
\tag{6.32}
$$

with

$$
\begin{aligned}
0 &= -d_0 \bar{S}_x(0) + v \bar{S}(0), \quad \bar{S}_x(L) = 0 \\
0 &= -d \bar{u}_x(0) + v \bar{u}(0), \quad \bar{u}_x(L) = 0 \\
0 &= \bar{w}_x(0) = \bar{w}_x(L).
\end{aligned}
\tag{6.33}
$$

The stability of washout is determined by a dominant eigenvalue but note that this eigenvalue problem decouples in the sense that the last two equations are independent of the first and therefore, we need to consider dominant eigenvalues both for this subsystem as well as for the full eigenvalue problem. The following result is proved in [45].

Lemma 1. *There exists a dominant real eigenvalue Λ of (6.32) and (6.33). That is, $\Re\lambda < \Lambda$ for all other eigenvalues λ of (6.32). The washout steady state is asymptotically stable if $\Lambda < 0$ and unstable if $\Lambda > 0$. Λ is related to the dominant eigenvalue Γ of the last two equations of (6.32) and (6.33) as follows. If $\Gamma > -\lambda_{d_0}$, which holds if $f_w(S^0)G(0) - k_w - \beta > -\lambda_{d_0}$, then*

$$f_w(S^0)G(0) - k_w - \beta < \Lambda = \Gamma < s(\tilde{A})$$

where

$$\tilde{A} = \begin{pmatrix} f_u(S^0) - k - \alpha & f_w(S^0)(1 - G(0)) + \beta \\ \alpha & f_w(S^0)G(0) - k_w - \beta \end{pmatrix} \tag{6.34}$$

and λ_{d_0} is defined in Theorem 2. If $\Gamma \leq -\lambda_{d_0}$, then $\Lambda = -\lambda_{d_0}$.

As a special case, note that

$$f_w(S^0)G(0) > k_w + \beta$$

implies $\Lambda > 0$ and hence the instability of the washout state.

As in the case when wall adherent bacteria are assumed to be immobile, one can give conditions for the global stability of the washout equilibrium (see Proposition 4.4 in [45]) but they are not sharp. We conjecture that $\Lambda < 0$ is sufficient for this.

The main result of [45] follows.

Theorem 5. *Assume $\Lambda > 0$. Then system (6.29)–(6.31) has at least one positive steady state:*

$$0 < S(x) < S^0 \,,\ 0 < u(x) \,,\ 0 < w(x) \leq w_{max} \,.$$

Furthermore, it is uniformly persistent in the sense that there exists $\eta > 0$ such that for any nonnegative continuous initial data (S_0, u_0, w_0), with at least one of $u_0(\cdot)$ and $w_0(\cdot)$ not identically zero, there exists $T_0 = T_0(S_0, u_0, w_0) > 0$ such that the solution (S, u, w) satisfies

$$S(x,t) \geq \eta \,,\quad u(x,t) \geq \eta \,,\quad w(x,t) \geq \eta \,,\quad x \in [0, L] \,,\quad t \geq T_0 \,.$$

6.5 Three-Dimensional Flow Reactor

The three dimensional model is substantially more complicated because the fluxes between adherent and planktonic compartments of the model appear as part of the (nonlinear) boundary conditions. Our treatment here follows Jones et al. [32]. Consider a cylindrical tube

$$\Omega = \{(x, y, z) \in \mathbb{R}^3 : 0 < x < L \,,\quad 0 \leq r^2 = y^2 + z^2 < R^2\}$$

under steady flow with velocity profile

$$v(r) = V_{max}[1 - (r/R)^2] \,.$$

The equations for nutrient density $S = S(x, y, z, t)$, planktonic biomass density $u = u(x, y, z, t)$ and the areal density of wall-attached cells $w = w(x, y, z, t)$ on the radial boundary $(r = R)$ for a single strain are given by:

$$S_t = d_x^S S_{xx} + d_r^S [S_{yy} + S_{zz}] - v(r)S_x - \gamma^{-1}uf_u(S)$$
$$u_t = d_x^u u_{xx} + d_r^u [u_{yy} + u_{zz}] - v(r)u_x + u(f_u(S) - k) \tag{6.35}$$

for $(x, y, z) \in \Omega$. Substrate diffusivity in the axial direction is denoted by d_x^S and in the radial direction by d_r^S. Planktonic bacteria are assumed to follow a random run and tumble motion which can be modeled by diffusion (Berg [9]). Motility coefficients d_x^u and d_r^u in the axial and radial directions are prescribed.

Wall-adherent cells are assumed to be immobile. Growth of bacteria on the wall $r = R$ is described by

$$w_t = w[f_w(S)G(W) - k_w - \beta] + \alpha u(1 - W) . \tag{6.36}$$

(The units of α are now lt^{-1})

Danckwerts' boundary conditions describe the interface conditions between up-stream and down-stream flow and the reactor. They are as follows:

at $x = 0$:
$$v(r)S^0 = -d_x^S S_x + v(r)S$$
$$0 = -d_x^u u_x + v(r)u , \tag{6.37}$$

at $x = L$:

$$S_x = u_x = 0 . \tag{6.38}$$

These conditions reflect the assumption that upstream flow brings sterile nutrient at concentration S^0 into the reactor at $x = 0$ and flushes out planktonic cells and unused nutrient at $x = L$. The radial boundary conditions at $r = R$ reflect important biological considerations:

$$-d_r^S S_r = \gamma^{-1}wf_w(S)$$
$$-d_r^u u_r = \alpha u(1 - W) - w[f_w(S)(1 - G(W)) + \beta] . \tag{6.39}$$

They describe the fluxes of nutrient and biomass between the fluid and wall environment. The first describes the flux of nutrient from the fluid to the wall environment due to consumption by wall-attached bacteria. The first term in the second equation represents the flux of biomass from the fluid to the wall due to passive attraction of planktonic cells to the wall; the second term represents flux in the opposite direction caused by a fraction of the progeny of wall-attached cells being forced into the fluid.

In addition, S, u, w satisfy (non-negative) initial conditions at $t = 0$:

$$S(x, y, z, 0) = S_0(x, y, z)$$
$$u(x, y, z, 0) = u_0(x, y, z) \tag{6.40}$$
$$w(x, y, z, 0) = w_0(x, y, z),$$

where S_0, u_0, w_0 are continuous. Existence of a unique weak solution of the system (6.35)–(6.40) and its Hölder continuity is established in [32].

It will also be of interest to allow the initial "charging" of the reactor with microbes to take place via the boundary condition at $x = 0$ by replacing zero on the left side of (6.37) by $v(r)u^0(t)$, where $u^0(t) \equiv u^0$, a constant, on $0 \leq t \leq t_0$ and $u^0(t) = 0, t \geq t_0$.

In order to clarify that the system (6.35)–(6.40) captures the intended mass transfer, we integrate the S equation over Ω and use cylindrical coordinates (r, θ, x), to obtain

$$
\frac{\partial}{\partial t} \int \int \int_\Omega S dV
$$

$$
= \int \int \int_\Omega \left([d_x^S S_{xx} - v(r)S_x] + d_r^S [\frac{1}{r}(rS_r)_r + \frac{1}{r^2} S_{\theta\theta}] \right) r dr dx d\theta
$$

$$
- \int \int \int_\Omega \gamma^{-1} u f_u(S) dV
$$

$$
= \int_0^{2\pi} \int_0^R (d_x^S S_x - v(r)S)|_{x=L} r dr d\theta - \int_0^{2\pi} \int_0^R (d_x^S S_x - v(r)S)|_{x=0} r dr d\theta
$$

$$
+ R d_r^S \int_0^{2\pi} \int_0^L S_r|_{r=R} dx d\theta - \int \int \int_\Omega \gamma^{-1} u f_u(S) dV
$$

$$
= - \int_0^{2\pi} \int_0^R v(r)S|_{x=L} dr d\theta + \int_0^{2\pi} \int_0^R v(r)S^0 dr
$$

$$
- R \int_0^{2\pi} \int_0^L \gamma^{-1} w f_w(S)|_{r=R} dx d\theta - \int \int \int_\Omega \gamma^{-1} u f_u(S) dV
$$

The rate of change of substrate in Ω is the flux of substrate into Ω at $x = 0$ minus the flux of substrate out of Ω at $x = L$ minus substrate consumed by wall-adherent organisms on the inside wall of the cylinder $r = R$ and minus substrate consumed by planktonic organisms in Ω.

Similarly, integrating the u equation leads to

$$
\frac{\partial}{\partial t} \int \int \int_\Omega u dV
$$

$$
= \int \int \int_\Omega \left([d_x^u u_{xx} - v(r)u_x] + d_r^u [\frac{1}{r}(ru_r)_r + \frac{1}{r^2} u_{\theta\theta}] \right) r dr dx d\theta
$$

$$
+ \int \int \int_\Omega u[f_u(S) - k] dV
$$

$$
= \int_0^{2\pi} \int_0^R (d_x^u u_x - v(r)u)|_{x=L} r dr d\theta - \int_0^{2\pi} \int_0^R (d_x^u u_x - v(r)u)|_{x=0} r dr d\theta
$$

$$
+ R d_r^u \int_0^{2\pi} \int_0^L u_r|_{r=R} dx d\theta + \int \int \int_\Omega u[f_u(S) - k] dV
$$

$$= -\int_0^{2\pi}\int_0^R v(r)u|_{x=L}rdrd\theta$$

$$+ R\int_0^{2\pi}\int_0^L (w[f_w(S)(1-G(W))+\beta]-\alpha u(1-W))|_{r=R}dxd\theta$$

$$+ \int\int\int_\Omega u[f_u(S)-k]dV$$

Planktonic biomass increases due to net growth, due to the sloughing of wall-adherent cells into the fluid, and due to the failure of a fraction of the daughter cells of wall attached bacteria to find wall sites and it decreases due to attachment of planktonic cells to the lateral surface and to washing out at $x = L$.

For brevity, we let L^S and L^u denote the differential operators for the S and u equations so they become:

$$S_t = L^S S - \gamma^{-1}uf_u(S)$$
$$u_t = L^u u + u(f_u(S) - k) .$$

Our aim now is to perform a linear stability analysis of the washout equilibrium:

$$S = S^0 , \ u = w = 0 \tag{6.41}$$

and to show the existence of a nontrivial steady state in which $u, w > 0$ under suitable conditions. The linear variational equation about the washout equilibrium is

$$S_t = L^S S - \gamma^{-1}uf_u(S^0)$$
$$u_t = L^u u + u[f_u(S^0) - k]$$
$$w_t = w[f_w(S^0)G(0) - k_w - \beta] + \alpha u$$

together with homogeneous boundary conditions $x = 0$ (formally, set $S^0 = 0$ in (6.37)) and $x = L$ and radial boundary conditions on $r = R$:

$$0 = d_r^S S_r + \gamma^{-1}wf_w(S^0) \tag{6.42}$$
$$0 = d_r^u u_r + \alpha u - w[f_w(S^0)(1 - G(0)) + \beta] .$$

The associated eigenvalue problem, obtained by seeking solutions $\hat{S} = e^{\lambda t}S(x, y, z)$ (and similarly for other variables), is

$$\lambda S = L^S S - \gamma^{-1}uf_u(S^0)$$
$$\lambda u = L^u u + u[f_u(S^0) - k] \tag{6.43}$$
$$\lambda w = w[f_w(S^0)G(0) - k_w - \beta] + \alpha u$$

together with the above boundary conditions.

We have the following from [32].

Proposition 1. *The eigenvalue λ^* of (6.43) with the largest real part is real, simple and satisfies $\lambda^* > f_w(S^0)G(0) - k_w - \beta$. It belongs to the interval with endpoints $f_w(S^0) - k_w$ and $f_u(S^0) - k - \frac{L}{V_{max}}\lambda$, where $-\lambda < 0$ is the principal eigenvalue of the (scaled $\bar{x} = x/L$, $\bar{r} = r/R$) eigenvalue problem:*

$$\lambda u = \theta_x u_{\bar{x}\bar{x}} - (1 - \bar{r}^2)u_{\bar{x}} + \theta_r \bar{r}^{-1}(\bar{r}u_{\bar{r}})_{\bar{r}} \, ,$$

$$0 = -\theta_x u_{\bar{x}} + (1 - \bar{r}^2)u \, , \quad \bar{x} = 0$$

$$0 = u_{\bar{x}} \, , \quad \bar{x} = 1$$

$$u_{\bar{r}} = 0 \, , \quad \bar{r} = 1 \, ,$$

where $\theta_x = (d_x^u/L^2)(L/V_{max})$, $\theta_r = (d_r^u/R^2)(L/V_{max})$. The washout state is stable in the linear approximation if $\lambda^* < 0$ and unstable if $\lambda^* > 0$.

As a result of the inequalities above, we see that $\lambda^* < 0$ and the washout state is stable if

$$f_w(S^0) - k_w < 0 \text{ and } f_u(S^0) - k - \frac{L}{V_{max}}\lambda < 0 \, ; \tag{6.44}$$

$\lambda^* > 0$ and the washout state is unstable if either

$$f_w(S^0)G(0) - k_w - \beta \geq 0 \tag{6.45}$$

or if

$$f_u(S^0) - k - \frac{L}{V_{max}}\lambda > 0 \text{ and } f_w(S^0) - k_w > 0 \, .$$

The reader will observe the similarity in the CSTR and PFR results. Obviously, $\frac{L}{V_{max}}\lambda$ is an effective dilution rate for PFR.

We conjecture that the washout state is globally stable if $\lambda^* < 0$ as in the case of the CSTR. It can be shown (Jones et al. [32]) that if (6.44) holds then the washout state is globally attracting in the sense that

$$\lim_{t\to\infty} \left(\int\int\int_\Omega u \, dV + \int_0^{2\pi}\int_0^L wR \, d\theta dx \right) = 0 \, .$$

The steady state equations are

$$0 = L^S S - \gamma^{-1} u f_u(S)$$
$$0 = L^u u + u[f_u(S) - k] \, , \quad \text{in } \Omega \tag{6.46}$$
$$0 = w[f_w(S)G(W) - k_w - \beta] + \alpha u(1 - W) \, , \quad \text{on } r = R \, ,$$

with boundary conditions (6.37)–(6.39). We summarize our main result for PFR below. It follows from Proposition 3.4 and Theorem 3.5 in [32] and Proposition 4.2(b) in [31].

Theorem 6. *If $\lambda^* < 0$, then the washout state is stable in the linear approximation; it is globally attracting if (6.44) holds. If $\lambda^* > 0$ and the non-degeneracy condition $b = f_w(S^0)G(0) - k_w - \beta \neq 0$ holds then there exists a radially symmetric steady state solution (S, u, w) of (6.46) satisfying (in cylindrical coordinates)*

$$0 < S(x, r) \leq S^0, \quad u(x, r) > 0, \quad \text{and } 0 < w(x) \leq w_{max}.$$

A persistence result similar to that described above for CSTR is also proved in [32] if $\lambda^* > 0$ and certain additional assumptions hold.

The initial conditions for wall-attached bacteria and planktonic bacteria are $S^0 = S_0$, $w_0 = 0$, and $u_0 = 0$, respectively. The reactor is charged with bacteria by replacing the original boundary condition on the reactor entrance with $v(r)u^0(t) = -d_x^u u_x + v(r)u$, where $u^0(t) = 10^{-9}U_5(t)$ and $U_5(t)$ is the step function of unit height turning to zero at $T = 5$. This mimics the introduction of an inoculum of planktonic cells into a bacteria-free PFR. Note the physical time scale is $L/V_{max} = 40\,\mathrm{h}$. In order to provide a steady-state profile, the equations were integrated to time $T = 1,500\,\mathrm{h}$ at which point no further change could be detected, as can be seen in Fig. 6.5 (left).

The PFR time series displayed in Fig. 6.7 (top) is roughly similar to that in Fig. 6.3 for CSTR but note the log–log scale used to accommodate the range of

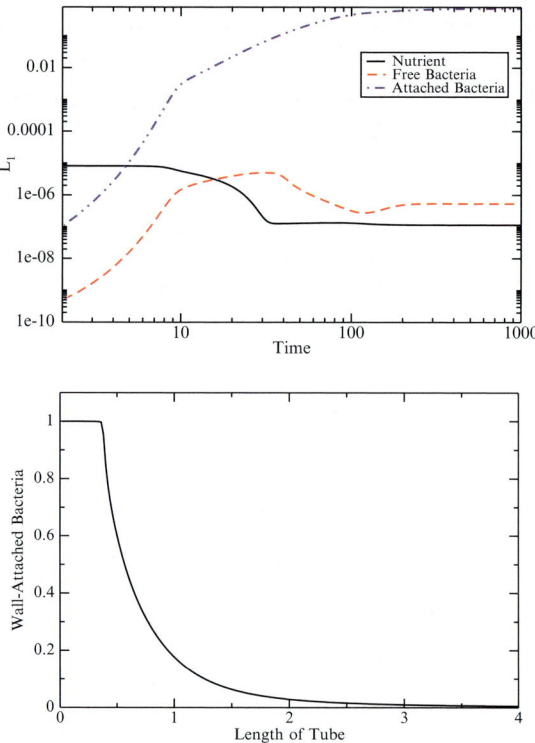

Fig. 6.7. *Top*: Time series plot of the total nutrient, free bacteria, and attached bacteria for (6.35)–(6.40). *Bottom*: Graph of steady-state, wall-attached bacteria

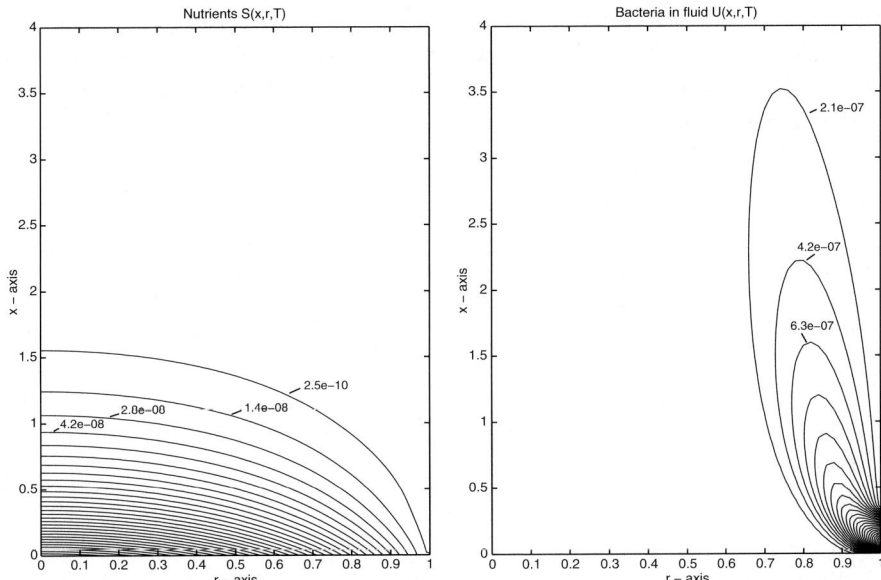

Fig. 6.8. *Left*: Contour plots of nutrient density at steady-state, ranging from 2.5×10^{-10} g ml^{-1} at *top* to 4.13×10^{-7} g ml^{-1} at *bottom*. *Right*: Contour plots of planktonic biomass density at steady state, ranging from near zero to 5.2×10^{-6} g ml^{-1}

the data. An initial period of unchanged total substrate where wall adherent and planktonic cells accumulate is followed by a rapid decline in substrate and a slowing of the rate of increase of adherent and planktonic cells. As the substrate levels off, the wall adherent population does too while the planktonic population level declines somewhat as steady-state is achieved (Fig. 6.8).

As noted in Sect. 2.2, wall adherent bacterial cells may have flagella and consequently have limited motility on the cylindrical surface $r = R$. Furthermore, including random cell motility on this surface should facilitate the realization of the biofilm system as a well-behaved semiflow in a suitable function space possessing a compact global attractor just as it did in Sect. 2.2. For that reason we are lead to introduce (6.35)–(6.40) where (6.36) is replaced by:

$$w_t = d^w[R^{-2}w_{\theta\theta} + w_{xx}] + w[f_w(S)G(W) - k_w - \beta] + \alpha u(1 - W)$$
$$w_x(x, \theta) = 0, \ x = 0, L \tag{6.47}$$
$$w(x, \theta) = w(x, \theta + 2\pi)$$

In future work we will extend the analysis in Sect. 2.2 to this system.

6.6 Mixed Culture

A mixed culture model treating competition for substrate and wall colonization sites in an advection-dominated flow reactor was considered by Jones and Smith in [35]. Here, we follow [31] where both advection and diffusion are included in a three dimensional flow reactor. A flow reactor amounts to a section of the cylindrical tube Ω in which a steady flow of fluid in the direction of increasing x is imposed. The flow carries fresh nutrient at concentration S^0 into the reactor across the $x = 0$ interface and carries unused nutrient and bacteria out of the reactor across the $x = L$ interface. The equations describing nutrient density $S = S(x, y, z, t)$ and biomass density $u^i = u^i(x, y, z, t)$ of bacteria of strain i in the fluid (i.e., in Ω) are given by:

$$S_t = d_x^S S_{xx} + d_r^S [S_{yy} + S_{zz}] - v(r)S_x - \sum_i \gamma_i^{-1} u^i f_{ui}(S) \qquad (6.48)$$

$$u_t^i = d_x^i u_{xx}^i + d_r^i [u_{yy}^i + u_{zz}^i] - v(r)u_x^i + u^i(f_{ui}(S) - k_i) \qquad (6.49)$$

for $(x, y, z) \in \Omega$. The areal density of wall-attached cells of strain i on the radial boundary $r = R$ of Ω, denoted by $w^i = w^i(x, y, z, t)$, satisfy the equations by

$$w_t^i = w^i[f_{wi}(S)G_i(W) - k_{wi} - \beta_i] + \alpha_i u^i(1 - W) . \qquad (6.50)$$

The specific growth rate of strain i in the fluid $f_{ui}(S)$ and on the wall $f_{wi}(S)$ are further described below. Constants k_i and k_{wi} represent cell death rates in fluid and wall environments, respectively, β_i is the rate of sloughing of wall-attached bacteria into the fluid, and α_i is the rate coefficient of adhesion to the wall surface for strain i. Axial and radial nutrient diffusion coefficients are given by d_x^S, d_r^S, respectively; axial and radial bacterial motility coefficients for the ith strain are d_x^i, d_r^i, respectively. A model assumption is that there is a maximum attainable areal density of wall-attached bacteria w_{max} and

$$W = \sum_i \frac{w^i}{w_{max}}$$

represents the occupation fraction. Danckwerts' boundary conditions describe the interface conditions between the up-stream and down-stream flow and the reactor. They are as follows:
at $x = 0$:

$$v(r)S^0 = -d_x^S S_x + v(r)S \qquad (6.51)$$
$$0 = -d_x^i u_x^i + v(r)u^i ,$$

at $x = L$:

$$S_x = u_x^i = 0 . \qquad (6.52)$$

The radial boundary conditions at $r = R$ are:

$$0 = d_r^S S_r + \sum_i \gamma_i^{-1} w^i f_{wi}(S) \qquad (6.53)$$

$$0 = d_r^i u_r^i + \alpha_i u^i(1 - W) - w^i[f_{wi}(S)(1 - G_i(W)) + \beta_i] .$$

In addition, S, u, w satisfy (non-negative) initial conditions at $t = 0$:

$$S(x, y, z, 0) = S_0(x, y, z)$$
$$u^i(x, y, z, 0) = u_0^i(x, y, z) \tag{6.54}$$
$$w^i(x, y, z, 0) = w_0^i(x, y, z) .$$

We assume that S_0, u_0^i, w_0^i are continuous.

It will also be of interest to allow the "charging" of the reactor with microbes to take place via the boundary condition at $x = 0$ by replacing zero on the left side of (6.51) by $v(r)u^{0i}(t)$, where $u^{0i}(t) \equiv u^{0i}$, a constant, on $0 \le t \le t_0$ and $u^{0i}(t) = 0, t \ge t_0$.

Keeping in mind that $0 \le w^i \le w_\infty$, the initial data and solutions S, u^i, w^i must satisfying:

$$S, u^i, w^i \ge 0 , \quad \sum_i w^i \le w_\infty . \tag{6.55}$$

Hereafter, we refer to these restrictions as the range conditions.

We stress here that the model equations describe competition among the n bacterial strains for limited wall colonization sites as well as for limited substrate. A strain may be a good competitor by being able to grow at low substrate concentrations and/or by a relatively strong ability for surface attachment.

For brevity, we let L^S and L^i denote the differential operators for the S and u^i equations so they become:

$$S_t = L^S S - \sum_i \gamma_i^{-1} u^i f_{ui}(S)$$
$$u_t^i = L^i u^i + u^i (f_{ui}(S) - k_i)$$

6.6.1 Eigenvalue Problems

A family of non-standard eigenvalue problems plays a central role in our analysis so we introduce them here. The adjoint operator to L^S (L^i) with homogeneous boundary conditions (6.51) with $S^0 = 0$, (6.52), and radial boundary condition $S_r = 0$ is denoted by L_S (L_i). L_S is given by:

$$L_S \phi = d_x^S \phi_{xx} + d_r^S [\phi_{yy} + \phi_{zz}] + v(r)\phi_x$$

with homogeneous boundary conditions

$$0 = d_x^S \phi_x + v(r)\phi , \quad x = L$$
$$0 = \phi_x , \quad x = 0$$
$$0 = \phi_r , \quad r = R$$

and similarly for L_i. It's as if the direction of flow through the reactor changed from left to right to right to left. Denote by $-\lambda^S$ the principal eigenvalue of the eigenvalue problem:

$$L_S \phi = \lambda \phi ,$$

together with the above boundary conditions. Then $-\lambda^S < 0$ and the corresponding eigenfunction ϕ satisfies $\phi > 0$ on $\overline{\Omega}$ and can be normalized by assuming that it attains a maximum of unity (see Appendix 6 in [32]). Let $-\lambda^i < 0$ be the principal eigenvalue of L_i subject to analogous homogeneous boundary conditions. It is well-known that $-\lambda^i < 0$ is also the principal eigenvalue of L^i corresponding to boundary conditions (6.51),(6.52), and $u_r = 0$ on $r = R$.

Another important pair of eigenvalue problems is the following.

$$\lambda u = L^i u + au , \quad \Omega$$
$$\lambda w = bw + \alpha u , \quad r = R$$
$$0 = d_r u_r + \alpha u - cw , \quad r = R \tag{6.56}$$
$$0 = -d_x u_x + v(r)u , \quad x = 0$$
$$0 = u_x , \quad x = L$$

The corresponding adjoint problem is given by:

$$\lambda u = L_i u + au , \quad \Omega$$
$$\lambda w = bw + cu , \quad r = R$$
$$0 = d_r u_r + \alpha u - \alpha w , \quad r = R \tag{6.57}$$
$$0 = d_x u_x + v(r)u , \quad x = L$$
$$0 = u_x , \quad x = 0$$

In order to see in what sense (6.57) is adjoint to (6.56) we make the following observation.

Proposition 2. *Let $u \in C^2(\Omega) \cap C^1(\overline{\Omega})$ satisfy the Danckwerts' boundary conditions at $x = 0, L$, $\hat{u} \in C^2(\Omega) \cap C^1(\overline{\Omega})$ satisfy the adjoint Danckwerts' boundary conditions at $x = 0, L$, u, w satisfy the inhomogeneous radial boundary condition*

$$h = d_r u_r + \alpha u - cw , \quad r = R$$

and $\hat{u}, \hat{w}$ satisfy the homogeneous adjoint radial boundary condition in (6.57). Then we have

$$\int_{\Omega} (L^i u)\hat{u}\,dV + \int_{r=R} (bw + \alpha u)\hat{w}\,dA$$
$$= \int_{\Omega} (L_i \hat{u})u\,dV + \int_{r=R} h\hat{u} + w(b\hat{w} + c\hat{u})\,dA \tag{6.58}$$

If $h \equiv 0$, then we obtain the adjoint relation of (6.56) and (6.57).

The proof, given in [31], boils down to the use of Green's identities.

One of our main tools is the following result, Theorem 3.3 in [32]. It can be generalized to non-constant coefficients; see [31].

Proposition 3. Principal Eigenvalue. *Let* $\alpha, c > 0$. *Then there exists a real simple eigenvalue* $\lambda^* > b$ *of* (6.56) *satisfying:*

$$b + c < \lambda^* \leq a - \lambda_i \ , \ if \ b + c < a - \lambda_i$$
$$b + c = \lambda^* \ , \qquad if \ b + c = a - \lambda_i$$
$$a - \lambda_i < \lambda^* < b + c \ , \ if \ b + c > a - \lambda_i$$

Corresponding to eigenvalue λ^* *is an eigenvector* $(\bar{u}, \bar{w})$ *satisfying* $\bar{u} > 0$ *in* $\overline{\Omega}$ *and* $\bar{w} > 0$ *in* $r = R$. *If* λ *is any other eigenvalue of* (6.56) *corresponding to an eigenvector* $(u, w) \geq 0$, *then* $\lambda = \lambda^*$ *and* $(u, w) = c(\bar{u}, \bar{w})$ *for some* $c > 0$. $\bar{u}, \bar{w}$ *are axially symmetric, i.e., in cylindrical coordinates* (r, θ, x), $\bar{u} = \bar{u}(r, x), \bar{w} = \bar{w}(x)$.

λ^* *is also an eigenvalue of* (6.57) *corresponding to an eigenvector* $(u, w) = (\psi, \chi)$. *Moreover,* (ψ, χ) *has the same uniqueness up to scalar multiple, positivity and symmetry properties as does* $(\bar{u}, \bar{w})$.

6.6.2 Estimates and Simulations

We begin by establishing that bacterial growth is limited by the supplied substrate. Let (ψ^i, χ^i) be the principal eigenvector corresponding to the eigenvalue $\overline{\lambda_i}$ of (6.57) in the case that $a = 0, b = -\beta_i, \alpha = \alpha_i, c = \beta_i, d_r = d_r^i, d_x = d_x^i$. Normalize (ψ^i, χ^i) by requiring $\psi^i, \chi^i \leq \phi \leq 1$ with equality holding at some point for each inequality. By Proposition 3 and the fact that $b + c = 0$, we have $\overline{\lambda_i} < 0$ and, by the second of equations (6.57), $\psi^i < \chi^i$ on $r = R$.

Our first result represents a significant improvement over Theorem 3.1 in [32] in the case of a single species since we obtain useful bounds even when $k_{wi} = 0$.

Theorem 7. *The following estimates hold for solutions of* (6.48)–(6.54):

$$\limsup_{t \to \infty} S(t, x, y, z) \leq S^0 \ ,$$

uniformly in $(x, y, z) \in \Omega$ *and*

$$\limsup_{t \to \infty} \left(\int_{\Omega} S\phi dV + \sum_i \gamma_i^{-1} \left[\int_{\Omega} u^i \psi^i dV + \int_{r=R} w^i \chi^i dA \right] \right)$$
$$\leq \frac{2\pi S^0 \int_0^R rv(r) dr}{M} \tag{6.59}$$

where

$$M = \min_{1 \leq j \leq n} \{ \lambda^S, -\overline{\lambda_j} + k_j, -\overline{\lambda_j} + k_{wj} \} \ .$$

Proof. It is easy to establish that $S \leq \tilde{S}$, where $\tilde{S}$ satisfies $S_t = L^S S$ with homogeneous radial boundary condition $S_r = 0$ and (6.51),(6.52), by a simple comparison argument. Furthermore, noting that $S = S^0$ is a steady state of this comparison equation, the linearization of which having a dominant negative eigenvalue, we conclude that $\tilde{S} \to S^0$ as $t \to \infty$ uniformly in $(x, y, z) \in \Omega$.

Now, corresponding to the normalized eigenfunctions ϕ, ψ^i, χ^i, define

$$X = \int_\Omega \phi S dV \ , \ Y^i = \int_\Omega \psi^i u^i dV \ , \ Z = \int_{r=R} \chi^i w^i dA \ .$$

We note the following, which follow by integration by parts and Green's third identity applied to the two dimensional Laplacian in y, z and using the boundary conditions satisfied by S:

$$\int_\Omega \phi L^S S dV = \int_\Omega S L_S \phi dV - \sum_i \gamma_i^{-1} \int_{r=R} \phi w^i f_{wi}(S) dA$$

$$+ S^0 \int_{r \le R} v(r) \phi(0, y, z) dy dz$$

$$= -\lambda^S X - \sum_i \gamma_i^{-1} \int_{r=R} \phi w^i f_{wi}(S) dA$$

$$+ S^0 \int_{r \le R} v(r) \phi(0, y, z) dy dz$$

Differention of X, Y^i, Z^i and using these relations leads to the following:

$$X_t = -\lambda^S X - \sum_i \gamma_i^{-1} \int_\Omega u^i \phi f_{ui}(S) dV - \sum_i \gamma_i^{-1} \int_{r=R} \phi w^i f_{wi}(S) dA$$

$$+ S^0 \int_{r \le R} v(r) \phi(0, y, z) dy dz$$

$$Y_t^i = \int_\Omega \psi^i L^i u^i dV + \int_\Omega u^i \psi^i [f_{ui}(S) - k_i] dV \qquad (6.60)$$

$$Z_t^i = \int_{r=R} \chi^i [-\beta_i w^i + \alpha_i u^i] dA + \int_{r=R} \chi^i w^i [f_{wi}(S) G_i(W) - k_{wi}] \quad (6.61)$$

$$- \chi^i \alpha_i u^i W dA$$

Now, using the adjoint relation (6.58), the eigenvalue problem satisfied by (ψ^i, χ^i), and $\psi^i < \chi^i$ on $r = R$, we find that

$$(Y^i + Z^i)_t = \int_\Omega \psi^i L^i u^i dV + \int_{r=R} \chi^i [-\beta_i w^i + \alpha_i u^i] dA$$

$$+ \int_\Omega (f_{ui}(S) - k_i) \psi^i u^i dV + \int_{r=R} \chi^i w^i [f_{wi}(S) G_i(W) - k_{wi}]$$

$$- \alpha_i \chi^i u^i W dA$$

$$= \int_\Omega (L_i \psi^i) u^i dV + \int_{r=R} \psi^i [\alpha_i u^i W + w^i f_{wi}(S)(1 - G_i(W))]$$

$$+ w^i [-\beta_i \chi^i + \beta_i \psi^i] dA + \int_\Omega f_{ui}(S) \psi^i u^i dV$$

$$+ \int_{r=R} \chi^i w^i f_{wi}(S) G_i(W) dA - k_i Y^i - k_{wi} Z^i$$

$$- \int_{r=R} \alpha_i \chi^i u^i W dA$$

$$= \overline{\lambda_i}(Y^i + Z^i) - k_i Y^i - k_{wi} Z^i + \int_\Omega f_{ui}(S) \psi^i u^i dV$$

$$+ \int_{r=R} w^i f_{wi}(S)[G_i(W)\chi^i + (1 - G_i(W))\psi^i] dA$$

$$+ \int_{r=R} \alpha_i u^i W(\psi^i - \chi^i) dA$$

$$\leq \overline{\lambda_i}(Y^i + Z^i) - k_i Y^i - k_{wi} Z^i + \int_\Omega f_{ui}(S) \psi^i u^i dV \qquad (6.62)$$

$$+ \int_{r=R} w^i f_{wi}(S)\chi^i dA$$

Let $Q = X + \sum_i \gamma_i^{-1}(Y^i + Z^i)$. Using $\chi^i, \psi^i \leq \phi \leq 1$, we find that

$$Q_t \leq -\lambda^S X + \sum_i \gamma_i^{-1}(\overline{\lambda_i} - k_i)Y^i + \sum_i \gamma_i^{-1}(\overline{\lambda_i} - k_{wi})Z^i$$

$$+ \sum_i \int_\Omega \gamma_i^{-1} u^i f_{ui}(S)[\psi^i - \phi] dV + \sum_i \int_{r=R} \gamma_i^{-1} w^i f_{wi}(S)[\chi^i - \phi] dA$$

$$+ S^0 \int_{r \leq R} v(r)\phi(0, y, z) dydz$$

$$\leq -\lambda^S X + \sum_i \gamma_i^{-1}(\overline{\lambda_i} - k_i)Y^i + \sum_i \gamma_i^{-1}(\overline{\lambda_i} - k_{wi})Z^i$$

$$+ S^0 \int_{r \leq R} v(r)\phi(0, y, z) dydz$$

$$\leq -\min_j\{\lambda^S, -\overline{\lambda_j} + k_j, -\overline{\lambda_j} + k_{wj}\}Q + 2\pi S^0 \int_0^R rv(r) dr.$$

Therefore,

$$\limsup_{t \to \infty} Q(t) \leq \frac{2\pi S^0 \int_0^R rv(r) dr}{\min_j\{\lambda^S, -\overline{\lambda_j} + k_j, -\overline{\lambda_j} + k_{wj}\}} \qquad (6.63)$$

This completes our proof. $\square$

As the numerator of the fraction on the right side of (6.59) is the net flux of nutrient into the reactor across $x = 0$, (6.59) says precisely that the output of organisms is limited by the input of substrate. Since $\psi^i > 0$ is continuous on $\overline{\Omega}$, (6.59) implies the existence of an a priori asymptotic estimate for $\int_\Omega u^i dV$ for each i.

The n-strain model is obviously less mathematically tractable than the pure culture case $n = 1$ and we can say very little about its asymptotic behavior via mathematical analysis. Aside from the washout state

$$S = S^0, \; u_i = 0, \; w_i = 0$$

there are the mono-culture equilibria E_i:

$$S = S_i > 0, \; u_i > 0, \; w_i > 0, \; u_j = 0, \; w_j = 0, \; j \neq i$$

which exist when a principal eigenvalue $\lambda_i^* > 0$ and a non-degeneracy condition are satisfied (see Theorem 6). The main interest focuses on coexistence equilibria in which one or more strains coexist. In [31] it was shown that the linearized stability of the mono-culture equilibrium E_i to invasion by strain j depends on the sign of a principal eigenvalue. However, because we have yet to show that system (6.48)–(6.54) defines a well-behaved semiflow, we have been unable to exploit the stability results so as to obtain persistence of multiple strains. We expect that the inclusion of wall-attached cell motility as in Sect. 4.1 may yield such a system.

Parameters for the simulations have been chosen following Freter et al. [24], modified for the different units used here (biomass density as opposed to cell density used by Freter) following Ballyk et al. [4]. The initial conditions for wall-attached bacteria and planktonic bacteria are $S^0 = S_0$, $w_0^i = 0$, and $u_0^i = 10^{-6}$ g ml^{-1}, respectively. In order to provide a steady-state profile, the equations were integrated to time $T = 35,000$ h at which point no further change could be detected.

The other parameters are as follows. We set $d_x^S = d_r^S = 0.2$ cm^2 h^{-1}, $d_x^{u_i} = d_r^{u_i} = 0.002$ cm^2 h^{-1} for $i = 1, 2, 3$. The concentration of the substrate feed, $S_0 = 2.09 \times 10^{-6}$ g ml^{-1}. The fluid in the center of the tube is $V_{max} = 5$ cm h^{-1}. The velocity is higher than suggested by the biology. However, coexistence does not seem possible at lower velocities (holding all other parameters fixed except for uptake functions, m_i, a_i). The larger velocity apparently provides a more uniform nutrient field near the tube wall allowing the microbes to persist. We use Monod uptake functions, $f_{w_i}(S) = f_{u_i}(S) = m_i S/(a_i + S)$ with $m_1 = 1.66$ h^{-1}, $a_1 = 9 \times 10^{-7}$ g ml^{-1}, $m_2 = .277$ h^{-1}, $a_2 = 1 \times 10^{-8}$ g ml^{-1}, and $m_3 = .45$ h^{-1}, $a_3 = 1.05 \times 10^{-7}$ g ml^{-1}.

Finally, the rate of adhesion is $\alpha_i = 500$ cm h^{-1}; the maximum areal biomass density of adherent bacteria is $w_\infty = 2.78 \times 10^{-6}$ g cm^{-2}; the yield constant $\gamma = 0.75$. The planktonic cell and adherent cell death rates are $k_i = k_{w_i} = .01$ h^{-1}.

Our simulations show that three populations can coexist in the flow reactor; these three populations compete for two limited resources, namely substrate and wall-attachment space (a refuge from washout). The three organisms differ only in their uptake functions (m_i, a_i); death rates, wall-affinities, sloughing rates, and yield constants are identical for the three populations. Figures 6.9 (top) and 6.11 (top) show transient oscillations in the

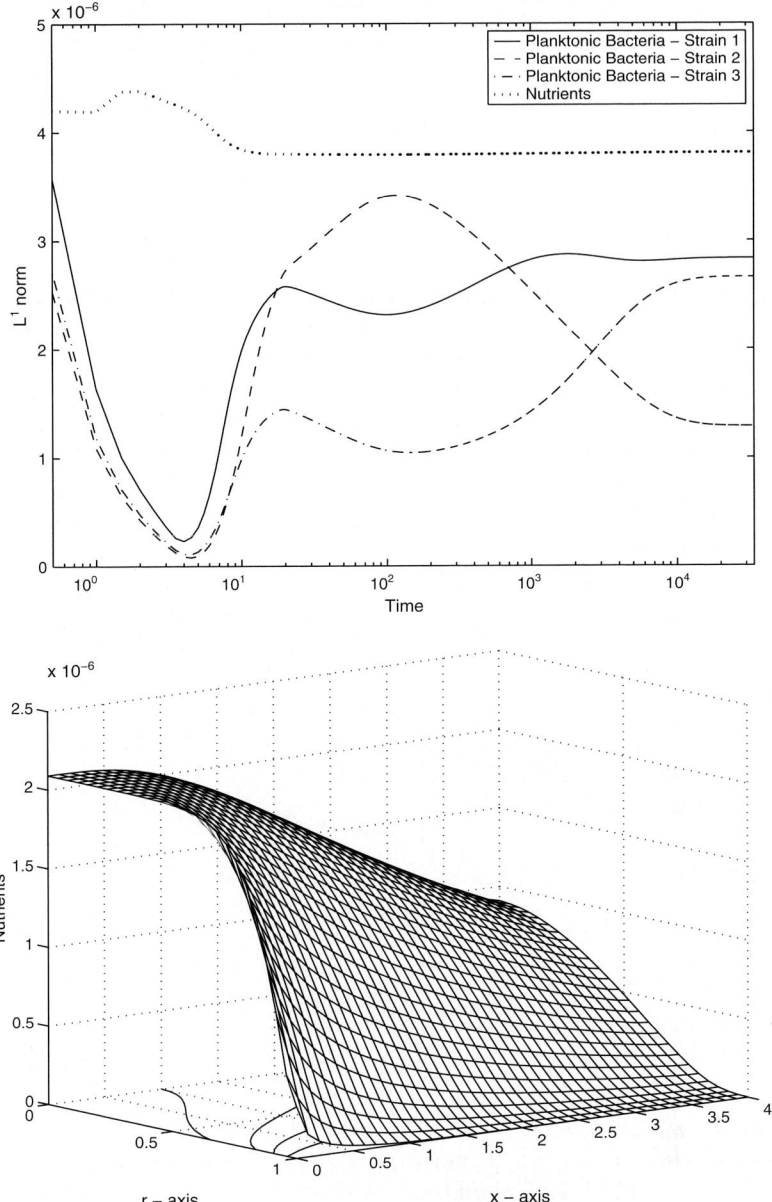

Fig. 6.9. L^1 norm versus time of the nutrients and free bacteria (*top*) and a surface plot of the nutrient density S (*bottom*)

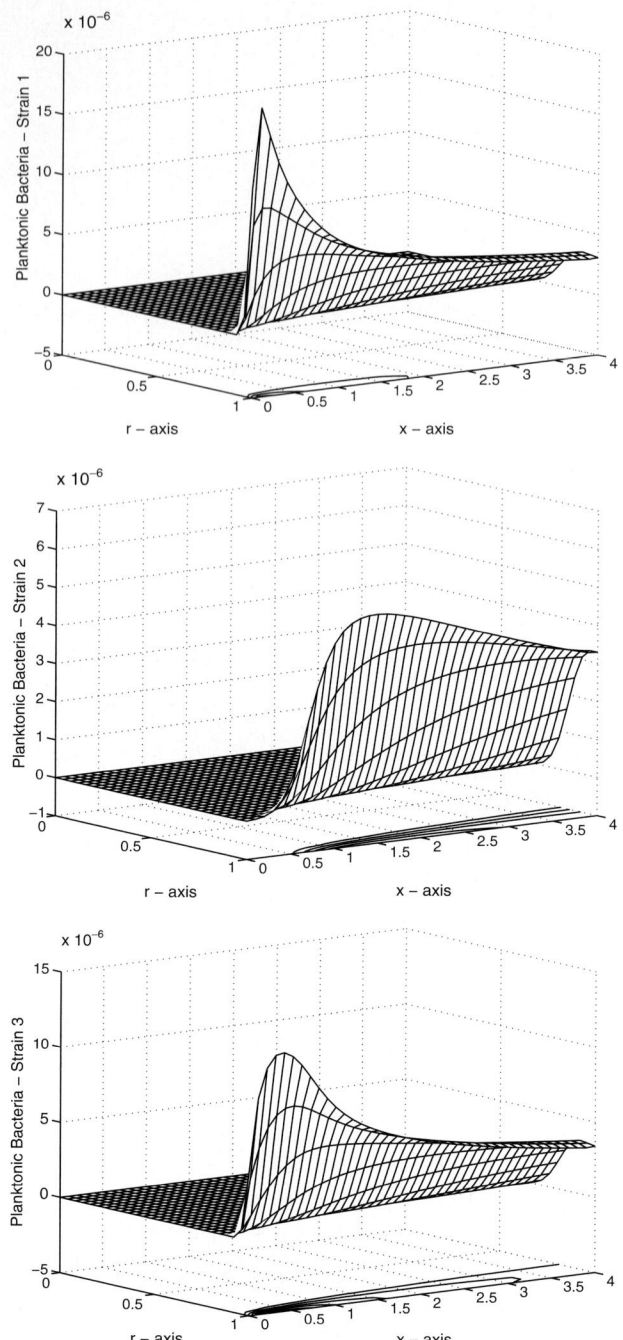

Fig. 6.10. Surface plots of the planktonic biomass density u^i- strain 1 (*top*), strain 2 (*center*), strain 3 (*bottom*)

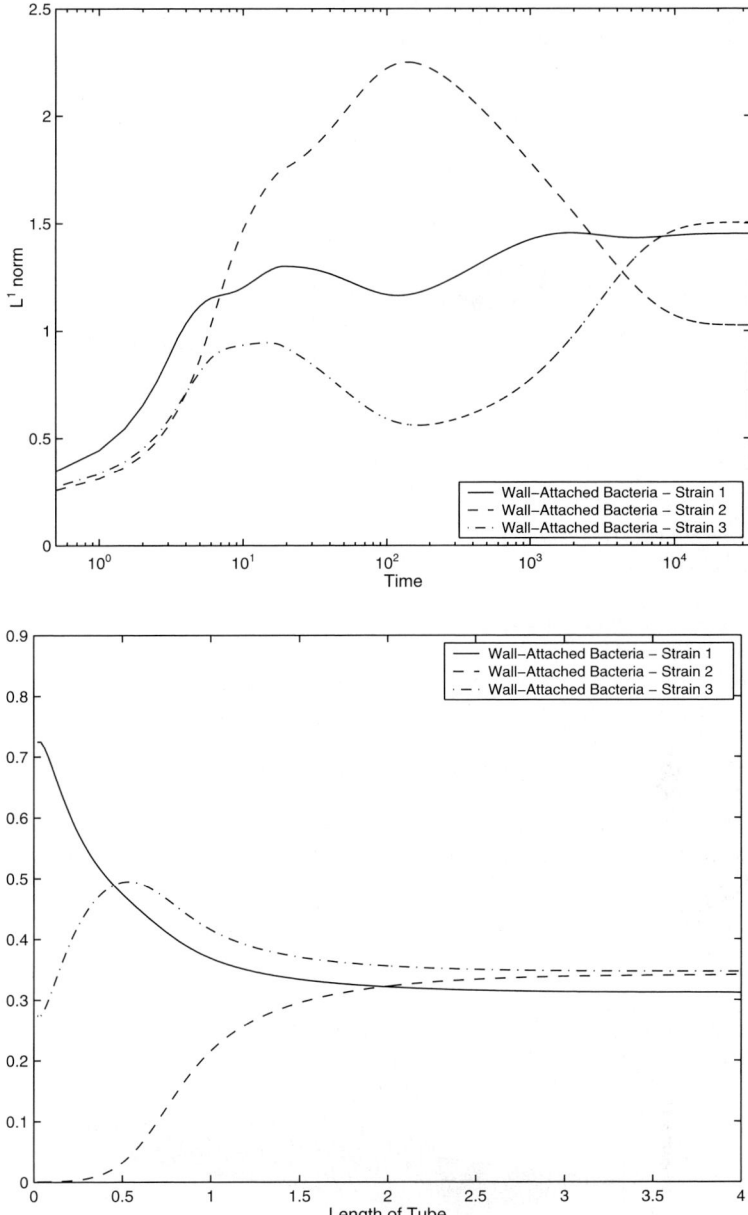

Fig. 6.11. L^1 norm versus time of the wall-attached bacteria (*top*) and a plot of the areal density of wall-attached bacteria w^i at a steady-state (*bottom*)

total planktonic and wall-attached populations, respectively, of each of the three organisms prior to reaching steady state. Coexistence is facilitated by a spatially inhomogeneous substrate steady state profile as depicted in Fig. 6.9 (bottom); the choice of substrate uptake functions gives each competitor an advantage over its rivals in a region of the bio-reactor. Equilibrium distributions of the three planktonic populations are shown in Fig. 6.10; their distributions on the reactor wall are shown in Fig. 6.11 (bottom). The latter form a pattern of segregation: roughly, one organism dominates the nutrient-rich upstream end, a second dominates an adjacent downstream segment and a third appears to share with the second organism the far downstream end. A similar segregation profile was shown in [4] for the analogous one-space dimensional system with three populations and in [35] for an approximate hyperbolic system with two populations. Although not shown here, populations one and two coexist in the absence of population three using the same parameters and initial data as above for the first two populations.

References

1. R. Aris: *Mathematical Modeling, a Chemical Engineer's Perspective* (Academic, New York 1999)
2. J. Bailey, D. Ollis: *Biochemical Engineering Fundamentals*, 2nd edn (McGraw Hill, New York 1986)
3. M. Ballyk, D. Jones, D. Le, H.L. Smith: Effects of random motility on microbial growth and competition in a flow reactor, SIAM J. Appl. Math. **59**, 2 (1998) pp 573–596
4. M. Ballyk, D. Jones, H.L. Smith: Microbial competition in reactors with wall attachment: a comparison of chemostat and plug flow models, Microb. Ecol. **41** (2001) pp 210–221
5. M. Ballyk, H.L. Smith: A Flow reactor with wall growth. In: *Mathematical Models in Medical and Health Sciences*, ed by M. Horn, G. Simonett, G. Webb (Vanderbilt University Press, Nashville, TN 1998)
6. M. Ballyk, H.L. Smith: A model of microbial growth in a plug flow reactor with wall attachment, Math. Biosci. **158** (1999) pp 95–126
7. B. Baltzis, A. Fredrickson: Competition of two microbial populations for a single resource in a chemostat when one of them exhibits wall attachment, Biotechnol. Bioeng. **25** (1983) pp 2419–2439
8. R. Bakke, M.G. Trulear, J.A. Robinson, W.G. Characklis: Activity of Pseudomonas aeruginosa in biofilms: steady state, Biotechnol. Bioeng. **26** (1984) pp 1418–1424
9. H. Berg: *Random Walks in Biology* (Princeton University Press, Princeton, NJ 1983)
10. A. Berman, R. Plemmons: *Nonnegative Matrices in the Mathematical Sciences* (Academic, New York 1979)
11. J. Bryers, ed: *Biofilms II, Process Analysis and Applications*, Wiley series in Ecological and Applied Microbiology (Wiley-Liss, NY 2000)
12. W. Characklis, K. Marshall (eds): *Biofilms*, Wiley Series in Ecological and Applied Microbiology (Wiley, New York 1990)

13. N.G. Cogan: Effects of persister formation on bacterial response to dosing, J. Theor. Biol. **238**, 3 (2006) pp 694–703
14. N.G. Cogan and J.P. Keener: The role of the biofilm matrix in structural development, Math. Med. Biol. **21** (2004) pp 147–166
15. J. Costerton: Overview of microbial biofilms, J. Indust. Microbiol. **15** (1995) pp 137–140
16. J. Costerton, P. Stewart, E. Greenberg: Bacterial biofilms: a common cause of persistent infections, Science **284** (1999) pp 1318–1322
17. J. Costerton, Z. Lewandowski, D. Debeer, D. Caldwell, D. Korber, G. James: Biofilms, the customized microniche, J. Bacteriol. **176** (1994) pp 2137–2142
18. J. Costerton, Z. Lewandowski, D. Caldwell, D. Korber, H. Lappin-Scott: Microbial biofilms, Annu. Rev. Microbiol. **49** (1995) pp 711–745
19. O. Diekmann, J. Heesterbeek: *Mathematical Epidemiology of Infectious Diseases, Model Building, Analysis and Interpretation* (Wiley, Chichester 2000)
20. R. Dillon, L. Fauci, A. Fogelson, D. Gaver: Modeling biofilm processes using the immersed boundary method, J. Comput. Phys. **129** (1996) pp 57–73
21. J. Dockery, I. Klapper: Finger formation in biofilm layers, SIAM J. Appl. Math. **62** (2001) pp 853–869
22. H.J. Eberl, D.F. Parker, M.C.M. van Loosdrecht: A new deterministic spatiotemporal continuum model for biofilm development, J. Theor. Med. **3** (3) (2001) pp 161–175
23. R. Freter: Interdependence of mechanisms that control bacterial colonization of the large intestine, Microecol. Ther. **14** (1984) pp 89–96
24. R. Freter: Mechanisms that control the microflora in the large intestine. In: *Human Intestinal Microflora in Health and Disease*, ed by D. Hentges (Academic, New York 1983)
25. R. Freter, H. Brickner, J. Fekete, M. Vickerman, K. Carey: Survival and implantation of *Escherichia coli in the intestinal tract*, Infect. Immun. **39** (1983) pp 686–703
26. R. Freter, H. Brickner, S. Temme: An understanding of colonization resistance of the mammalian large intestine requires mathematical analysis, Microecol. Ther. **16** (1986) pp 147–155
27. C.A. Fux, J.W. Costerton, P.S. Stewart, and P. Stoodley, Survival strategies of infectious biofilms, Trends Microbiol., **13** (2005) pp 34–40
28. D. Herbert, R. Elsworth, R. Telling: The continuous culture of bacteria; a theoretical and experimental study, J. Can. Microbiol. **14** (1956) pp 601–622
29. M. Imran, D. Jones, H.L. Smith: Biofilms and the plasmid maintenance question, Math. Biosci. **193** (2005) pp 183–204
30. M. Imran, H.L. Smith: A mathematical model of gene transfer in a biofilm. In: *Mathematics for Ecology and Environmental Sciences*, Vol.1 (Springer, Berlin Heidelberg New York 2006)
31. D. Jones, H. Kojouharov, D. Le, H.L. Smith: Bacterial wall attachment in a flow reactor: mixed culture, Can. Appl. Math. Q. **10** (2004) pp 111–138
32. D. Jones, H. Kojouharov, D. Le, H.L. Smith: Bacterial wall attachment in a flow reactor, SIAM J. Appl. Math. **62** (2002) pp 1728–1771
33. D. Jones, H. Kojouharov, D. Le, H.L. Smith: Microbial Competition for Nutrient in a 3*D* Flow Reactor, Dynamics of Continuous, Discrete Impulsive Dynamical Syst. **10** (2003) pp 57–67
34. D. Jones, H. Kojouharov, D. Le, H.L. Smith: The Freter model: a simple model of biofilm formation, J. Math. Biol. **47** (2003) pp 137–152

35. D. Jones, H.L. Smith: Microbial competition for nutrient and wall sites in plug flow, SIAM J. Appl. Math. **60** (2000) pp 1576–1600

36. J.-U. Kreft, C. Picioreanu, J. Wimpenny, M. van Loosdrecht: Individual-based modelling of biofilms, Microbiology **147** (2001) pp 2897–2912

37. C.M. Kung, B. Baltzis: The growth of pure and simple microbial competitors in a moving distributed medium, Math. Biosci. **111** (1992) pp 295–313

38. C.S. Laspidou and B.E. Rittmann: Modeling the development of biofilm density including active bacteria, inert biomass, and extracellular polymeric substances, Water Res. **38** (2004) pp 3349–3361

39. D. Noguera, S. Okabe, C. Picioreanu: Biofilm modeling: present status and future directions, Water Sci. Technol. **39** (1999) pp 273–278

40. D. Noguera, G. Pizarro, D. Stahl, B. Rittman: Simulation of multispecies biofilm development in three dimensions, Wat.Sci. Tech. **39** (1999) 123–130

41. C. Picioreanu, M.C.M. van Loosdrecht, J. Heijnen: Mathematical Modeling of biofilm structure with a hybrid differential-discrete cellular automaton approach, Biotechnol. Bioeng. **58** (1998) pp 101–116

42. S. Pilyugin and P. Waltman: The simple chemostat with wall growth, SIAM J. Appl. Math. **59** (1999) pp 1552–1572

43. H.L. Smith: A semilinear hyperbolic system, *Proceedings of the Mathematics Conference*, ed by S. Elyadi, F. Allan, A. Elkhader, T. Muhgrabi, M. Saleh (World Scientific, Singapore 2000)

44. H.L. Smith, P. Waltman: *The Theory of the Chemostat* (Cambridge University Press, New York 1995)

45. H.L. Smith, X.-Q. Zhao: Microbial growth in a plug flow reactor with wall attachment and cell motility, JMAA **241** (2000) pp 134–155

46. E. Stemmons, H.L. Smith, H.L.: Competition in a chemostat with wall attachment, SIAM J. Appl. Math. **61** (2000) pp 567–595

47. P. Stewart, G. Mcfeters, C.-T. Huang: Biofilm control by antimicrobial agents, Chapter 11. In: *Biofilms II: Process Analysis and Applications*, ed by J. Bryers (Wiley-Liss, New York 2000)

48. S.M. Hunt, M.A. Hamilton, P.S. Stewart: A 3D model of antimicrobial action on biofilms, Water Sci. Technol. **52**, 7 (2005) 143–148

49. H. Topiwala and G. Hamer: Effect of wall growth in steady-state continuous cultures, Biotechnol. Bioeng. **13** (1971) pp 919–922

50. C.Y. Wen, L.T. Fan: Models for flow systems and chemical reactors. In: *Chemical Processing and Engineering*, vol. 3 (Marcel Dekker, New York 1975)

51. J.W.T. Wimpenny, R. Colasanti: A unifying hypothesis for the structure of microbial biofilms based on cellular automaton models, FEMS Microb. Ecol. **22** (1997) pp 1–16

Lecture Notes in Mathematics

For information about earlier volumes
please contact your bookseller or Springer
LNM Online archive: springerlink.com

Vol. 1803: G. Dolzmann, Variational Methods for Crystalline Microstructure – Analysis and Computation (2003)
Vol. 1804: I. Cherednik, Ya. Markov, R. Howe, G. Lusztig, Iwahori-Hecke Algebras and their Representation Theory. Martina Franca, Italy 1999. Editors: V. Baldoni, D. Barbasch (2003)
Vol. 1805: F. Cao, Geometric Curve Evolution and Image Processing (2003)
Vol. 1806: H. Broer, I. Hoveijn. G. Lunther, G. Vegter, Bifurcations in Hamiltonian Systems. Computing Singularities by Gröbner Bases (2003)
Vol. 1807: V. D. Milman, G. Schechtman (Eds.), Geometric Aspects of Functional Analysis. Israel Seminar 2000-2002 (2003)
Vol. 1808: W. Schindler, Measures with Symmetry Properties (2003)
Vol. 1809: O. Steinbach, Stability Estimates for Hybrid Coupled Domain Decomposition Methods (2003)
Vol. 1810: J. Wengenroth, Derived Functors in Functional Analysis (2003)
Vol. 1811: J. Stevens, Deformations of Singularities (2003)
Vol. 1812: L. Ambrosio, K. Deckelnick, G. Dziuk, M. Mimura, V. A. Solonnikov, H. M. Soner, Mathematical Aspects of Evolving Interfaces. Madeira, Funchal, Portugal 2000. Editors: P. Colli, J. F. Rodrigues (2003)
Vol. 1813: L. Ambrosio, L. A. Caffarelli, Y. Brenier, G. Buttazzo, C. Villani, Optimal Transportation and its Applications. Martina Franca, Italy 2001. Editors: L. A. Caffarelli, S. Salsa (2003)
Vol. 1814: P. Bank, F. Baudoin, H. Föllmer, L.C.G. Rogers, M. Soner, N. Touzi, Paris-Princeton Lectures on Mathematical Finance 2002 (2003)
Vol. 1815: A. M. Vershik (Ed.), Asymptotic Combinatorics with Applications to Mathematical Physics. St. Petersburg, Russia 2001 (2003)
Vol. 1816: S. Albeverio, W. Schachermayer, M. Talagrand, Lectures on Probability Theory and Statistics. Ecole d'Eté de Probabilités de Saint-Flour XXX-2000. Editor: P. Bernard (2003)
Vol. 1817: E. Koelink, W. Van Assche (Eds.), Orthogonal Polynomials and Special Functions. Leuven 2002 (2003)
Vol. 1818: M. Bildhauer, Convex Variational Problems with Linear, nearly Linear and/or Anisotropic Growth Conditions (2003)
Vol. 1819: D. Masser, Yu. V. Nesterenko, H. P. Schlickewei, W. M. Schmidt, M. Waldschmidt, Diophantine Approximation. Cetraro, Italy 2000. Editors: F. Amoroso, U. Zannier (2003)
Vol. 1820: F. Hiai, H. Kosaki, Means of Hilbert Space Operators (2003)
Vol. 1821: S. Teufel, Adiabatic Perturbation Theory in Quantum Dynamics (2003)
Vol. 1822: S.-N. Chow, R. Conti, R. Johnson, J. Mallet-Paret, R. Nussbaum, Dynamical Systems. Cetraro, Italy 2000. Editors: J. W. Macki, P. Zecca (2003)
Vol. 1823: A. M. Anile, W. Allegretto, C. Ringhofer, Mathematical Problems in Semiconductor Physics. Cetraro, Italy 1998. Editor: A. M. Anile (2003)
Vol. 1824: J. A. Navarro González, J. B. Sancho de Salas, $\mathscr{C}^\infty$ – Differentiable Spaces (2003)
Vol. 1825: J. H. Bramble, A. Cohen, W. Dahmen, Multiscale Problems and Methods in Numerical Simulations, Martina Franca, Italy 2001. Editor: C. Canuto (2003)
Vol. 1826: K. Dohmen, Improved Bonferroni Inequalities via Abstract Tubes. Inequalities and Identities of Inclusion-Exclusion Type. VIII, 113 p, 2003.

Vol. 1827: K. M. Pilgrim, Combinations of Complex Dynamical Systems. IX, 118 p, 2003.
Vol. 1828: D. J. Green, Gröbner Bases and the Computation of Group Cohomology. XII, 138 p, 2003.
Vol. 1829: E. Altman, B. Gaujal, A. Hordijk, Discrete-Event Control of Stochastic Networks: Multimodularity and Regularity. XIV, 313 p, 2003.
Vol. 1830: M. I. Gil', Operator Functions and Localization of Spectra. XIV, 256 p, 2003.
Vol. 1831: A. Connes, J. Cuntz, E. Guentner, N. Higson, J. E. Kaminker, Noncommutative Geometry, Martina Franca, Italy 2002. Editors: S. Doplicher, L. Longo (2004)
Vol. 1832: J. Azéma, M. Émery, M. Ledoux, M. Yor (Eds.), Séminaire de Probabilités XXXVII (2003)
Vol. 1833: D.-Q. Jiang, M. Qian, M.-P. Qian, Mathematical Theory of Nonequilibrium Steady States. On the Frontier of Probability and Dynamical Systems. IX, 280 p, 2004.
Vol. 1834: Yo. Yomdin, G. Comte, Tame Geometry with Application in Smooth Analysis. VIII, 186 p, 2004.
Vol. 1835: O.T. Izhboldin, B. Kahn, N.A. Karpenko, A. Vishik, Geometric Methods in the Algebraic Theory of Quadratic Forms. Summer School, Lens, 2000. Editor: J.-P. Tignol (2004)
Vol. 1836: C. Năstăsescu, F. Van Oystaeyen, Methods of Graded Rings. XIII, 304 p, 2004.
Vol. 1837: S. Tavaré, O. Zeitouni, Lectures on Probability Theory and Statistics. Ecole d'Eté de Probabilités de Saint-Flour XXXI-2001. Editor: J. Picard (2004)
Vol. 1838: A.J. Ganesh, N.W. O'Connell, D.J. Wischik, Big Queues. XII, 254 p, 2004.
Vol. 1839: R. Gohm, Noncommutative Stationary Processes. VIII, 170 p, 2004.
Vol. 1840: B. Tsirelson, W. Werner, Lectures on Probability Theory and Statistics. Ecole d'Eté de Probabilités de Saint-Flour XXXII-2002. Editor: J. Picard (2004)
Vol. 1841: W. Reichel, Uniqueness Theorems for Variational Problems by the Method of Transformation Groups (2004)
Vol. 1842: T. Johnsen, A. L. Knutsen, K$_3$ Projective Models in Scrolls (2004)
Vol. 1843: B. Jefferies, Spectral Properties of Noncommuting Operators (2004)
Vol. 1844: K.F. Siburg, The Principle of Least Action in Geometry and Dynamics (2004)
Vol. 1845: Min Ho Lee, Mixed Automorphic Forms, Torus Bundles, and Jacobi Forms (2004)
Vol. 1846: H. Ammari, H. Kang, Reconstruction of Small Inhomogeneities from Boundary Measurements (2004)
Vol. 1847: T.R. Bielecki, T. Björk, M. Jeanblanc, M. Rutkowski, J.A. Scheinkman, W. Xiong, Paris-Princeton Lectures on Mathematical Finance 2003 (2004)
Vol. 1848: M. Abate, J. E. Fornaess, X. Huang, J. P. Rosay, A. Tumanov, Real Methods in Complex and CR Geometry, Martina Franca, Italy 2002. Editors: D. Zaitsev, G. Zampieri (2004)
Vol. 1849: Martin L. Brown, Heegner Modules and Elliptic Curves (2004)
Vol. 1850: V. D. Milman, G. Schechtman (Eds.), Geometric Aspects of Functional Analysis. Israel Seminar 2002-2003 (2004)
Vol. 1851: O. Catoni, Statistical Learning Theory and Stochastic Optimization (2004)
Vol. 1852: A.S. Kechris, B.D. Miller, Topics in Orbit Equivalence (2004)
Vol. 1853: Ch. Favre, M. Jonsson, The Valuative Tree (2004)

Vol. 1854: O. Saeki, Topology of Singular Fibers of Differential Maps (2004)

Vol. 1855: G. Da Prato, P.C. Kunstmann, I. Lasiecka, A. Lunardi, R. Schnaubelt, L. Weis, Functional Analytic Methods for Evolution Equations. Editors: M. Iannelli, R. Nagel, S. Piazzera (2004)

Vol. 1856: K. Back, T.R. Bielecki, C. Hipp, S. Peng, W. Schachermayer, Stochastic Methods in Finance, Bressanone/Brixen, Italy, 2003. Editors: M. Fritelli, W. Runggaldier (2004)

Vol. 1857: M. Émery, M. Ledoux, M. Yor (Eds.), Séminaire de Probabilités XXXVIII (2005)

Vol. 1858: A.S. Cherny, H.-J. Engelbert, Singular Stochastic Differential Equations (2005)

Vol. 1859: E. Letellier, Fourier Transforms of Invariant Functions on Finite Reductive Lie Algebras (2005)

Vol. 1860: A. Borisyuk, G.B. Ermentrout, A. Friedman, D. Terman, Tutorials in Mathematical Biosciences I. Mathematical Neurosciences (2005)

Vol. 1861: G. Benettin, J. Henrard, S. Kuksin, Hamiltonian Dynamics – Theory and Applications, Cetraro, Italy, 1999. Editor: A. Giorgilli (2005)

Vol. 1862: B. Helffer, F. Nier, Hypoelliptic Estimates and Spectral Theory for Fokker-Planck Operators and Witten Laplacians (2005)

Vol. 1863: H. Führ, Abstract Harmonic Analysis of Continuous Wavelet Transforms (2005)

Vol. 1864: K. Efstathiou, Metamorphoses of Hamiltonian Systems with Symmetries (2005)

Vol. 1865: D. Applebaum, B.V. R. Bhat, J. Kustermans, J. M. Lindsay, Quantum Independent Increment Processes I. From Classical Probability to Quantum Stochastic Calculus. Editors: M. Schürmann, U. Franz (2005)

Vol. 1866: O.E. Barndorff-Nielsen, U. Franz, R. Gohm, B. Kümmerer, S. Thorbjønsen, Quantum Independent Increment Processes II. Structure of Quantum Lévy Processes, Classical Probability, and Physics. Editors: M. Schürmann, U. Franz, (2005)

Vol. 1867: J. Sneyd (Ed.), Tutorials in Mathematical Biosciences II. Mathematical Modeling of Calcium Dynamics and Signal Transduction. (2005)

Vol. 1868: J. Jorgenson, S. Lang, $Pos_n(R)$ and Eisenstein Series. (2005)

Vol. 1869: A. Dembo, T. Funaki, Lectures on Probability Theory and Statistics. Ecole d'Eté de Probabilités de Saint-Flour XXXIII-2003. Editor: J. Picard (2005)

Vol. 1870: V.I. Gurariy, W. Lusky, Geometry of Müntz Spaces and Related Questions. (2005)

Vol. 1871: P. Constantin, G. Gallavotti, A.V. Kazhikhov, Y. Meyer, S. Ukai, Mathematical Foundation of Turbulent Viscous Flows, Martina Franca, Italy, 2003. Editors: M. Cannone, T. Miyakawa (2006)

Vol. 1872: A. Friedman (Ed.), Tutorials in Mathematical Biosciences III. Cell Cycle, Proliferation, and Cancer (2006)

Vol. 1873: R. Mansuy, M. Yor, Random Times and Enlargements of Filtrations in a Brownian Setting (2006)

Vol. 1874: M. Yor, M. Émery (Eds.), In Memoriam Paul-André Meyer - Séminaire de Probabilités XXXIX (2006)

Vol. 1875: J. Pitman, Combinatorial Stochastic Processes. Ecole d'Eté de Probabilités de Saint-Flour XXXII-2002. Editor: J. Picard (2006)

Vol. 1876: H. Herrlich, Axiom of Choice (2006)

Vol. 1877: J. Steuding, Value Distributions of L-Functions (2007)

Vol. 1878: R. Cerf, The Wulff Crystal in Ising and Percolation Models, Ecole d'Eté de Probabilités de Saint-Flour XXXIV-2004. Editor: Jean Picard (2006)

Vol. 1879: G. Slade, The Lace Expansion and its Applications, Ecole d'Eté de Probabilités de Saint-Flour XXXIV-2004. Editor: Jean Picard (2006)

Vol. 1880: S. Attal, A. Joye, C.-A. Pillet, Open Quantum Systems I, The Hamiltonian Approach (2006)

Vol. 1881: S. Attal, A. Joye, C.-A. Pillet, Open Quantum Systems II, The Markovian Approach (2006)

Vol. 1882: S. Attal, A. Joye, C.-A. Pillet, Open Quantum Systems III, Recent Developments (2006)

Vol. 1883: W. Van Assche, F. Marcellàn (Eds.), Orthogonal Polynomials and Special Functions, Computation and Application (2006)

Vol. 1884: N. Hayashi, E.I. Kaikina, P.I. Naumkin, I.A. Shishmarev, Asymptotics for Dissipative Nonlinear Equations (2006)

Vol. 1885: A. Telcs, The Art of Random Walks (2006)

Vol. 1886: S. Takamura, Splitting Deformations of Degenerations of Complex Curves (2006)

Vol. 1887: K. Habermann, L. Habermann, Introduction to Symplectic Dirac Operators (2006)

Vol. 1888: J. van der Hoeven, Transseries and Real Differential Algebra (2006)

Vol. 1889: G. Osipenko, Dynamical Systems, Graphs, and Algorithms (2006)

Vol. 1890: M. Bunge, J. Funk, Singular Coverings of Toposes (2006)

Vol. 1891: J.B. Friedlander, D.R. Heath-Brown, H. Iwaniec, J. Kaczorowski, Analytic Number Theory, Cetraro, Italy, 2002. Editors: A. Perelli, C. Viola (2006)

Vol. 1892: A. Baddeley, I. Bárány, R. Schneider, W. Weil, Stochastic Geometry, Martina Franca, Italy, 2004. Editor: W. Weil (2007)

Vol. 1893: H. Hanßmann, Local and Semi-Local Bifurcations in Hamiltonian Dynamical Systems, Results and Examples (2007)

Vol. 1894: C.W. Groetsch, Stable Approximate Evaluation of Unbounded Operators (2007)

Vol. 1895: L. Molnár, Selected Preserver Problems on Algebraic Structures of Linear Operators and on Function Spaces (2007)

Vol. 1896: P. Massart, Concentration Inequalities and Model Selection, Ecole d'Été de Probabilités de Saint-Flour XXXIII-2003. Editor: J. Picard (2007)

Vol. 1897: R. Doney, Fluctuation Theory for Lévy Processes, Ecole d'Été de Probabilités de Saint-Flour XXXV-2005. Editor: J. Picard (2007)

Vol. 1898: H.R. Beyer, Beyond Partial Differential Equations, On linear and Quasi-Linear Abstract Hyperbolic Evolution Equations (2007)

Vol. 1899: Séminaire de Probabilités XL. Editors: C. Donati-Martin, M. Émery, A. Rouault, C. Stricker (2007)

Vol. 1900: E. Bolthausen, A. Bovier (Eds.), Spin Glasses (2007)

Vol. 1901: O. Wittenberg, Intersections de deux quadriques et pinceaux de courbes de genre 1, Intersections of Two Quadrics and Pencils of Curves of Genus 1 (2007)

Vol. 1902: A. Isaev, Lectures on the Automorphism Groups of Kobayashi-Hyperbolic Manifolds (2007)

Vol. 1903: G. Kresin, V. Maz'ya, Sharp Real-Part Theorems (2007)

Vol. 1904: P. Giesl, Construction of Global Lyapunov Functions Using Radial Basis Functions (2007)

Vol. 1905: C. Prévôt, M. Röckner, A Concise Course on Stochastic Partial Differential Equations (2007)
Vol. 1906: T. Schuster, The Method of Approximate Inverse: Theory and Applications (2007)
Vol. 1907: M. Rasmussen, Attractivity and Bifurcation for Nonautonomous Dynamical Systems (2007)
Vol. 1908: T.J. Lyons, M. Caruana, T. Lévy, Differential Equations Driven by Rough Paths, Ecole d'Été de Probabilités de Saint-Flour XXXIV-2004 (2007)
Vol. 1909: H. Akiyoshi, M. Sakuma, M. Wada, Y. Yamashita, Punctured Torus Groups and 2-Bridge Knot Groups (I) (2007)
Vol. 1910: V.D. Milman, G. Schechtman (Eds.), Geometric Aspects of Functional Analysis. Israel Seminar 2004-2005 (2007)
Vol. 1911: A. Bressan, D. Serre, M. Williams, K. Zumbrun, Hyperbolic Systems of Balance Laws. Cetraro, Italy 2003. Editor: P. Marcati (2007)
Vol. 1912: V. Berinde, Iterative Approximation of Fixed Points (2007)
Vol. 1913: J.E. Marsden, G. Misiołek, J.-P. Ortega, M. Perlmutter, T.S. Ratiu, Hamiltonian Reduction by Stages (2007)
Vol. 1914: G. Kutyniok, Affine Density in Wavelet Analysis (2007)
Vol. 1915: T. Bıyıkoğlu, J. Leydold, P.F. Stadler, Laplacian Eigenvectors of Graphs. Perron-Frobenius and Faber-Krahn Type Theorems (2007)
Vol. 1916: C. Villani, F. Rezakhanlou, Entropy Methods for the Boltzmann Equation. Editors: F. Golse, S. Olla (2008)
Vol. 1917: I. Veselić, Existence and Regularity Properties of the Integrated Density of States of Random Schrödinger (2008)
Vol. 1918: B. Roberts, R. Schmidt, Local Newforms for GSp(4) (2007)
Vol. 1919: R.A. Carmona, I. Ekeland, A. Kohatsu-Higa, J.-M. Lasry, P.-L. Lions, H. Pham, E. Taflin, Paris-Princeton Lectures on Mathematical Finance 2004. Editors: R.A. Carmona, E. Çinlar, I. Ekeland, E. Jouini, J.A. Scheinkman, N. Touzi (2007)
Vol. 1920: S.N. Evans, Probability and Real Trees. Ecole d'Été de Probabilités de Saint-Flour XXXV-2005 (2008)
Vol. 1921: J.P. Tian, Evolution Algebras and their Applications (2008)
Vol. 1922: A. Friedman (Ed.), Tutorials in Mathematical BioSciences IV. Evolution and Ecology (2008)
Vol. 1923: J.P.N. Bishwal, Parameter Estimation in Stochastic Differential Equations (2008)
Vol. 1924: M. Wilson, Littlewood-Paley Theory and Exponential-Square Integrability (2008)
Vol. 1925: M. du Sautoy, L. Woodward, Zeta Functions of Groups and Rings (2008)
Vol. 1926: L. Barreira, V. Claudia, Stability of Nonautonomous Differential Equations (2008)
Vol. 1927: L. Ambrosio, L. Caffarelli, M.G. Crandall, L.C. Evans, N. Fusco, Calculus of Variations and Non-Linear Partial Differential Equations. Cetraro, Italy 2005. Editors: B. Dacorogna, P. Marcellini (2008)
Vol. 1928: J. Jonsson, Simplicial Complexes of Graphs (2008)
Vol. 1929: Y. Mishura, Stochastic Calculus for Fractional Brownian Motion and Related Processes (2008)
Vol. 1930: J.M. Urbano, The Method of Intrinsic Scaling. A Systematic Approach to Regularity for Degenerate and Singular PDEs (2008)

Vol. 1931: M. Cowling, E. Frenkel, M. Kashiwara, A. Valette, D.A. Vogan, Jr., N.R. Wallach, Representation Theory and Complex Analysis. Venice, Italy 2004. Editors: E.C. Tarabusi, A. D'Agnolo, M. Picardello (2008)
Vol. 1932: A.A. Agrachev, A.S. Morse, E.D. Sontag, H.J. Sussmann, V.I. Utkin, Nonlinear and Optimal Control Theory. Cetraro, Italy 2004. Editors: P. Nistri, G. Stefani (2008)
Vol. 1933: M. Petkovic, Point Estimation of Root Finding Methods (2008)
Vol. 1934: C. Donati-Martin, M. Émery, A. Rouault, C. Stricker (Eds.), Séminaire de Probabilités XLI (2008)
Vol. 1935: A. Unterberger, Alternative Pseudodifferential Analysis (2008)
Vol. 1936: P. Magal, S. Ruan (Eds.), Structured Population Models in Biology and Epidemiology (2008)
Vol. 1937: G. Capriz, P. Giovine, P.M. Mariano (Eds.), Mathematical Models of Granular Matter (2008)
Vol. 1938: D. Auroux, F. Catanese, M. Manetti, P. Seidel, B. Siebert, I. Smith, G. Tian, Symplectic 4-Manifolds and Algebraic Surfaces. Cetraro, Italy 2003. Editors: F. Catanese, G. Tian (2008)
Vol. 1939: D. Boffi, F. Brezzi, L. Demkowicz, R.G. Durán, R.S. Falk, M. Fortin, Mixed Finite Elements, Compatibility Conditions, and Applications. Cetraro, Italy 2006. Editors: D. Boffi, L. Gastaldi (2008)
Vol. 1940: J. Banasiak, V. Capasso, M.A.J. Chaplain, M. Lachowicz, J. Miękisz, Multiscale Problems in the Life Sciences. From Microscopic to Macroscopic. Będlewo, Poland 2006. Editors: V. Capasso, M. Lachowicz (2008)
Vol. 1941: S.M.J. Haran, Arithmetical Investigations. Representation Theory, Orthogonal Polynomials, and Quantum Interpolations (2008)
Vol. 1942: S. Albeverio, F. Flandoli, Y.G. Sinai, SPDE in Hydrodynamic. Recent Progress and Prospects. Cetraro, Italy 2005. Editors: G. Da Prato, M. Röckner (2008)
Vol. 1943: L.L. Bonilla (Ed.), Inverse Problems and Imaging. Martina Franca, Italy 2002 (2008)
Vol. 1944: A. Di Bartolo, G. Falcone, P. Plaumann, K. Strambach, Algebraic Groups and Lie Groups with Few Factors (2008)

Recent Reprints and New Editions

Vol. 1702: J. Ma, J. Yong, Forward-Backward Stochastic Differential Equations and their Applications. 1999 – Corr. 3rd printing (2007)
Vol. 830: J.A. Green, Polynomial Representations of GL_n, with an Appendix on Schensted Correspondence and Littelmann Paths by K. Erdmann, J.A. Green and M. Schoker 1980 – 2nd corr. and augmented edition (2007)
Vol. 1693: S. Simons, From Hahn-Banach to Monotonicity (Minimax and Monotonicity 1998) – 2nd exp. edition (2008)
Vol. 470: R.E. Bowen, Equilibrium States and the Ergodic Theory of Anosov Diffeomorphisms. With a preface by D. Ruelle. Edited by J.-R. Chazottes. 1975 – 2nd rev. edition (2008)
Vol. 523: S.A. Albeverio, R.J. Høegh-Krohn, S. Mazzucchi, Mathematical Theory of Feynman Path Integral. 1976 – 2nd corr. and enlarged edition (2008)

Printing: Krips bv, Meppel, The Netherlands
Binding: Stürtz, Würzburg, Germany